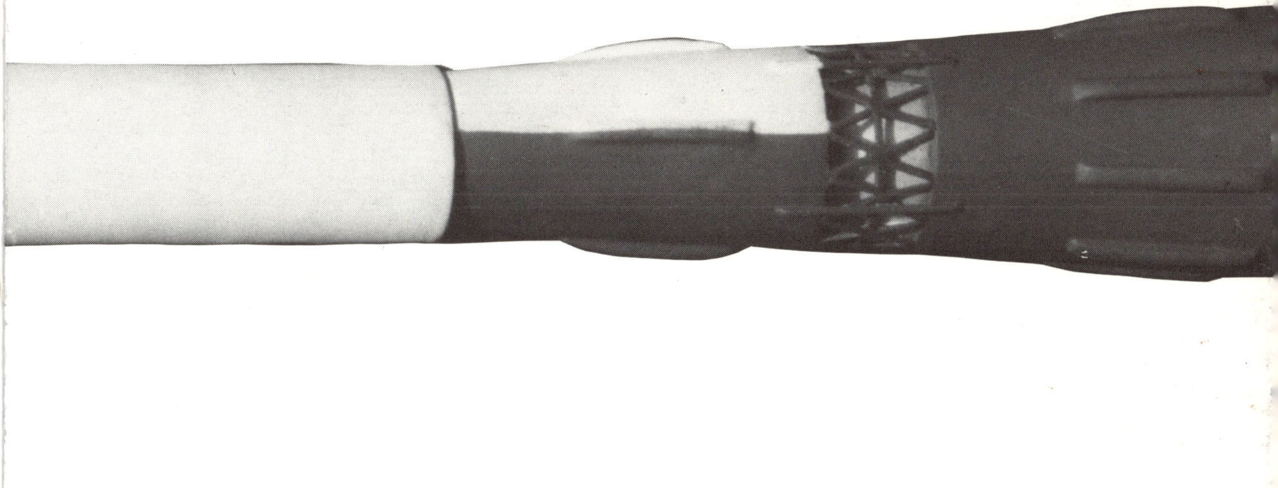
UNITED STATES
USA
USA

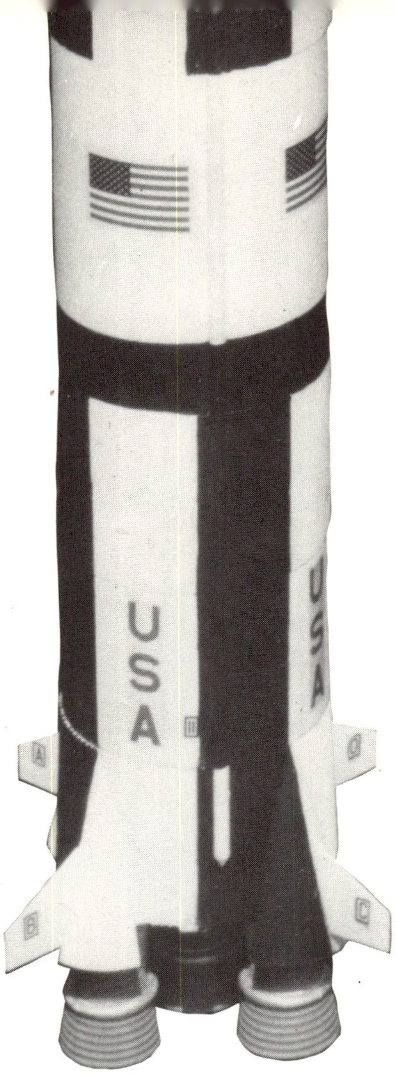
USA
USA

The Buzz Aldrin's Race Into Space Companion

The Buzz Aldrin's Race Into Space Companion

Fritz Bronner

Osborne **McGraw-Hill**

Berkeley New York St. Louis San Francisco
Auckland Bogotá Hamburg London Madrid
Mexico City Milan Montreal New Delhi Panama City
Paris São Paulo Singapore Sydney
Tokyo Toronto

Osborne **McGraw-Hill**
2600 Tenth Street
Berkeley, California 94710
U.S.A.

For information on translations or book distributors outside of the U.S.A., please write to Osborne **McGraw-Hill** at the above address.

The Buzz Aldrin's Race Into Space Companion

1234567890 DOC 99876543

ISBN 0-07-881938-5

To Laika, the first explorer of the heavens

Acquisitions Editor	William Pollock
Associate Editor	Emily Rader
Technical Editor	Ivan Luk
Project Editor	Mark Karmendy
Copy Editor	Mark Karmendy
Proofreader	Mick Arellano
Computer Designer	Marcela Hancik
Illustrators	Marla Shelasky
	Steve Stipp
	Margie Beeson
Cover Designer	Ken Hodges
Historical Writing Contributions	Robert Reeves
	Erik Anderson
	Wes Kowato
	Ken Wheeler
Gameplay Writing Contributions	Ivan Luk
	Erik Anderson
	Michael McCarty
Game Strategy Contributions	Erik Anderson
	Wes Kowato
	Ken Wheeler
	Ivan Luk

Contents

Part II:
Installation and Game Strategies

Part III:
Debriefing

Acknowledgments

There are numerous individuals to whom I am grateful for their assistance and support in the writing of this strategy book. Thanks go to Robert Reeves, Erik Anderson, Wes Kowato, and Ken Wheeler for their historical research and writing contributions. Additional thanks to Ivan Luk for his advice and gameplay writing contributions. For gameplay strategies, I thank Erik Anderson, Wes Kowato, and Ken Wheeler.

At Strategic Visions, a hearty thanks to Michael McCarty for his technical support and expertise. Thanks also to Rusty Buchert and Bruce Schlickbernd at Interplay, for their support and assistance.

At Osborne McGraw-Hill, thanks to Emily Rader for her suggestions and persistence. Thanks to Mark Karmendy for his wry humor and grammatical expertise. Thanks also to Bill Pollock and his vision for this book.

I thank the entire staffs at Strategic Visions, Interplay, and Osborne McGraw-Hill. I also thank Buzz Aldrin for his interest and assistance in this endeavor.

A special thanks to my wife Margie for supporting another long and involved project.

Foreword

I have often looked up to the night sky with wonder and amazement. The celestial knowledge that we have discovered in the last century is nothing short of monumental. The more we search and challenge the mysteries of our universe, the more new questions arise. As portions of this gigantic puzzle unravel, mankind's eternal questions still remain unanswered. Who are we? Where do we come from, and where are we going? It is this thirst for truth that drives mankind on a quest for answers.

Earth's companion, the Moon, has always reflected humanity's current knowledge. For eons during man's past, the Moon was a spirit or god. More recently, it represented a challenge of prestige and technology. Interestingly, nearly twenty-five years after *Apollo 11,* the Moon is now rarely discussed. The general public rarely talks of the Moon other than to reminisce an event that occurred long ago. Sadly, most have forgotten that we ever set foot there.

Critics of the space race and of space exploration have often complained about the huge price tag that accompanies such an endeavor. There are arguments that the money could be better spent on more pressing needs at home.

What is usually ignored by the critics is that if the money were indeed cut from the space program, there would be no guarantee of these funds ever going to special needs. It is a fact that there are great wastes in government and in society. The entire Apollo-Moon program cost less than what Americans spent on cosmetics in the same time period. What Americans spend over the next ten years on smoking could easily fund a space station and a manned mission to Mars. Motivation is more important than money.

Solutions will arrive when goals are made. But it all starts with a dream.

May we rediscover the dream....

Introduction

The beginning of man's initial exploration of the heavens is recounted in *The Buzz Aldrin's Race into Space Companion.* In the first part of this book, there is startling new information about the space program of the former Soviet Union; thanks to glasnost, secret data has finally been unveiled. Also, after years of neglect, forgotten plans of America's space program are now being remembered. In the second part of the book, veteran players of the game offer winning tips and strategies on how to more effectively direct your space program. Step-by-step procedures on how to maneuver within your space program are detailed as well. Also revealed in this book are secrets on how the game was designed, with clues on how your computer opponent's artificial intelligence operates.

While playing "Buzz Aldrin's Race into Space," I suggest that you step outside and take a fresh look at our nighttime companion. The Moon might appear to you now as slightly different. Perhaps you might recall vague memories of dancing, ghost-like images on the surface of a distant world that lies somewhere between fantasy and the past. After playing this game, you might view that old Moon in a new light. You'll have re-experienced man's exploration of space with all its glory, folly, tragedy, and excitement. May this stir up desire in another generation of explorers so that we may journey farther into space.

How This Book is Organized

Part I:

- Chapter 1 offers a brief history of the development of rocketry and an explanation of the relation of lunar myths and religion to science fiction and space technology. There are also biographies of Goddard, von Braun, and Korolev.

- Chapter 2 is an account of the Cold War, especially the confrontations between the superpowers during the space race. The art of propaganda and how the Soviets excelled at it are described.
- Chapter 3 reveals the machines, animals, men, and women that explored the frontiers of space. There are also biographies of some of the more famous American and Soviet space explorers.
- Chapter 4 is a history of the close calls and tragedies that occurred in the space race. With every step of progress in space exploration, there is the risk of disaster and catastrophe.
- Chapter 5 reveals new data on what the Soviets were secretly attempting in the space race. Also included are some of the forgotten American proposals for space exploration.
- Chapter 6 offers Buzz Aldrin's insights into the space race and his own views on the Soviet competitors.
- Chapter 7 is a chronology of space that outlines the major accomplishments and moments in the space race.

Part II:

- Chapter 8 describes all of the hardware configurations and system requirements for playing "Buzz Aldrin's Race Into Space." Information is provided on how to install the game and what to look for in sound cards.
- Chapter 9 defines overall goals and victory conditions that you should know as director of the space program. Over 20 different manned lunar landing approaches are described and outlined.
- Chapter 10 explains how to schedule and launch all of the major types of space missions in the game. It also gives strategy and advice on what the player should do during the three phases of the game.
- Chapter 11 provides recommendations for efficiently operating your spaceport and buildings. It also offers details on what is inside each building and how to find your way around the spaceport.
- Chapter 12 gives a specific breakdown and description of how to utilize the various buildings for scheduling and launching missions.

- Chapter 13 offers various winning strategies for both the Soviets and the Americans from game experts of "Buzz Aldrin's Race into Space." It also reveals numerous tips, hints, and play styles.

Part III:

- Appendix A reveals secrets on how artificial intelligence operates and describes its effect in the game. It also gives a list of all space events so you can avoid potential pitfalls and prepare for bonuses.
- Appendix B is a brief history and overview of how "Buzz Aldrin's Race into Space" became a reality. From the conception of the board game to the development and publication of the computer game, the game's designer, Fritz Bronner, gives various insights into and comments on both the game's design and the board/computer game industry.

A SPECIAL OFFER TO ORDER
LIFTOFF!
The Original Race to the Moon Board Game

Created by Fritz Bronner, Designer of the Board Game and the Computer Game,**BUZZ ALDRIN'S RACE INTO SPACE!**

STELLAR REVIEWS!

"LIFTOFF! is technically accurate and educationally stimulating...a game for the nineties!"

—National Space Society

"(LIFTOFF!) is an excellent game of strategy and planning, and quite educational as well. An excellent choice for starting kids on gaming."

—Fire & Movement Magazine

"(LIFTOFF!) offers new challenges...refreshing change of pace...it's fun!... simpler than most war games...unusual mechanics...excellent production quality....in the spirit of movies such as MAROONED!...I like games such as LIFTOFF!...LIFTOFF! is out of this world!"

—Wargamer Magazine

"We are fascinated with this game!"

—Los Angeles Daily News

"(LIFTOFF!) Absorbing!...Rings true!...is probably the way NASA managers feel!

—Final Frontier Magazine

ONE TO FOUR PLAYERS CAN PLAY

PART I

Space History

CHAPTER 1

The Evolution of Rocketry

Today, powerful rockets routinely send enormous payloads of hardware into space. Long before mankind had the technological means for space travel, it was imagination that supplied the spacecraft and energy. Perhaps it was tens of thousands of years ago, when early man observed sparks from the cooking fire rising into the night sky, that the seed of traveling to the heavens was planted. This dream would not lie dormant forever...

A Brief History of Rocketry

Rockets, in their crudest form, have existed for nearly a millennium. The first historical records about rockets are found in ancient Chinese writings. These early records also show that rockets have been used for warfare virtually as long as their concept has existed. When the Mongols laid siege to the town of K'ai-fung-fu in the year 1232, the defenders used crude gunpowder-filled tubes attached to arrows to repel the invaders.

Early rocket technology migrated to the Middle East where in the year 1280, Syrian writer al-Hassan-al-Rammah described the making of gunpowder and rockets in "The Book of Fighting on Horseback and With War Engines." He referred to these missiles as "Chinese arrows." Italian records from the year 1379 also describe the military adaptation of rocketry when during the siege of Chiozzia a black powder rocket was used to incinerate the town's defending tower.

> *"The wise man looks into space and does not regard the small as too little, nor the great as too big, for he knows there is no limit to dimensions."*
>
> — Lao Tse, Ancient Chinese Philosopher

Western civilization largely ignored the rocket as a weapon because gun technology soon outstripped the rocket in both range and accuracy. The British, however, encountered an effective use of rocketry in the eighteenth century when they clashed with Indian ruler Hyder Ali at the Battles of Seringapatam in 1792 and 1799. The Indians used thousands of troops whose specialty was launching five-pound (three-kilogram) bombardment rockets.

Some of these Indian rockets were carried to England where in 1804, artillery officer William Congreve examined them and improved their design. Within a year, Congreve produced an 11-pound (seven-kilogram) bombardment rocket with a range of nearly one mile (1,800 meters), and soon after, a 15-pound (nine-kilogram) rocket with a range of just over one and one-half miles (2,700 meters). These were successfully used against the French during the Napoleonic Wars and against the Americans in the War of 1812. After this, rocket artillery battalions were a standard part of European armies.

Rocket research saw its birth in 1810 when Englishman W. Moore published his mathematical studies titled, "On the Motion of Rockets." Another Englishman, William Hale, improved on the Congreve rocket by introducing spin stabilization in 1843. Hale eliminated the stabilizing stick and passed the rocket's exhaust through vents drilled at slight angles. By 1865, Hale spun his rockets by introducing curved vanes into the exhaust stream.

The earliest concepts of manned rocket travel were devised in the late nineteenth century. The Russian inventor Nikolai Kibalchich conceived of a man-carrying craft propelled by a

> *"Ideals are like stars, you will not succeed in touching them with your hands. But like seafaring men on the desert of waters, you choose them as your guides, and following them you will reach your destiny."*
>
> — Carl Shurz, European Immigrant 1859

series of gunpowder explosions. Unfortunately, the young inventor never saw later adaptations of his design as when he created them in 1881, he was in prison awaiting execution for the murder of Czar Alexander II.

> *"This is a very difficult business...it will require knowledge, devotion, and will, and many long years, maybe a whole lifetime."*
>
> — Konstantin Tsiolkovsky, regarding the development of spaceflight

Another visionary Russian foresaw the true future of rocketry. As early as 1883, Konstantin Tsiolkovsky wrote articles on spaceflight, accurately describing the state of weightlessness and artificial Earth satellites. He also published mathematical evaluations of the principles of rocket flight in 1898. Later, he realized the advantages of liquid fuels and the need for multi-stage rockets. Tsiolkovsky, shown in Figure 1-1, even foresaw liquid hydrogen as the most practical rocket fuel. Tsiolkovsky's theoretical work predicted all the principles used in modern day space boosters.

The United States

In the United States during World War 1, Robert Goddard perfected a small tube-launched artillery rocket which eventually became the very effective bazooka.

Figure 1-1. Russian rocket visionary Konstantin Tsiolkovsky

> *"It is difficult to say what is impossible, for the dream of yesterday is the hope of today, and the reality of tomorrow."*
>
> —Robert Goddard 1904

A photograph of Goddard is shown in Figure 1-2. In 1919, he published his theoretical work, "A Method of Reaching Extreme Altitudes." Two years later, he began pioneering development on liquid-fueled rocket engines. On March 16, 1926, from a field not far from his home in Auburn, Massachusetts, Goddard launched the world's first liquid-fueled rocket. Goddard's design used liquid oxygen and gasoline. It reached a speed of 62 miles (100 kilometers) per hour and landed 170 feet (56 meters) away from the launch stand. No one realized it, but that first small step would lead to the giant Saturn 5 Moon rocket in only 41 short years.

Goddard continued his rocket research from a launch site in New Mexico where in 1935, his gyroscopically stabilized missiles reached near supersonic speeds. In the course of his rocket development, Goddard applied for over 200 patents related to rocket design. When *Apollo 11* lifted off for the Moon three decades later, it incorporated every one of Goddard's patented design principles.

Figure 1-2. Robert Goddard and his liquid-fueled rocket

Europe

In Europe, Hermann Oberth, a Rumanian living in Germany, was also attracted to the concept of liquid-fueled rockets. After he published his pioneering "Rocket Into Planetary Space," Oberth's followers founded the German Society For Space Travel, known as the "VfR," and began experimenting with rocket technology. The Society's Johannes Winkler became the second person to launch a liquid-fueled rocket with a short three-meter test hop on February 21, 1931. Three weeks later, the same liquid methane powered rocket attained a height of 2,000 feet (600 meters). Joining in the experiments was an 18-year-old named Wernher von Braun.

> *"I want to help turn the wheels of progress."*
>
> —Wernher von Braun, as a small boy

Launching their creations from the "Raketenflugplatz," or the Rocket Flying Field in Berlin, the VfR soon achieved altitudes in excess of one-half mile (one kilometer). However, as the world-wide economic depression dried up funds and forced the experiments to cease, von Braun and his associates approached the German Army about sponsoring their rocket experiments. This effort changed the course history.

After reviewing von Braun's experiments, army officials realized rocketry fell outside the scope of the Treaty of Versailles which prevented Germany from rearming after World War I. A special section of the Army Weapons Department was created under the guidance of Captain Walter Dornberger for the development of large liquid-fueled rockets. Wernher von Braun, who was still studying for his doctoral physics degree, was placed in charge of rocket development. By 1934, von Braun's small research team was firing advanced design A-2 rockets to altitudes of one and one-half miles (2.5 kilometers).

These initial tests led to army approval for the development of the six-ton (12,700-kilogram), 46-foot (14-meter) long V-2 ballistic missile capable of lofting a one-ton warhead 200 miles (320 kilometers). Von Braun's research team was moved in 1937 to the new secret German rocket base at Peenemunde on the Baltic coast. Here, his fellow rocket researchers from the VfR joined him and their efforts produced the first successful V-2 launch on October 3, 1942. The V-2 rocket is shown in Figure 1-3. The missile was the first rocket to exceed the speed of sound as it ultimately reached an altitude of 53 miles (85 kilometers). After the success, a jubilant Dornberger proclaimed, "Today, the spaceship was born!"

Figure 1-3. The much feared World War II V-2 rocket lacked accuracy

Soviet Union

Rocket research also flourished in the Soviet Union. In the 1920s Friedrich Tsander, a follower of Tsiolkovsky, developed primitive liquid-fueled rocket engine designs. In 1930, he perfected a working motor which ran on gasoline and compressed air. Tsander established the Group for the Study of Rocket Propulsion (GIRD), where he was joined by a young engineer named Sergei Korolev. With Korolev's assistance, the first liquid-fueled engine using liquid oxygen and gasoline was developed. Just days after Tsander's death in 1933 from typhoid fever, GIRD launched the first Soviet all-liquid-fueled rocket, attaining an altitude of almost two miles (three kilometers).

> *"...The Earth is the cradle of the mind, but man cannot live in the cradle forever."*
>
> — Konstantin Tsiolkovsky 1895

Like the Germans, the Soviet military also became interested in the development of rocket technology and took over the amateur rocket society. Creating the Rocket Science Research Institute (RNII), the Soviet military employed the brightest rocket engineers such as Korolev and Valentin Glushko to develop weapons systems.

In the Soviet Union, however, the development of rocket technology went terribly wrong in 1937. The RNII came under the wing of the Soviet Chief of Armaments, Marshal Mikhail Tukhachevsky. Through German sabotage and the dictator Stalin's paranoia, Tukhachevsky was falsely accused of being a Nazi spy and was executed along with his entire staff. The RNII, being associated with Tukhachevsky, also became suspect by Stalin, and all its members were arrested and imprisoned. From 1937 to 1945, the bulk of Soviet aeronautical and research engineers remained imprisoned by Stalin, including Sergei Korolev, the man who a decade later would develop the first Soviet intercontinental ballistic missile (ICBM).

> *"We must master the highest technology or be crushed!"*
>
> —Vladimir Lenin 1919

In Germany meanwhile, von Braun's research team continued rocket development as best they could under wartime conditions. Von Braun himself had to endure such obstacles as arrest by the SS on trumped up treason charges, while Peenemunde came under fierce bombardment by British bombers.

The End of World War II

In 1945, as the Red Army approached Peenemunde and its capture was imminent, von Braun polled his engineers to see whom they wanted to surrender to. With only one dissenting vote, all decided to seek the Americans. Von Braun, along with five thousand employees and family members left Pennemunde for the safety of Bleicherode in the Hartz Mountains. At one point, the rocket scientists even feared for their lives as they thought the SS would liquidate them rather than allow their capture. The Soviets occupied Pennemunde and captured 12,000 tons of von Braun's technical equipment, but the brain power behind the missile development program surrendered to the Americans in May of 1945.

> *"My name is Magnus von Braun. My brother invented the V-2. We want to surrender."*
>
> —Magnus von Braun, May 2, 1945

After von Braun's team was in American custody, an amazing clandestine operation was begun which would forever change the course of rocket history. On July 19, 1945, an American Intelligence project known as "Overcast" was created. This involved the selection of 127 top German rocket scientists to continue their research in the United States. As the cards of those selected were marked with a paper clip, the effort gained the now famous name of "Operation Paperclip."

By September of 1945, von Braun was in the United States. The rest of his Paperclip team joined him at Ft. Bliss, Texas, by December. The basis of future American rocketry was then established at the nearby White Sands Proving Grounds in New Mexico. From here, captured V-2s were fired for military and scientific tests over the next six years.

> *"We can see no more clearly all the utility and implications of spaceships than the Wright Brothers could see air transports circling the globe."*
>
> —RAND Corporation space outlook report May 2, 1946

The Cold War

The Korean War pushed American rocket development into high gear. As tensions with Communist countries heightened, von Braun and most of his research team moved to a new missile development center at the Redstone Arsenal in Huntsville, Alabama. Here, the team perfected the Army's first large ballistic missile, the Redstone, which first flew from Cape Canaveral, Florida, on August 20, 1953.

In 1955, von Braun proposed using the Redstone as the basis for a program to orbit a small satellite. However, his Project Orbiter was turned down and the task of developing the first American satellite was given to the Navy's Vanguard satellite project.

In the late 1950s, further Cold War escalations raised the priority of long range missile programs. The Army and Air Force gave their Atlas, Jupiter, Thor, and Titan missile development a near crash program status.

Meanwhile, the Soviet Union was not idle. Rocket designer Sergei Korolev, after being freed from prison, led the perfection of postwar Soviet rocketry. Although the Soviets had the services of hundreds of former German rocket engineers, they relied primarily on their own designs for missile development.

Korolev quickly improved on the design and range of the V-2, producing an all-Soviet version of the rocket called the "Pobeda," or Victory. In the early 1950s, he also completed the first Soviet medium range ballistic missile and gained approval for the development of the first Soviet ICBM, the R-7. Flight testing of the R-7 began in early 1957, with the first full intercontinental range flight coming in August of that year.

> *"Rockets are weapons and science."*
>
> —Sergei Korolev

With the performance of the R-7 proven in the summer of 1957, Korolev received permission from Soviet Premier Nikita Khrushchev to attempt the launch of an Earth satellite using the booster. By October 4, 1957, all hardware was ready and *Sputnik 1*, the first artificial satellite of the Earth, was launched from the ICBM test range at Baikonour. With the Russian satellite circling overhead, the Space Race between the U.S. and USSR was born.

> *"The conquering of space has begun. We can be proud that it was begun by our country."*
>
> —Sergei Korolev, After the launch of *Sputnik 1* October 4, 1957

The United States, fully expecting to be the first to orbit a satellite with its Vanguard program, was shocked at the appearance of a Soviet satellite in the sky. *Vanguard*, America's response to *Sputnik*, is shown in Figure 1-4. When *Vanguard* vanished in a spectacular explosion during liftoff in December of 1957, public outcry spurred a congressional inquiry about the status of the American missile and space program.

The job of quickly launching an American satellite to answer the Sputnik challenge fell to von Braun. Reactivating his mothballed Project Orbiter equipment, he promised to have an American satellite in orbit within 90 days. On January 31,

Figure 1-4. *Vanguard*: The press called it "Flopnik"

1958, he delivered the *Explorer 1* satellite to a grateful America. The Redstone rocket and its payload, *Explorer1*, are displayed in Figure 1-5.

The Soviets continued to press ahead with a series of space exploration spectaculars which culminated with the first manned orbital flight on April 12, 1961. The American answer to Yuri Gagarin's pioneering trip around the globe came in May 1961, when President Kennedy proclaimed that the United States would land a man on the Moon and return him safely by the end of the decade.

America took up Kennedy's bold lunar challenge in the 1960s with a politically driven "Beat-the-Russians" national fervor. The Apollo lunar landing project grew from that effort as did the development of the family of huge Saturn rocket boosters. The resources of the entire nation were harnessed as NASA aimed for the Moon.

Thus, in the summer of 1969, *Apollo 11* stood ready at the Kennedy Space Center in Florida to take men from Earth to the surface of another world. In the 24 short years since von Braun's final V-2 rocket had ascended from Pennemunde, the dreams of generations of visionary thinkers were at last fulfilled on July 20, 1969, when Neil Armstrong and Buzz Aldrin set foot upon the Moon.

Figure 1-5. *Explorer 1*: At the launchpad

How Dreamers Created a Science

For eons mankind watched the night sky. One object, the Moon (shown in Figure 1-6), always brightly stood out against the black velvet star field. Early societies attempted to explain this companion as a great spirit or a powerful god. The lure of the Moon deity persisted until humanity began to question the universe and their relationship within it.

During mankind's quest for knowledge, questions arose about our place in the universe—deep questions about who we are and what our purpose is. In an effort to understand ourselves, the universe we live in first had to be identified and defined. These observations would turn into patterns of behavior which are assembled into rules that accurately predict results. A series of rules form the basis of science. Science and fiction always moved in harmony with each other throughout the ages. New discoveries quickly become routine devices in fictional stories. Imaginative writers often foresaw the path of science.

Figure 1-6. The Moon, the focus of mankind's attention throughout the ages

The Ancients

The concept of space travel could not even be conceived until certain discoveries were made. While the Babylonians and the Chinese precisely predicted the movements of the Moon and the planets, they could only see these images as powerful gods. The concept of planets revolving around the Sun was beyond their own beliefs. Any ideas opposed to the status quo would not bode well within these societies.

Even the ancient Greeks had the same notion of the night sky. The Greek word "planet" meant wanderer. It wasn't until 640 BC on the island of Cos, a school of astronomy was founded and the cycles, "saros," of the Moon were taught. These saros could predict eclipses and quickly changed the perception of the Moon from a spirit to that of a mere object. In the tale of the Odyssey, the Moon was described as a shield that reflected the land and water of the Earth.

New generations of Greek Astronomers, such as Pythagoras of Samos, identified the ecliptic and developed the concept that the Earth floats in space. In 360 BC, Heraclides of Pontus (a disciple of Plato) predicted that Mercury and Venus revolved around the Sun. By 280 BC, Aristarchus of Samos combined all of these concepts into what is now called the Copernican system, a solar system of planets revolving around a sun. A few decades later, Eratosthenes of Cyrene accurately measured the size of the Earth.

In a few thousand years, mankind had redefined his universe and was ready to explore it. In 160 AD, the Greek satirist Lucian of Samosata wrote a story about a voyage to the Moon. His book, *Vera Historia* (True History), told of seafaring travelers caught in a storm and lifted out of the sea. For seven days and nights, the ship traveled to a destination unknown to its occupants. On the eighth day, they landed in a great county in the air, similar to a shining island. It was the Moon. The travelers were met by warriors atop three-headed giant vultures. The story tells of preparations that are being made for a great war between the inhabitants of the Moon and the Sun.

Lucian wrote another Moon adventure in which Icaromenippus intentionally travels to the Moon. By combining the wings of an eagle and a vulture, the explorer learns to fly and reaches the Moon. After exploring the Moon, he travels to heaven and the stars. The Gods resent the intrusion and order Mercury to escort Icaromenippus back to Earth. His wings are taken away, so that he can never visit the Moon and the stars again.

These notions of the Earth as a mere satellite was disturbing to some of the Greek philosophers. Other than observations, there were measurements or rules. Kepler's and Newton's laws would not be realized for nearly two thousand years. An alternate view was developed which put the Earth back at the center of the

Universe. It was Aristotle's great influence and belief that no other worlds existed which froze any further discussion or any scientific progress challenging this belief.

Past the Greeks and through the rise and fall of the Roman Empire, the "rules" of the universe remained chiseled in stone. For the next fourteen centuries, the new Christian philosophy would forbid any alternative thinking as heretical. For centuries, observations of the night sky were noted and cataloged by men and women of all faiths. No one dared to question the inconsistencies.

The Age of Discovery

Three books and the telescope began what was called the astronomical revolution. Nicholas Copernicus wrote *On the Revolutions of Celestial Orbs* in 1543. The second book, entitled *On the Motions of Mars*, was by Johannes Kepler and was published in 1609. Galileo Galilei invented the telescope and recorded observations of the heavens, which resulted in his book *Messenger of the Stars*. Each book was placed on the Index Liborum Prohibitorum (Index of Forbidden Books).

Copernicus' and Kepler's books explained the mechanism of the solar system while Galileo's telescope visually confirmed the new order. Through his eyes, Galileo traveled across space and discovered detailed features of the Moon. He described the lunar surface as "rough and uneven... vast protuberances, deep chasms." Kepler and Galileo corresponded with each other about these new discoveries. Kepler suggested the possibility of traveling to the Moon in "celestial boats with golden sails." Kepler wrote "Somnium" (Dream), a fictional story of a voyage to the Moon. When the Earth's shadow crosses the Moon, demons can transport humans in the tunnel of darkness to the Moon. To survive in the vacuum of space, humans are given an anesthetic and breath through moist sponges.

The seventeenth century was time of migration to the New World, the Americas. Tales of new civilizations and new lands to be explored gave speculation to the possibilities of intelligent societies on the Moon and other planets. New methods of travel would have to be devised in order to visit these worlds.

Francis Goodwin (who eventually became a bishop in England) wrote "The Man in the Moon" (1638), a tale of Gonsales, a Spaniard who inadvertently traveled via bird power to the lunar surface. Another future bishop, John Wilkens, wrote a story based on the scientific facts of the era. He surmised that if one could attain an altitude greater than twenty miles away from the Earth's influence, the remainder of the journey to the Moon would be relatively easy. Ironically, this simplistic view is quite accurate. The most difficult part of traveling to the Moon is leaving the Earth's gravity well. Wilkens believed that someday a "flying chariot" would be constructed.

It was none other than the Frenchman Cyrano de Bergerac, a satirist, playwright, philosopher, and frequent swordsman who wrote the next major work of space travel, *Histoire Comique des Etats et Empires de la Lune*. His method for leaving the Earth is vials of captured morning dew. The more vials of dew, the faster and higher one travels. To lower himself, the pilot simply breaks a vial. When too many vials are broken, the poor pilot ends up in French-speaking Canada.

In 1686, *Conversations on the Plurality of Worlds*, a popular astronomy book by Bernard de Fontenelle, took Europe by storm. It offered a fascinating view of life conforming to the features or surroundings of the planet. Fontenelle had descriptions of inhabitants on every planet except on the Moon, due to the rarefied air.

The Age of Invention

The authors of this era delved into more technical detail, the modes of space travel along with measurements, distances, speeds, methods of construction, and materials.

Voyage to the Moon (1703) by David Russen included Moon-springs as the mode of travel. With the use of well-tempered steel, springs, cords, and pulleys, he proposed that a man could safely travel to the Moon. Another writer, Daniel Defoe, describes a ship in the shape of a chariot, with two vast wings fifty yards in length, feathers pressed so tight that air cannot pass. The ship was driven by flames and fire.

The philosopher Voltaire satirizes the insignificance of man by having a giant 120,000-foot-tall extraterrestrial named Micromegas, who travels on comets, sunbeams, and by gravitation through the universe and visits our solar system. After a philosophical discussion with the head scientist of Saturn, the two explore the rest of the solar system and are astonished with the discovery that there is intelligent life on Earth.

Louis-Guillaume de la Follie wrote *Philosophe sans Prétention* (1775), a story of an invention by an inhabitant on Mercury. A young scientist constructs a flying machine powered by electricity, but is denied admission to the Science Academy on the grounds that it is not possible to fly. To prove his point, he flies to the planet Earth. Unfortunately, an irreplaceable part of the machinery is broken and the inventor is trapped on the third planet.

The Age of Steam

The nineteenth century was the birth of the Industrial Revolution. Science and technical know-how were rapidly increasing and changing the world.

American author George Tucker's *A Voyage to the Moon with some Account of the Manners and Customs, Science and Philosophy of the People of Morosofia and other Lunarians* (1827), described a six-cubed copper vessel with its edges rounded off, propelled by the rare metal lunarium which behaved like anti-gravity. Eight years later, a student of Mr. Tucker, Edgar Allan Poe, also wrote of a lunar flight in a cigar-shaped balloon in "The Unparalleled Adventure of One Hans Pfaall." Heavily in debt (similar to the author), the character flees to the Moon in order to avoid his creditors.

The greatest and most successful scientific sham was published in 1835. Richard Locke of the New York Sun had the largest readership in the world. The paper had a circulation of over 19,000 readers, two thousand more than the London Times. He wrote a series of articles falsely reporting that the noted astronomer Sir John Herschel had observed abundant plant life and a bat-like civilization on the Moon. When the "Moon-Hoax" was revealed, the general readership was not amused.

By the mid-1850s, the vacuum conditions of space were generally well known. Creative authors utilized current technologies to bridge the void. With the invention of submersibles or submarines, fictional spacecraft designs included airtight cabins with compressed air tanks. Shielding and insulation were created to protect occupants from the cold temperatures and meteorites. Experience, borrowed from Arctic expeditions, included food and provisions compressed into cans for long duration and more efficient storage.

> *"The chess board is the world, the pieces are the phenomena of the universe, the rules of the game are what we call the laws of nature."*
>
> —Thomas Huxley,
> English Biologist.

In the year 1865, five novels of Moon voyages were published. The most memorable and famous was *From the Earth to the Moon* by the French author Jules Verne. Verne approached his writings with gritty realism. His characters are not romantics, but ambitious achievers who would create a better world. The setting of the story is in the United States with all of the main characters as former Civil War gunnery experts who form the Gun Club. They devise a plan to send a man around the Moon. The method of spaceflight is also striking. A giant cannon, the Columbiad, fires a twelve-by-nine foot bullet-shaped cabin, made of solid aluminum. The interior, shown in Figure 1-7, provides all of the comforts including gas lighting, ventilated air, preserved foods, and reclined seat cushions to protect the crew from the acceleration during launch. To ensure safety, the designers even perform a test launch with animals in a scale-down version of the ship.

Figure 1-7. The Jules Verne spaceship

Five years later, Verne wrote a sequel, *Round the Moon*. The launch takes off from the Florida coast. Three astronauts are accelerated into space towards to Moon. During launch, the crew loses consciousness due to the G-forces. None of the crew recalls hearing the sound of the blast until they realize that they were traveling faster than the speed of sound. After avoiding a meteorite and landing on the Moon, the spaceship returns to a splashdown in the Pacific Ocean. The USS *Susquehanna*, an American Corvette, recovers the three-man crew. Verne's vision of the future was astonishingly accurate, as you can see in Figure 1-8. Only a hundred years later, American astronauts would routinely liftoff from the shores of the Florida coast into space and splashdown for recovery in the Pacific Ocean.

In 1877, Giovanni Schiaperelli, an Italian astronomer, turned all attention away from the Moon to the red planet Mars. He announced that he had seen "canali," which translates to "grooves" or "channels." Throughout the world, the press announced that canals had been seen on Mars. Canals implied constructed waterways by sentient creatures.

Three decades followed with enthusiasm and speculation of the life on Mars. American Percival Lowell wrote numerous articles sensationalizing a dying civilization on the red planet. Soon newspapers and magazines caught the bug and published hundreds of serialized articles on life on Mars.

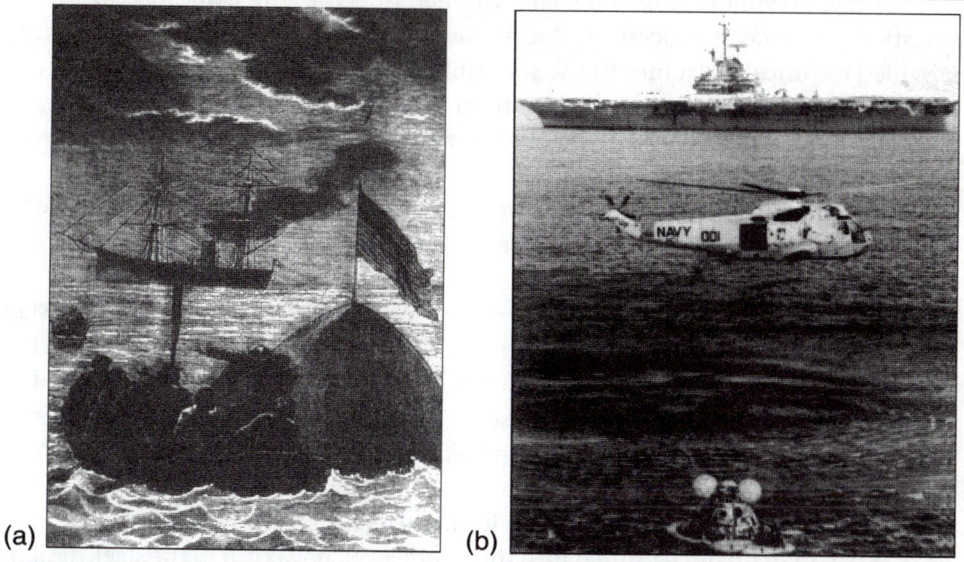

Figure 1-8. An accurate vision of the future: (a) a Jules Verne recovery; (b) an Apollo recovery

The Twentieth Century

H.G. Wells's *First Men in the Moon* (1901) created a sensation about an insect society that lives under the lunar surface. The method of travel was by anti-gravity.

In 1912, American writer Edgar Rice Burroughs continued the fascination with Mars with serialized installments about John Carter of Mars. This lead to numerous imitators in the pulp magazines.

After World War I, American Robert Goddard began testing and launching liquid-fueled rockets. With solid mechanical skills and many failures, Goddard patiently constructed and launched the world's first liquid-propellant rocket on October 16, 1926. The ten-pound rocket flew 184 feet. This small achievement was the beginning of the age of rockets.

The late twenties saw the publication of two competitive sci-fi magazines, *Amazing Stories* (1928) and *Astounding Stories* (1930). The success of these magazines eventually led to hardcover printings of the works of Edwin Balmer, Philip Wylie, H.G. Wells, and Edgar Rice Burroughs. This was the beginning of the golden age of science fiction.

In the 1929 movie *Frau im Mond*, German director Fritz Lang was the first to realistically portray a spaceship. Rocket designer Hermann Oberth meticulously provided technical assistance in the spaceship design and how the crew would live and eat in zero-G conditions. A scene from the film, which translates to "Woman on the Moon," is depicted in Figure 1-9. A whole slew of motion pictures followed suit with various rocket designs.

The thirties saw the success of the Flash Gordon series and comic book hero Buck Rodgers on the big screen. Both flew spaceships with art deco egg-shaped designs.

After World War II, the V-2 rocket design forced a revision of spacecraft design in the movies and science fiction. The roar of a V-2 launch dramatically opens the movie *Destination Moon*. The descendant of the Sanger space plane is portrayed in the movie *When Worlds Collide*. After a series of successful articles by von Braun and stunning artwork of Chesley Bonestell in *Collier's* magazine on "How to Get to Mars," von Braun collaborated with Walt Disney on a visually imaginative film series. This raised the public's excitement over the future of space travel.

When man finally ventured into space, however, it was not in the sleek designs of winged spaceships. The Mercury and Vostok spacecraft, on the other hand, were efficient and functional in design. The movies also began to follow suit.

Figure 1-9. Walking on the Moon: A scene from *Frau im Mond*

Probably the most influential film to realistically portray future space travel was Stanley Kubrick's *2001: A Space Odyssey*. The science fiction writer Arthur C. Clarke and Kubrick co-wrote the screenplay and worked closely with NASA designers and futurists on look and design of space stations, lunar habits, and interstellar spacecraft.

Once again, science fiction had momentarily leaped ahead of the present. Nearly all of the technology envisioned in the film *2001* exists today. Now, only the leadership is needed to fulfill these goals. Who knows what future dreams will someday turn into realities?

The Rocket Men

Throughout history from the ancient Chinese to Galileo and Congreve, there were countless individuals who advanced the knowledge of rocketry. At the turn of the century, there were the practical designs of rockets by Tsiolkovsky. In the 1920s Oberth made additional refinements and actually constructed rockets. But it was not until Goddard, von Braun, and Korolev, that rockets would be transformed from small devices into awesome giants capable of carrying tons of payloads. These three men developed the modern rocket into a powerful weapon of war and a transport to the stars. Still today, rockets are a double-edged sword of progress: a weapon of death and a useful tool for science and exploration.

Robert H. Goddard: Father of Modern Rocketry

On March 16, 1926, a new age began with the successful launch of the world's first liquid-propellant rocket. At his Aunt Effie's farm in Auburn, Massachusetts, designer Robert Goddard (shown in Figure 1-10) saw the ten-pound rocket reach an unimpressive height of 184 feet. What was more important than altitude, however, was that his liquid-propellant design had worked. Altitude records would be set on another day.

As a boy, Goddard dreamed of space travel. His interests and desires lead him to become a physics professor and to study the practicality of rockets. From the ancient Chinese up to World War I, all rockets had been propelled by powder fuels. Goddard tested the efficiency of the U.S. Navy's current inventory of rockets which used a double-based mixture of shotgun powder.

As early as 1909, Goddard foresaw the possibilities of liquid-propellant rockets but lacked the necessary funds to begin experiments. He continued developing his rocket design skills with cartridge-fed solid-propellant rockets.

Figure 1-10. Robert Goddard

After many misfires, Goddard was determined to pursue a propellant design with liquid oxygen and gasoline.

Obtaining a small grant from the Smithsonian, Goddard constructed fuel and oxidizer tanks, combustion chambers, and pumps. This was unknown territory; no one had ever dealt with the problems of how to keep rocket motors cool or how to properly store liquid oxygen. Goddard's patience and persistence finally paid off with the successful launch in 1926.

A few years earlier in 1919, he published a paper entitled "A Method of Reaching Extreme Altitudes." When the press got hold of it, it created a small sensation. The Goddard paper was very technical and cautious, but the press summarized a portion of it regarding the theoretical possibility of a rocket reaching the lunar surface and exploding flash paper to announce its arrival. Goddard was quickly labeled the "Moon-man," which might help explain his secretive behavior of zealously guarding his results. It was not until ten years later in 1936 that Goddard would published a paper on the results of his first historic launch.

Soon his rockets were reaching several hundred feet, and causing a commotion with local residents with the occasional small grass fires in the surrounding fields. With the help of Charles Lindberg, Goddard was able to obtain additional financial assistance through the Guggenheim Fund. In 1930, he closed up shop and moved his small team to Roswell, New Mexico. The climate there offered year-round good weather for launchings, and the sparse population ensured better security.

He continued testing for several years, carefully improving the designs of his rockets. He developed the gyroscope-stabilizer which greatly increased the rocket's altitude performance to the height of 4800 feet. Unfortunately, the United States paid very little attention to Goddard's results. Unknown to Goddard, most of his patents were meticulously studied by the German rocket program and assisted its scientists in the development of the V-2. It wasn't until after Pearl Harbor that the Americans sought Goddard's services. Ironically, he wasn't involved in rocket weapons development but was assigned to the Navy to assist in the development of take-off rockets for seaplanes. On August 10, 1945, Goddard tragically died after a throat operation.

Goddard just missed witnessing the fruition of his life's work. Had he lived a few years longer, he would have seen the beginning of the space age. After the war, only a few miles away from his test facility in New Mexico, the Army test-launched captured German V-2s. Most of the major components of all rockets used today are based on the patents and designs of Robert H. Goddard.

Wernher von Braun: Designer of the Moon Rocket

The man destined to be the guiding spirit that lifted the American space program to prominence after the initial shock of *Sputnik*, was born into German aristocracy on March 23, 1912. Wernher von Braun was raised on the ancestral family estate in Silesia. Young Wernher took an interest in music and the serious study of languages, but curiously did poorly in mathematics. After switching to a more challenging school, he soon became a top scholar.

Von Braun's fascination with space travel was triggered after his mother gave him an astronomical telescope as a Lutheran confirmation present. Wernher soon read a magazine article which described an imaginary trip to the Moon. Von Braun remembered, "It filled me with a romantic urge. Interplanetary travel! Here was a task worth dedicating one's life to!"

In high school, von Braun heard about the rocket experiments of Fredriech Tsander and Robert Goddard. Upon graduation he worked as an apprentice at the locomotive shop of the Borsig Engineering Works in Berlin. There he met Willy Ley, a fellow rocket enthusiast. In 1929, Ley introduced von Braun to Hermann Oberth, the head of the Society for Space Travel. Von Braun soon became a dedicated member of the Society, engaging in the first crude rocket experiments.

By 1932, the Society's experiments attracted the interest of the German military. That year, von Braun became a civilian employee of the German Army. He used military facilities for experimental work necessary in his 1934 Ph.D. thesis on rocket propulsion. World War I treaties limited the German Army's ability to rearm after the war, but said nothing about rocketry! Thus in 1936, Adolph Hitler, through the Army, assigned the group headed by von Braun to create the V-2 ballistic missile.

As World War II engulfed Europe, progress continued on the development of the V-2 at the German rocket base at Peenemunde. By October 1942, von Braun had successfully launched test versions of the V-2. The Army ordered the rocket into mass production, and in 1944, it began raining warheads down upon Antwerp and London.

Although von Braun succeeded in perfecting the V-2 for the German military, life in a dictatorship proved harsh. Among other false charges, von Braun was accused of treason for clinging to his ideas of space exploration using his rocket creations. Only direct intervention by Hitler saved von Braun from execution by the SS!

Following the German defeat, von Braun and his rocket team surrendered en mass to the American Army. Operation Paperclip reunited von Braun and his colleagues with V-2s in the deserts of New Mexico. He and his team were transported to the United States where they were employed for American missile research and eventually offered citizenship. The first American V-2 was launched on June 28, 1946. The preparation for that launch is shown in Figure 1-11.

Initially, von Braun was an illegal alien as he was brought to the United States without State Department approval. To remedy this sticky legal problem, von Braun was taken to Mexico, where he reentered the United States from Juarez. When asked by border agents if he was from Mexico, von Braun answered, "Ja!"

In 1947, von Braun was allowed to return to Bavaria to marry Maria Quistorp, his second cousin. The von Brauns settled at the newly developed Rocket Research

Figure 1-11. The first American V-2 is prepared for launch

and Development Center at Ft. Bliss near El Paso, Texas. Here, von Braun and his team engaged in test launches of his V-2 for scientific investigations.

At one point, von Braun's German war background was the focus of an FBI investigation. However, J. Edgar Hoover was convinced von Braun was not a Nazi and determined he was vital to the security of the United States in light of the escalating Cold War with Russia.

Although he had developed the V-2 war rocket for Hitler, von Braun's unyielding devotion to space exploration won him many admirers. Indeed, in 1949, in the country where 1,100 V-2s caused thousands of casualties, von Braun was extended the honor of becoming an Honorary Fellow of the British Interplanetary Society.

Politics and the Korean Conflict found von Braun transferred to the Army's Redstone Arsenal in Huntsville, Alabama, in 1950. Here, he engaged in the design of the Redstone ballistic missile which would later be pivotal in launching the first American satellites and initial manned spaceflights. This work would be the cornerstone of his large Moon rocket designs just a few years later. The team at the Army Ballistic Missile Agency in 1956 in Huntsville, Alabama, is shown in Figure 1-12.

The dream of spaceflight never left von Braun's thoughts, and he never overlooked a chance to preach the gospel of interplanetary travel. His dream was the conquest of Mars. Trips to the Moon were to be merely stepping stones to the planets. The dashingly handsome, aristocratic von Braun was a natural salesman

Figure 1-12. Von Braun and the army team: clockwise from foreground, Hermann Oberth, Eberhard Rees, Major General Tuftoy, Ernst Stuhlinger, and Wernher von Braun

for the exploration of space. His articles became popular in national magazines and his authoritative appearance on television inspired the spaceflight dream in younger generations.

Von Braun was also a natural leader and encouraged a team spirit in any enterprise. He made a special effort to make everyone, from theoretician to wrench-turner, feel needed in the project. In his eyes, no one was "too high" or "too low" to carry out the needed work. In that light, von Braun himself even made some special wrenches in a machine shop during a funding shortage.

Von Braun called for the orbiting of an artificial Earth satellite as early as 1955 using existing Army Redstone missile components. His plan was defeated, however, by the political decision to use the civilian Vanguard rocket to orbit as America's first satellite. When the Soviets politically humiliated America by orbiting a satellite first, von Braun demanded to be allowed to use his Army missile components to launch an American answer to *Sputnik* within 90 days. Approval was given and von Braun delivered America's first satellite, *Explorer 1*, on January 31, 1958.

After the creation of NASA, the Redstone Arsenal became the site of the George C. Marshal Space Flight Center in 1960, with von Braun as its director. Here, von Braun led the team that developed the family of Saturn rockets, the culmination of which was the Saturn 5 manned Moon rocket (shown in Figure 1-13), which carried Neil Armstrong and Buzz Aldrin to the Moon.

Figure 1-13. Von Braun's Moon rocket

After the Apollo manned flights to the Moon ended in 1972, von Braun left NASA to take a position with the Fairchild Corporation. Von Braun died of cancer on June 23, 1977. Through his career, Wernher von Braun had seen his childhood dream come true: the birth of rocketry, the perfection of spaceflight, and ultimately, manned travel to another world.

Sergei Korolev: Conqueror of the Cosmos

The creator of the Soviet space program remained totally unknown in the West until after his death in 1966. Known only as the mysterious "Chief Designer," the Soviet government kept him isolated and anonymous as Premier Khrushchev absurdly warned of CIA "abductions" or "assassination attempts." Only after his death was the world told that the builder of Sputnik, Luna, Venera, and Vostok was really Academician Sergei Pavlovich Korolev.

Korolev was born on December 30, 1907, in the Ukrainian town of Zhitomir. Later, as a boy living in Odessa, he became fascinated with aviation while watching seaplanes take off from the city harbor.

After graduating from the Bauman Higher Technical School, Korolev had an aviation career as both a designer and a test pilot. He soon became interested in the concept of spaceflight and at age 25, he met the "Father of Spaceflight," Konstantin Tsiolkovsky.

In 1931, Korolev assisted in the creation of the Group for the Study of Rocket Propulsion (GIRD). Here, Korolev helped develop the first successful Soviet rocket, the GIRD-09 which reached an altitude of one-quarter mile (400 meters) in August 1933.

In 1933, when GIRD was merged by the military into the Rocket Science Research Institute (RNII), Korolev was appointed deputy director with a military commission equal to that of a major general. During the 1930s, Korolev produced the largest Russian rocket yet, the ten-foot (three-meter) long, 660-pound (300-kilogram) thrust "Aviavnito." He also designed flying bombs and rocket-propelled gliders and planes.

> *"Your pyrotechnics and fireworks are not only unnecessary to our country but are actually dangerous!*
>
> —Secret Police interrogator to Sergei Korolev, 1937

In 1938, Korolev was falsely arrested for the crime of being an "accomplice of a spy." He was sentenced to life at hard labor in the Kolyma gold mines in Magadan, but with the arrival of World War II, the Soviet government had him reassigned to a special concentration camp for aircraft designers. After seven years in prison, he was released.

In 1946, Korolev was one of the Soviet observers sent to the German rocket development base at Peenemunde. While there, he stayed at Bleicherode in the abandoned villa of Wernher von Braun. Korolev revived the German V-2 production facilities and improved the V-2 by lengthening the missile ten feet (three meters) and increasing its thrust from 25 to 32 tons.

After developing a series of Soviet military rockets by 1957, Korolev had perfected the R-7 ICBM. However, Korolev remained more interested in spaceflight than in war rockets. Korolev knew the R-7 was a clumsy ICBM, but the rocket would be a marvelous space launcher. Eventually, the R-7 came to be known as the SL-1 and SL-3 space boosters, and also became the lower stages of the SL-4 and SL-6 boosters which are still in constant use today after 35 years!

> *"She really flew! She really flew! We didn't do all that work for nothing!"*
>
> —Sergei Korolev, after the first rocket success, August 17, 1933.

In August of 1957, Korolev was given permission use the R-7 to launch the first artificial satellite of the Earth. After six weeks of intense preparation Korolev's rocket and satellite sprang to life at 10:28 P.M., local Moscow time, on the night of October 4, 1957. Ninety-five minutes later after circling the globe, *Sputnik* passed over the launch site for the first time and the famous "beep-beep-beep" of its signals was heard.

After the huge political success of the first Sputnik, Korolev was charged with creating further space spectaculars for Premier Nikita Khrushchev's propaganda machine. In January 1959, *Mechta*, the first of Korolev's Moon probes, successfully left the Earth's gravity forever. By September of that year, Korolev's efforts had struck the Moon when *Luna 2* impacted upon Earth's natural satellite. A mere three weeks later, his *Luna 3* circled behind the Moon and relayed the first views of the hidden side.

On April 12, 1961, Korolev's dream of manned spaceflight came true when Yuri Gagarin rode his Vostok spacecraft into the history books by becoming the first man to orbit the Earth.

Korolev also designed the Soyuz manned spacecraft, which after a shaky start, would eventually prove so sound as to provide the basis for all Soviet manned spaceflight for the next 25 years.

At the peak of his career, Korolev's health failed him. While undergoing a surgical operation in January 1966, he suffered severe hemorrhaging and died on the operating table. He was 58 years old.

To honor Korolev's space achievements, he was given a state funeral and buried in the highest place of Soviet honor, the Kremlin wall. In addition, a large crater on the Moon's far side is named in his honor.

Tragically, Korolev's last great project, the N-1 manned Moon rocket, never achieved success. Four test launchings, beginning three years after his death, all ended in failure. Had fate been kinder to Korolev, the outcome of the political race to the Moon may have been in his favor. He possessed the gift of hypnotic persuasion and had a great air of authority. While he was not infallible, the progress of the Soviet space program between 1957 and 1965 is due in great part to Korolev's dedication to his ideal, the exploration of space.

CHAPTER 2

The Cold War And Space

After a long and bloody conflict, the Allied powers were victorious over Nazi Germany and the Japanese. Over a hundred million died from the devastation of World War II. As the final months of the war came to a close, a chill of mistrust developed.

Mistrust Between Allies

As the Allies approached victory at the end of World War II, the old habits of Soviet/American distrust began to reappear. When the Americans exploded two atomic bombs over Japan (see Figure 2-1), a dark cloud emerged from the Kremlin. At first, the Soviets applauded the invention of the bomb and its effectiveness on the Japanese. Only a few weeks went by, however, before there was criticism of the nuclear monopoly. The Soviets of course had been secretly trying to develop their own atomic bomb since 1942. Stalin would not tolerate a potential atomic weapons advantage by the United States. To the Americans facing a Red Army of nearly three hundred divisions, what appeared as a comfortable stalemate in fact created insecurity for Soviet dictator Josef Stalin.

The highest priority was given by Stalin to his scientists to close the nuclear gap. This was the beginning of the technological race between the two superpowers.

Figure 2-1. Atomic explosion

It consumed enormous resources, talent, and wealth on both sides. It was also a spiraling, intoxicating addiction to destruction and power. It is amazing that a portion of this struggle became a demilitarized race into space, emphasizing political prestige instead of weapons. But still, the specter of space will always loom in military minds.

In August 1949, the Soviets successfully triggered an atomic bomb. By now, the Americans had a huge bomber force that could deliver these weapons deep into the Russian motherland. This spurred and accelerated the growth of a rocket program by the Soviets. With the development of intercontinental ballistic missiles (ICBMs), the Kremlin could counter the American bomber threat and unleash nuclear terror on the American heartland.

> *"The launching of a stratoplane to the Moon and the creation of an artificial satellite of the Earth is a real possibility."*
>
> —A.N. Nesmeyanov, November 27, 1953

The United States responded with an increase in defense spending and an accelerated program for a fusion, or hydrogen bomb. By June 1950, with the Korean War underway, the ante was raised for both sides and the Kremlin responded by increasing troop strength to nearly six million armed combatants. In 1951, Presi-

> *"American Imperialists, who are possessed to be a mania of world domination, are threatening the entire universe and making extravagant plans for transforming the entire solar system into an American colony."*
>
> —Excerpt from Soviet magazine *Znanie Sila*, 1954

dent Truman approved the costly Air Force Atlas ICBM Program. The Soviets followed suit with the quick development of their hydrogen bomb. The Americans exploded their first test of an H-Bomb in November 1952, and Soviets tested theirs in August 1953.

The War Against Communism

In Korea, after General Douglas MacArthur successfully lead United Nations forces past the 38th parallel, he pushed his forces to within a few miles of the Chinese border. China intervened with 300,000 "volunteers" and overran the multi-national Allied troops and forced a hasty retreat. General MacArthur suggested to Truman the dropping of thirty to fifty nuclear bombs on China's industrial centers. Truman rejected that idea and fired MacArthur after the General campaigned his plan openly. General Omar Bradley had responded, "The wrong war, at the wrong place, at the wrong time, with the wrong enemy." Another factor might have been China's mutual assistance pact with the Soviet Union.

Anti-communist hysteria began to sweep the United States in the early 1950s, cresting on the wave of McCarthyism. The nation became paranoid about communists and citizens questioned each others' loyalty. About the same time, Stalin was considering a repeat of his infamous political purges of the 1930s, but Stalin's death stymied those plans.

Following the Second World War, an audience was building of newly formed nations. Many of the old empires and colonies were crumbling and new nations were opening as spheres of influence for the superpowers. Instead of just trying to impress their opponents, the superpowers played to a world stage. This made the stakes even higher. Vietnam moved away from French imperialism, only to be later dominated by the Americans. Soviet influence likewise loomed over the Suez Crisis and caused the French to withdraw. Encouraged by the Soviets, Egypt and Syria experimented with the United Arab Republic. Colonial Africa emerged as a patchwork of independent Third World nations.

By 1955, a standoff was reached in weapons. Eastern Europe saw the formation of the Warsaw Pact by eastern European communist nations as a response to the North Atlantic Treaty Organization (NATO) created by the Allies. The great superpower competition would move into the Third World, which was ripe for the taking. The method of winning influence was by proving technological superiority over one's ideological opponent. The violent suppression of the Hungarian Revolt in 1956 by the Soviets was a display to the world of the Soviet way of doing business.

Artificial Satellites: New Technolgies, New Horizons

Just as Truman was kept in the dark by Franklin Roosevelt about the development of the atomic bomb with the Manhattan Project, Stalin's successor, Nikita Khrushchev, also was unaware of the Soviet rocket program until Stalin's death. It was only when chief rocket scientist Sergei Korolev briefed the new Secretary General did Khrushchev learn about the new technology and the potential of sending aloft a satellite. It was this double-edged sword, both a deliverer of weapons and a scientific program with its accompanying propaganda that convinced Khrushchev to support Korolev's rocket program. By August 1957, successful launchings allowed the Kremlin to proudly claim it had a working ICBM. Only after this claim did Korolev have approval to develop a satellite program.

America was aware of the potential of Earth satellites long before the revelation to Khrushchev by Korolev. A forgotten study released by the RAND Corporation in 1946 gave an assessment on technology in the future (see Table 2-1). The report summarized rocketry as having a major scientific impact, and that the public reaction of a satellite would be as startling as the atomic bomb. Sadly, the political leadership of the United States quietly forgot the RAND report.

As early as 1948, the Pentagon was looking for military uses for Earth satellites. The problem with satellites, in the military's view, was that they were not weapons and would not add any value to national security. It was also determined that the launching of a satellite would not be secret since all the world could view it. It was also postulated that the Soviet reaction could be hostile and might consider a satellite launch as an invasion of its territory. Besides, it would have to be announced early or the public would view it with suspicion, and then it of course wouldn't be a secret anymore.

> *"The insane plans of the imperialists are doomed to failure. The science of our country, having spoken the first word on interplanetary flights, will also be the first to realize the long cherished dream of mankind."*
>
> —Excerpt from Soviet magazine *Znanie Sila*, 1954

Although technologically capable of launching a satellite as early as 1955, the United States military completely missed the practical considerations and stumbled into a technological Pearl Harbor after the launch of *Sputnik*. Like speculation about Franklin Roosevelt's knowledge of the impending attack on

	Soviets	U.S.
Scientific Earth Satellites	1957	1957
Lunar Satellites or Lunar Probes	1958-59	1958-59
Mars Probe	1958	1960
Lunar Soft Landing Probes	1959-60	1960
Venus Probe	1959	1961
Manned Recoverable Vehicles		
Capsule	1959-60	
Glide Vehicles	1960-61	1961-63
Manned Circumlunar Pass	1961-62	1962-65
Manned Lunar Landing	1965	1968

Table 2-1. The 1946 RAND Corporation Report of Future Space Technology (from the book *The Heavens and the Earth* by Walter A. McDougal)

Pearl Harbor as an impetus of America's entry into World War II, the Pentagon could be accused of wearing blinders towards Earth satellites.

Satellites versus Spy Planes

The Pentagon had other more interesting and less technically challenging pet projects, like the U-2 reconnaissance aircraft, shown in Figure 2-2. The U-2 was developed in less than 180 days to ferret the secrets hidden for a decade behind the Iron Curtain. The obvious asset was that a spy plane could be kept secret. By June 1956, clandestine flights of the U-2 had already crossed the boundaries of the USSR. The dangerous part of the U-2 was the potentiality that it could be shot down and possibly start a war. However, the security needs of the free world during the Cold War dictated that the U-2 be developed and flown, regardless of the risks. Operational reconnaissance satellites were still years away from perfection, and the political pressures of the Cold War dictated that a reliable intelligence gathering system be developed quickly.

The risks of overflying Soviet territory with the U-2 eventually caught up with the Americans. On May 1, 1960, the U-2 was caught spying, red-handed so to speak, by the Soviets when pilot Francis Gary Powers was shot down by a missile 1,300 miles (2,100 kilometers) inside Soviet territory. Repercussions of the U-2 Incident

Figure 2-2. U-2 spy plane

caused the May 15, 1960 Paris Summit meeting between Khrushchev and President Eisenhower to be an embarrassing disaster for the Americans. Tried as a spy in a show trial, Powers was convicted and sentenced to ten years in a Soviet prison. However, in 1962, he was released in an exchange for Soviet spy Colonel Rudolf Abel.

> *"We now have all the rockets we need, long range rockets, intermediate range rockets and close range rockets. The age of the bomber is over!"*
>
> —Nikita Khrushchev

The International Geophysical Year (IGY) gave the Americans the window of opportunity to initiate a satellite launch under a peaceful guise. Using the IGY banner, the United States could have benevolent claims as its satellite flew over Soviet airspace. There was one problem, however. The White House wanted to emphasize the scientific aspects of the satellite and did not want the military tainting it in any way. Thus the initial American satellite project was to be a purely civilian venture, although it was to be managed by the Navy.

All of the armed services wanted their own rocket program to be the satellite delivery system. Although Wernher von Braun's Jupiter Program was moving along well, there was probably some jitters over the idea of putting America's first satellite on a descendant of the V-2. The Air Force and Navy eagerly pitched their own programs but were also turned down. So Vanguard was created, a "civilian" satellite program with assistance by the Department of Defense (DOD) instead of a DOD project with civilian involvement. Unfortunately, the military treated Vanguard as a second class project and there were many delays. What should have been a position of strength, Vanguard in fact was a low priority project compared to the military's own pet missile projects.

Terror in the Sky!

On the night of October 4, 1957, the whole world could confirm what the Soviets had boasted, that the first satellite, *Sputnik* (shown in Figure 2-3), was proof of Soviet ICBMs. The event not only proved to the West the existence of Soviet technology, but it also caused a political panic among Allies. There was a fear that the Soviets could now influence the world just with the threat of ICBMs, resulting in a change in the balance of power. The Soviet rocket threat was emphasized again only a month later when on November 3, 1957, a 1,121-pound (508-kilogram) satellite called *Sputnik 2*, attached to nearly seven tons of booster rocket, reached orbit carrying the live dog, Laika.

> *"The conquering of space has begun. We can be proud that it was begun by our country."*
>
> —Sergei Korolev after the successful launch of *Sputnik 1*

In December 1957, the Americans were ready to respond with the launch of their satellite, *Vanguard*, shown in Figure 2-4. What was supposed to be a development test flight was billed by the White House as an operational flight, ready for an orbital attempt. The world press viewed this event and saw the rocket rise four feet, then explode into a huge fireball. *Vanguard* became known as "Flopnik." In Europe, opinion polls placed Soviet military technology far ahead of the United States and expected that in 20 years, the Soviet Union would prevail over the United States even without a military conflict.

> *"That week after the first Sputnik was one prolonged nightmare...people...were dashing in and out of the President's office. Each new visitor had a longer face than the one before."*
>
> —Anonymous Presidential advisor

Figure 2-3. *Sputnik,* the world's first satellite on display

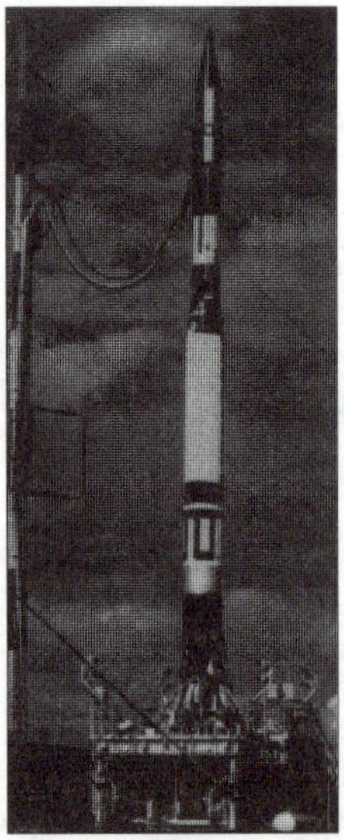

Figure 2-4. *Vanguard,* before the explosion

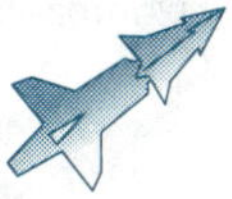

> *"For God's sake, turn us loose and let us do something. We can put up a satellite in sixty days, Mr. Mcelroy! Just give us the green light and sixty days!"*
>
> —Wernher von Braun,
> October 4, 1957

After the *Vanguard* embarrassment, the Army and Navy competition for the honor of the first U.S. satellite was so severe that by early 1958, inter-service relations became testy. On January 31, 1958, the Army used von Braun's Jupiter-C booster and successfully launched the first American satellite, *Explorer 1*. Two months later, the Navy evened the score with *Vanguard 1* on March 17, 1958.

To Race Or Not To Race

> *"I can't understand why the American people have got so worked up over this thing... this small ball...it's certainly not going to drop on their heads."*
>
> —President Dwight Eisenhower,
> October 4, 1957, after the launch of *Sputnik 1*

> *"We could make no greater mistake than to brush off this event as a scientific stunt...We have had a grim and timely reminder of a truth we must never overlook, that the Soviet Union has developed a scientific and industrial capacity of great magnitude."*
>
> —Richard Nixon, October 1957

In the late 1950s, the Eisenhower Administration and the Pentagon grappled with U.S. space policy. The Air Force wanted the whole manned space program under its jurisdiction. They even had plans for a fifty-man military base on the Moon by 1971. This militarization of space made many observers uncomfortable and helped lead the way for the formation of a civilian agency called the National Aeronautics and Space Administration (NASA). The new agency was immediately given the goal of, at the earliest date, achieving the orbital flight and successful recovery of a manned satellite. The project, called Mercury, had the political purpose of achieving a man in space before the Soviets and preventing another Sputnik-like propaganda defeat.

The 1960 American presidential election was a bitterly fought, extremely close battle. Both sides used the perceived U.S./Soviet "space gap" as a campaign weapon. Neither candidate, Richard Nixon nor John Kennedy, had any real space agenda at the time. In 1960, a top secret study by the Pentagon concluded there would be "...Shocking impact...drastic influence of first man in space," and propaganda predominance with the first man in space. Thus, the international stage was set and another race was about to begin (see Figure 2-5).

> *"If half the engineering efforts and public interests that go into the research on the American female bosom had gone into our guided missile program, we would now be running hot dog stands on the Moon."*
>
> —Al Capp, creator of *Li'l Abner*, 1959.

Just a few weeks after Kennedy's inauguration, the chimpanzee Ham was successfully recovered after a suborbital Mercury test flight. It was only a matter of time before a man would make his first flight into space. On February 3, 1961, Alan Shepard's manned suborbital Mercury flight was delayed for two months in order to schedule one more unmanned test flight to wring the bugs out of the Mercury-Redstone booster. This delay gave the honors of first man in space to Soviet cosmonaut Yuri Gagarin. On April 12, 1961, Sergei Korolev gave the orders that launched the young Soviet pilot into history. Gagarin was only in space for one orbit, but the feat easily dwarfed NASA's suborbital

Figure 2-5. Technicians at work on a Mercury capsule under white-room conditions

attempts. The news was devastating to the White House. The Americans were second again.

President Kennedy wanted to bring in fresh ideas to revitalize America's international image, but things were not going well for his new administration. The humiliation of Laos in Southeast Asia and the Cuban Bay of Pigs invasion, topped by Gagarin's flight, created a crisis. Kennedy needed something to boost American morale and prestige. He called upon his advisors for ideas and they recognized the political and scientific potential of space. Kennedy assigned Vice President Lyndon Johnson to evaluate ways of beating the Soviets in what was popularly known as the "space race." Johnson's report came back with several recommendations: the Soviets were ahead in the space race, and this lead would continue for at least several years. A goal with outstanding propaganda value was needed, one that would require several years of technological development. This time would give the United States a chance to catch up. Johnson's proposal was an expedition to send men to the Moon!

> *"Now is the time for this nation to commit itself... in achieving the goal, of landing a man on the Moon and returning him safely to the Earth."*
>
> —Excerpt from President Kennedy's Moon speech to Congress

Experts gave the Americans a fifty-fifty chance of winning the envisioned manned lunar effort. In a few months, the Administration's policy of battling the Soviets in the Third World, and suffering embarrassing setbacks, had changed to the exciting challenge of space. Kennedy delayed his announcement about the lunar plans until he had the momentum of good news, that is, until after the success of Alan Shepard's suborbital trip into space (see Figure 2-6).

In a May 25, 1961, speech before Congress, Kennedy (shown in Figure 2-7) challenged the Soviets to a race of landing a man on the Moon and returning him safely to the Earth before the decade was out. Khrushchev's initial reaction to Kennedy's Moon speech was the belief that the American president was bluffing. Unfortunately, instead of pursuing the space challenge, the tone of world politics in 1961 took a downturn as Khrushchev erected the Berlin Wall. By using concrete, barbed wire, and guns, he sealed off his "worker's paradise" from the West.

> *"...it is capitalism that makes United States Moon rockets fall into the ocean."*
>
> —Nikita Khrushchev, 1959

Western relations with the Soviet Union continued to degenerate toward the real threat of world war. In mid-1962, the Central Intelligence Agency (CIA) reported that there were over

Figure 2-6. Recovery of Alan Shepard's capsule

Figure 2-7. Kennedy presents his Moon mandate to Congress

5,000 technicians and 65 ballistic missile sites under construction in Cuba. American military activity reached a frenzy in South Florida as the tension level soared between the superpowers.

> *"I want to see our country mobilized to a wartime basis, because we are at war!"*
>
> —Victor Anfuso (D., NY)
> April 1961

In the fall of 1962, Kennedy announced a naval shipping quarantine around Cuba and demanded removal of the missiles. The antiseptic word "quarantine" was nothing more than a naval blockade. The prevention of shipping was provocative, if not an outright act of war. The world literally held its breath and waited for a week as Khrushchev sailed his missile-laden ships to Cuba. It was a classic case of political brinkmanship; neither side was going to back down. Cuban leader Fidel Castro proudly informed his Soviet benefactors that he was ready to go the whole way. Fortunately, in the eyeball-to-eyeball confrontation, the Soviets blinked first and backed down in the face of American resolve.

The Soviets Secretly Join the Lunar Race

By 1963, superpower relations began to improve slightly. Kennedy even surprised the Soviets by offering cooperation on a joint manned lunar trip. Khrushchev remained very noncommittal to the idea. Eventually, the U.S. Congress demanded that a joint U.S./Soviet lunar landing must have their approval, so the effort to reach the Moon was to remain a race between nations.

With American momentum in space growing, Khrushchev finally demanded a Soviet manned lunar agenda, to be kept as a state secret of course! But to start a program of this magnitude and complexity in late 1963, would be too little and too late.

Khrushchev continued to pressure Korolev for more politically appealing space spectaculars. The Soviet decision of sending the first woman into space was a huge success in contrast to American military test pilots. Since the Americans were developing a two-man capsule to follow the Mercury flights, Khrushchev demanded that Korolev upstage the Americans again! This disrupted the design and production of the three-man Soyuz moonship. To quickly produce a working three-man spacecraft, all safety features had to be stripped from the Vostok spacecraft in order to cram three men into a one-man capsule. The resulting craft, named *Voskhod,* offered no scientific advancement. It was nothing more than a celestial circus stunt. Shrouded in secrecy, the three-man *Voskhod* flight appeared as a sophisticated

Apollo-like design, and falsely implied the Soviets were at least four years ahead of the Americans.

Eventually, the Soviet Politburo grew tired of Khrushchev's political short sightedness and propaganda excesses and finally overthrew him in October 1964. The new man in charge, Leonid Brezhnev, was also attracted to the high prestige of space and gave approval for continuing the manned lunar program with the goal of beating the Americans. However, the sudden death of designer Korolev in early 1966 created a vacuum in visionary leadership. Soviet strategy became bogged down with the petty rivalries and egos of competing designers. Meanwhile, selected cosmonauts continued their arduous training schedules, including vertical helicopter landings, to acclimate themselves in preparation for the real lunar missions.

> *"A corrupt Capitalist society, by its very nature, is extremely capable of turning the greatest peaceful achievements of mankind into the total means of destruction for mankind."*
>
> —*Pravda* editorial, August 27, 1961

Tragedy on Both Sides

The tenth anniversary of spaceflight in 1967 promised to be an exciting year of great space achievements, climaxing with a manned flight around the Moon! Who would be first, the Soviets or the Americans? It was anyone's guess. But from the start, tragedy occurred on both sides. For the Americans, *Apollo 1*, the maiden flight of the lunar spacecraft, erupted in flames during a ground test, killing astronauts Grissom, Chaffee, and White. The Soviets, perhaps a bit overconfident like the Americans were before their loss, also meet death head on. Vladimir Komarov, a veteran and talented cosmonaut, died aboard the *Soyuz* (see Figure 2-8) during a failed emergency landing. This was a horrible setback for both programs. It would be eighteen months before another manned mission would be attempted by either side.

As the lunar race continued, politics kept a firm hand on the tiller. The Soviet manned lunar agenda was kept a state secret. That way, if they lost to the Americans, it could be denied that they ever intended to go to the Moon in the first place. In 1968, the publicly distasteful Vietnamese Tet Offensive began, escalating the Vietnam War. Social changes were exploding all over the world from China to France and from Chicago to Czechoslovakia. Space was almost becoming a distraction to the clashes that were occurring around the world.

Figure 2-8. The *Soyuz* on the launchpad

> *Man...Moon...Decade!!!"*
>
> *There were only three sacred specs, and they were man, Moon, and Decade. "If those are the real three things you've gotta do, then everything else can be traded off underneath."*
>
> —Apollo Program director Shea (*The Race to the Moon,* by Charles Murray and Catherine Bly Cox, Simon & Schuster 1989)

Last Attempts for Victory

In 1968, the race to the Moon was nearing the final two laps. It was too close to predict which side would be first to fly men around the Moon. As the year ended, an apparent rocket failure delayed the secret Soviet manned circumlunar attempt and gave *Apollo 8* the laurels (see Figure 2-9).

America was in front now and the Soviets were getting desperate. The military arms race and the Moon race was eating up a large percentage of their budget. An unmanned lunar soil sample return mission was designed as a contingency plan if the manned lunar program failed. This way, the Soviets would still recover lunar rock samples ahead of the Americans. Cold War politics had made a fistful of Moon dust the most expensive and sought after geological sample in history! The test flights of giant Soviet Moon rockets exploded four consecutive times, ending the Soviet manned lunar program. The unmanned Soviet robot

Figure 2-9. NASA art of *Apollo 8* heading towards the Moon

probe, *Luna 15*, crashed silently on the Moon just hours after *Apollo 11* (Figure 2-10) made its historic landing.

In the aftermath of the lunar race, Brezhnev still received his Citizen's Medal (the highest award in the Soviet Union) and 12 brave Americans walked on the Moon. The Soviet leader soon lost all interest in the Moon and withdrew his support. By the whim of the Soviet leader, the toil and sacrifices of thousands of Soviet manned lunar program engineers was erased and all evidence of the program quietly covered up. The remaining hardware was disassembled and parts were destroyed or whisked away into hidden storage.

> *"I am only sorry that we didn't manage to send a man to the Moon during Korolev's lifetime."*
>
> —Nikita Khrushchev

In the early 1970s, American lunar landings became seemingly routine and the public lost interest. It is ironic that in order to keep interest and support alive for their space programs, the superpowers agreed to the international Apollo-Soyuz Test Project joint mission (see Figure 2-11). In spite of criticism at the time of giving away secrets to the enemy, the joint space mission was probably a great boost to ending the adversary relationship between the superpowers.

Figure 2-10. *Apollo 11*, the first expedition to the Moon

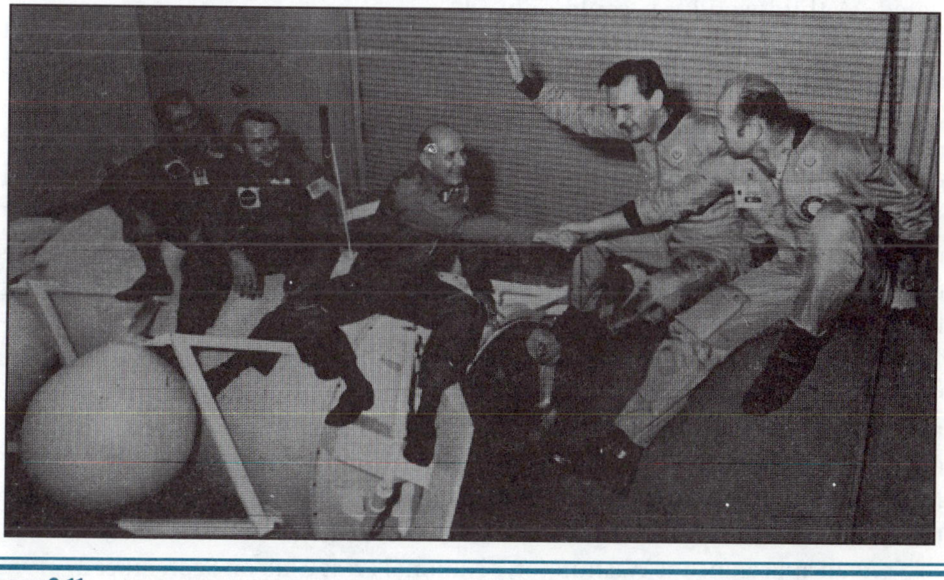

Figure 2-11. Detente in Space: ASTP (Apollo-Soyuz Test Project), the crew of *Apollo 8-Soyuz* spacecraft greet each other

By the late 1980s, thanks to *glastnost*, official Soviet versions of history and state space secrets are becoming unraveled and the truth is beginning to surface. Perhaps now that *glastnost* has blossomed, and hard line Communist rule has ended in the Soviet Union, future joint missions will occur and the costs and benefits will be shared by all.

Propaganda: How To Influence the World

The Soviet political propaganda machine went into high gear with the launching of the world's first satellite, *Sputnik*. Articles appearing in *Pravda*, *Izvestia*, and other periodicals showed a verbose ability to publish volumes of words on the subject but really told nothing about the technical aspects of the satellite. *Pravda* further hyped the event by stating that the brilliant achievement of *Sputnik* was possible only "under the wise leadership of the Communist Party." Other periodicals carried the Party line. A typical example was, "We Russians, a backward people ourselves less than a lifetime ago, can now do even more spectacular things than the rich and pompous West—thanks to Communism."

Flush with the reaction to *Sputnik*, Soviet Premier Nikita Khrushchev extended his space and missile rhetoric and began to tell outright lies. He stated, "A third world war could only end in the collapse of Capitalism...We now have all the rockets we need—long range rockets, intermediate range rockets, and close range rockets. The age of the bomber is over!" Khrushchev further stated that Soviet missiles were in "serial production" and were being produced "like sausages." Khrushchev gambled the world would believe his false claims of missile superiority on the strength of the *Sputnik* space spectacular.

> *"The Americans don't launch Sputniks, they hop up and fall down in the ocean."*
>
> —Nikita Khrushchev, July 1961, after the suborbital flight of *MR 4*

As Khrushchev began to realize the political power inherent in the prestige of space successes, he became a fixture in the world press, always using space and rocketry to push his Communist ideals. Two days after the launch of *Luna 1* into orbit around the Sun, Khrushchev declared his country had "outstripped the U.S. in science and technology," and that *Luna 1* was "another weapon in the hands of the Soviet man to contradict the slanders of the enemy."

At an appearance at Washington's National Press Club on September 27, 1959, Khrushchev chided the Americans for their lack of a scientific educational system.

He boasted, "It is the graduates of Soviet schools...who astonished the world with the first Sputniks. We are proud that the Russian word 'Sputnik' is understood all over the world without translation."

Khrushchev also claimed in early 1960, "Our flag is flying on the Moon. Is that not enough to prove the superiority of Communism over Capitalism?"

> *"A road has thus been blazed from the Earth to the Moon...a pennant bearing the national emblem of the Soviet Union is now on the Moon. Our Earth has now become somewhat lighter while the Moon has gained several hundred kilograms."*
>
> Nikita Khrushchev, September 13, 1959, after *Luna 2* impacts the Moon.

Always claiming that Soviet space efforts were purely scientific, Khrushchev defended his space program at the 1960 Paris Summit Conference when President Charles De Gaulle of France questioned whether *Sputnik 4* was photographing French territory. Khrushchev raised both hands and replied, "As God is my witness, our hands are clean and our soul is pure. We will not put a camera in the Sputnik!" Two years later, *Cosmos 4*, the first Soviet reconnaissance satellite was launched.

Two days after his pioneering orbital flight, Yuri Gagarin appeared with Khrushchev and taunted the West at a gigantic Moscow celebration by gloating, "Now let the other countries try to catch us!"

Khrushchev reflected on Vostok by saying Gagarin's flight marked "a stage in the scientific and technical development of our country and reveals its mighty upward flights from backwardness to progress. In this flight are reflected the heroic accomplishments of the Soviet people, the working class, collective farm peasantry, working intelligentsia, and our wonderful scientists, who have created a technical modern-day miracle, the cosmic ship Vostok, this greatest triumph of Lenin's ideas."

After Gagarin's orbital flight, Khrushchev remarked about the Project Mercury suborbital flight of *Liberty Bell 7* carrying Gus Grissom, "The Americans don't launch any Sputniks, they hop up and fall down in the ocean. We are glad that the American flier did not drown."

By late 1961, Khrushchev was responding to questions about the Soviet man-to-the-Moon program. He replied that "there was not yet any fixed schedule for the landing of a Soviet citizen on the Moon. It is not a question of mooning him but of de-mooning him. Our national emblem is already on the Moon, but we don't want to place a coffin beside it...We can fly to the Moon, but the difficulty is getting him away from there."

Khrushchev insisted that flashy space spectaculars coincide with major Soviet political events on Earth. His recurring theme was "Soviet superiority in space implied Socialist superiority on Earth." Titov's one-day *Vostok 2* flight just before the erection of the Berlin Wall, and the group flight of *Vostok 3* and *4* as Soviet nuclear missiles were being deployed to Cuba were examples of this.

In spite of his space bravado, Khrushchev lost the Moon race in 1961 by delaying any action on a landing program for several years after President Kennedy announced the American lunar goals. Khrushchev believed the Soviet heavy booster advantage would allow him to eventually win a Moon race and delayed action on a Moon landing program. Figure 2-12 is a fictional depiction of a cosmonaut and his lander on the Moon.

Khrushchev completely underestimated the American industrial and technical ability to quickly master a major engineering challenge and build up a leadership, development, and manufacturing infrastructure capable of developing the necessary lunar program. By the time Khrushchev realized the American resolve in 1964, it was too late to catch up. Apollo had an insurmountable lead.

Figure 2-12. History would have been very different if this event had occurred: Soviets landing on the Moon first!

CHAPTER 3

Machines, Animals, and Man: The First Explorers

Before man would venture into the cosmos, he was preceded by plucky mechanical satellites and trusting and trained animals. These early pioneers would explore the risks and dangers. This chapter is an account of the first unmanned satellites, and the animals and humans that explored space.

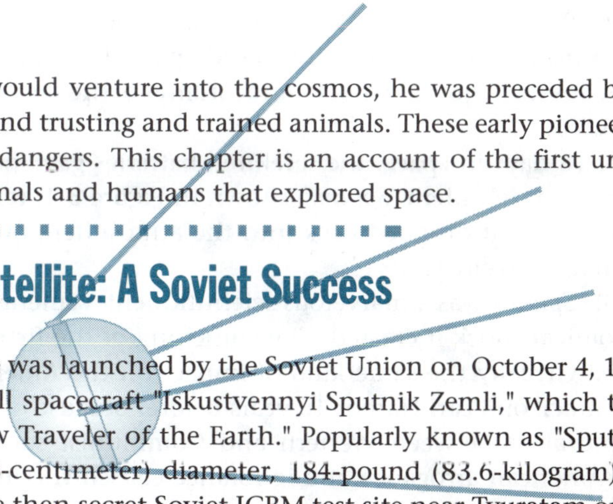

The First Satellite: A Soviet Success

The first Earth satellite was launched by the Soviet Union on October 4, 1957. The Soviets called the small spacecraft "Iskustvennyi Sputnik Zemli," which translates into English as "Fellow Traveler of the Earth." Popularly known as "Sputnik," the spherical two-foot (58-centimeter) diameter, 184-pound (83.6-kilogram) satellite was launched from the then secret Soviet ICBM test site near Tyuratam on an SL-1 booster. Its initial orbit ranged from 141 to 588 miles (228 to 947 kilometers) altitude, inclined 65.1 degrees to the equator, with a period of 96.2 minutes.

When the weight of *Sputnik* was announced, comparisons with the light-weight, yet-to-be-launched American *Vanguard* satellite led some people to think that a decimal point had been misplaced. Some observers erroneously thought that in order to launch a satellite ten times heavier than that planned by America, the Soviet Union had developed exotic fuels or atomic power for their rocket boosters. In reality, the launching of *Sputnik* was made possible by a Soviet prototype military rocket developed during the mid-1950s by Sergei Korolev. Known later in the West as the SS-6 "Sapwood" ICBM, this booster used a cluster of 20 rocket nozzles designed by Valentin Glushko. The SS-6 was designed to deliver heavy Soviet H-bombs over intercontinental distances. For this reason the SS-6, with its more than one-million-pound (half-million-kilogram) lift-off thrust, had great lifting capability compared to the seemingly puny American *Vanguard* booster.

> *"We have been projected into the Space Age by world forces almost beyond our control. A dream of the ages has become a necessity of our times."*
>
> —Isidor Rabi, Nobel Laureate, 1959

Sputnik's only instrument was a thermometer that minutely changed the frequency of the satellite's twin chemical battery-powered radio transmitters. The frequencies alternated between the two transmitters with a duration of .3 seconds with a .3-second break between, resulting in the famous "beep-beep-beep." The one-watt signal was transmitted at frequencies of 20.005 and 40.002 megahertz through four metal rod antennae ranging from eight to nine and one-half feet (2.4 to 2.9 meters) long, which were folded back at launch but hinged out once in space. The signals from *Sputnik* were easily picked up by HAM radio operators all over the world. The Soviets requested that tape recordings of the signal be mailed to "Sputnik, Moscow, USSR."

After 18 days in space, the satellite's batteries gave out and on October 22, 1957, the radios fell silent. *Sputnik*, shown in Figure 3-1, remained in orbit a total of 94 days before it dropped back into the atmosphere on January 4, 1958, and burned up upon reentry.

While *Sputnik* was a marvelous scientific achievement, it is best remembered for the political shock it created. The launching of the first satellite by the Soviet Union had virtually the same kind of impact on the American public as did the Japanese attack on Pearl Harbor 16 years before. Because of the Cold War political climate prevalent between Western and Communist societies after the Korean Conflict, American arrogance and ignorance caused the mistaken belief that the

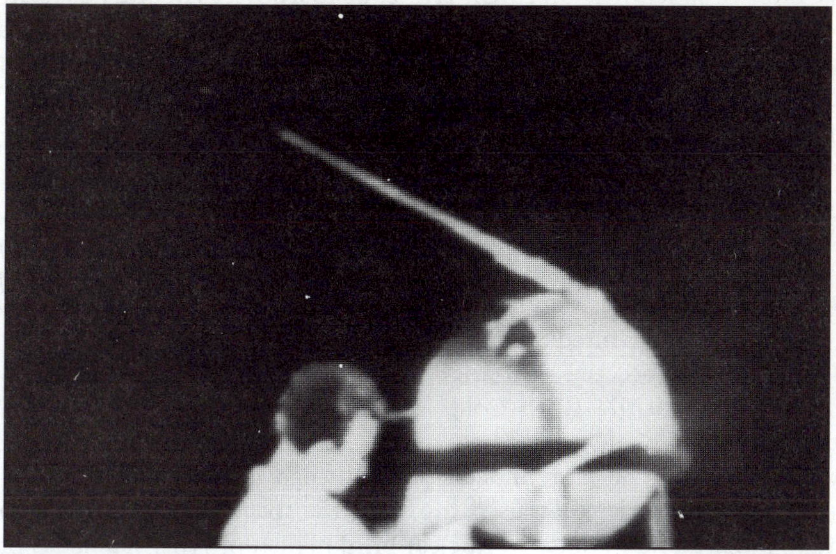

Figure 3-1. *Spunik 1*, Man's first artificial satellite

Soviet Union was incapable of performing such a complex scientific and technical feat as launching an Earth satellite.

The very idea that the Soviets were planning the launching of *Sputnik* was a complete surprise to most Americans. In reality, the Soviets openly predicted as early as August 2, 1955, that the USSR would launch an Earth satellite as part of its participation in the International Geophysical Year (IGY). At that time Academician Leonid Sedov, with the USSR Academy of Sciences, publicly predicted the first Soviet space satellite would be launched within two years. He went so far as to accurately state the size of the satellite's planned orbit. Throughout 1957, the Soviet newspaper *Pravda* and Radio Moscow made continued announcements about the upcoming Soviet satellite launch. Soviet news media accurately described the weight, size, orbital inclination, radio transmission frequencies, and the scientific studies to be carried out. *Sputnik* was a surprise to the Western world simply because it chose to ignore Soviet predictions about their planned satellite and regarded them as empty propaganda.

Even Soviet Premier Nikita Khrushchev was unprepared for the magnitude of the world reaction to *Sputnik*. He did not expect the huge political outcry from the American failure to place the first satellite in orbit, but instead expected a stunned paralysis reaction where he could score a propaganda triumph over the West. Khrushchev used *Sputnik* to begin his familiar theme: Soviet and Socialist success in space implied superiority on Earth. While the first Soviet satellite was technically crude and simple, it started the myth of Soviet space superiority over the West. The successful launching of the first satellite even made a new addition to our language. The very name "Sputnik" became synonymous with any satellite in orbit.

The act of sending a small metal ball into orbit around the Earth was not the greatest shock imposed by *Sputnik*. Instead, it was the realization by the Western world that if Russian rockets could orbit a satellite, they could loft H-bombs into the United States as well. This realization became a chilling fear in the climate of the 1950s Cold War confrontation that existed between the West and the East. Indeed, with his ego inflated by the propaganda success of *Sputnik*, Khrushchev used the scientific space achievement of the first satellite to pursue a new course in nuclear brinkmanship. With *Sputnik* as proof of the ability of his rockets, Khrushchev threatened a missile attack on any nation that allowed American military bases on its soil near the Soviet Union. Khrushchev was able to play this bluff because Soviet secrecy left the Western public ignorant of the fact that Soviet missile production at the time could barely support scientific space launches, let alone massive ICBM production. Khrushchev used this nuclear bluff to influence world politics for several years after the launch of *Sputnik*.

> *"In the open West, you learn to live closely with the sky. It is part of your life. But now, somehow, in some new way, the sky seemed almost alien."*
>
> —Senator Lyndon Johnson,
> October 4, 1957

When the initial political storm created by *Sputnik* subsided, the world came to realize that the primary lesson of *Sputnik* was not scientific or political, but rather that the Space Age itself had arrived. *Sputnik* showed that space travel was more than just a science fiction fantasy and was to become a common part of the twentieth century.

Animals in Space

As soon as rocket flight was perfected after World War II, it was realized that before man could risk venturing into the unknown environment of outer space, questions about the physiological effects of spaceflight on humans had to be answered. The search for these answers consumed a major part of the early American and Soviet space exploration effort, and in many ways today still occupies a large portion of in-orbit research by astronauts and cosmonauts.

In the 1950s, ground testing using centrifuges and isolation chambers with human subjects could not answer the ultimate question of what prolonged weightlessness and exposure to cosmic radiation would do to a living organism. Before man could be exposed to these potential dangers, animals would have to pave the way for humans and prove there were no hidden dangers awaiting a space traveler.

The initial American space biology experiments began in June 1948, when Dr. James Henry and a group of Air Force scientists launched a rhesus monkey named Albert to an altitude of 37 miles (60 kilometers) above the White Sands Proving Grounds in New Mexico using a captured German V-2 rocket. Although radio telemetry relayed Albert's ECG, respiration, and blood pressure showing he survived the flight in good shape, his cabin was not recoverable and he sacrificed all for the birth of American bioastronautics.

Dr. Henry continued his space biological research and by September 1951, successfully recovered a monkey and several mice which had been lofted to 45 miles (72 kilometers) altitude by an Aerobee-2 sounding rocket. His initial research indicated that the several minutes of weightlessness endured by the test subjects caused them no ill effects.

By April 1958, animal astronauts had reached an altitude of 1,391 miles (2,240 kilometers) when the mouse "Wickie" was launched from Cape Canaveral during the Mouse-in-Able (MIA) program. Although the animals aboard the MIA flights were not recovered, telemetered physiological data gave researchers confidence that humans could successfully fly in space.

The greatest boost for the hopes of future human flight in space came in May 1959, when the monkeys Able and Baker were recovered in a Jupiter missile nose cone after a flight 298 miles (480 kilometers) into space. Although Able died several days later from an anesthesia accident during the removal of biosensors from her body, Baker lived in good health for many more years and effectively was a milestone in the quest to send American astronauts into space.

As the American manned space program progressed with Project Mercury, monkeys were chosen to precede man into space during spacecraft test flights. Although the Mercury astronauts realized the medical wisdom of sending animals into the unknown first, they were less than pleased, being seasoned high performance test pilots, to have a monkey blaze the space trail for them.

The first significant Project Mercury animal test flight was the Mercury Redstone-2 mission in February 1961, in which the chimpanzee HAM (so named for the Holloman Aero Medical division where he was trained) flew 155 miles (250 kilometers) into space. HAM's flight helped validate the Mercury spacecraft in preparation for Alan Shepard's historic first American manned spaceflight. HAM is shown during his flight in Figure 3-2.

Another chimpanzee, Enos, preceded man into orbit. To provide a final test of the Mercury spacecraft's life support system, Enos was launched aboard the Mercury Atlas-5 mission on November 29, 1961. Although a spacecraft malfunction shortened the flight by one orbit, Enos survived 183 minutes of weightlessness with

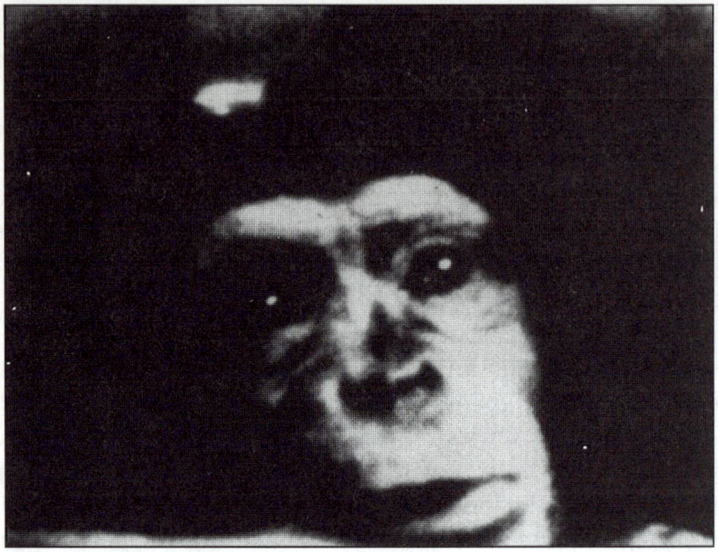

Figure 3-2. Chimpanzee HAM in space

no difficulty, clearing the way for John Glenn's first American manned orbital spaceflight three months later.

Soviet space doctors chose to use dogs instead of monkeys as specimens on sounding rocket flights between 1951 and 1957. The similarity between the blood vessels of dogs and man, as well as the even temperament of the dogs were noted as reasons for their selection. The dogs used on the flights were mongrels secured from the Moscow area. All had white hair so they could be observed and photographed easily, and were female to simplify the collection of wastes in the flight cabin.

Using a Soviet improvement of the German V-2 rocket, researchers Galkin, and later Burgov, initially launched dogs to altitudes reaching 68 miles (110 kilometers) above the Kapustin Yar missile test range. In some instances, the dogs were carried in unpressurized cabins and wore pressure suits. Later flights in 1957 using an improved rocket reached altitudes in excess of 124 miles (200 kilometers). All the dogs were successfully recovered.

The first living creature to orbit the Earth was the Russian dog Laika (Little Barker). Launched aboard the *Sputnik 2* satellite on November 3, 1957, the female black and white fox terrier was carried in a hermetically sealed cylindrical cabin. Automatic equipment provided Laika with air, temperature control, and a food supply. Laika, shown in Figure 3-3, was fitted in a special harness that monitored her pulse, respiration, and blood pressure, as well as recorded electrocardiograms.

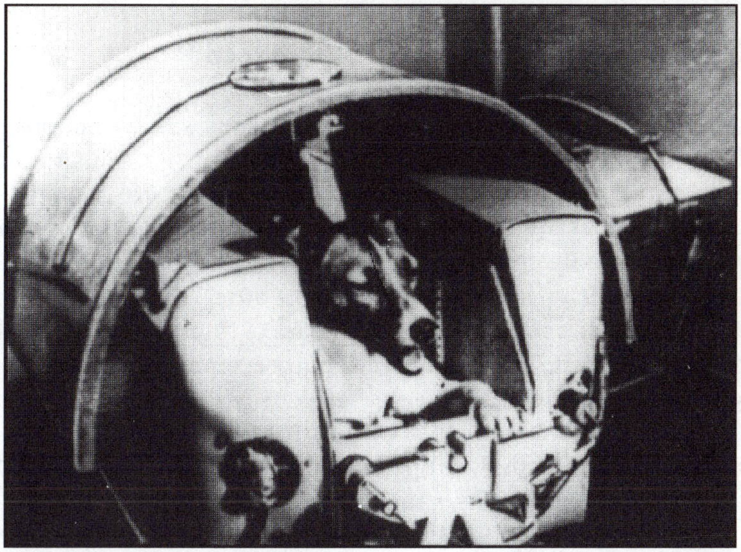

Figure 3-3. Canine Laika, the first creature in space

Food was provided in paste form and wastes removed via a rubber reservoir attached to the dog.

During launch, Laika's pulse rate rose to three times normal. It quickly stabilized after entering orbit. After several days in orbit, Laika's heart rate actually dropped below her normal rate while on Earth. Laika survived the launch and subsequent space travel very well and considerable data was returned. No recovery was planned for Laika or *Sputnik 2*, so after one week in orbit, Laika was euthanised by lethal injection.

Later test flights of the Soviet Vostok manned spacecraft, launched as the *Sputnik 5, 6, 9,* and *10* satellites, all carried dogs as orbital test subjects. The animals aboard *Sputnik 5, 9,* and 10 were recovered in good health after spending up to 24 hours in orbit. A spacecraft malfunction during the reentry of *Sputnik 6* caused the incineration of the spacecraft and its two canine passengers, Pchelka and Mushka. The two dogs became the first living creatures to die while returning from orbit.

The early spaceflight tests with animals showed there were no hidden biological or psychological dangers awaiting humans in space. The growing legions of space travelers today owe a debt to the animal pioneers who preceded them into the then unknown void of space.

The Race to Explore the Solar System

Travel to other planets has been a dream of astronomers, science fiction enthusiasts, and finally rocket engineers for 300 years. Some the first serious thoughts about interplanetary travel were published in Cyrano de Bergerac's 1649 novel, *The Voyage to the Moon*. Later literary efforts such as Jules Verne's 1865 classic, *From The Earth to the Moon*, remain an inspiration even to modern day space farers.

The development of large military rockets after World War II provided the means of finally fulfilling the centuries old dream of interplanetary travel. After the creation of NASA in 1958, American space engineers quickly designed adaptations of Air Force rockets capable of lofting instruments toward Earth's planetary neighbors. At the same time in the Soviet Union, rocket designer Sergei Korolev also pursued his spaceflight dream by proposing that his newly developed R-7 rocket be used to boost scientific instruments toward other planets.

The Cold War political rivalry between the U.S. and USSR in the 1950s provided fertile ground for a scientific interplanetary competition between the two countries. The sting of losing the honor of the first Earth satellite to the Soviet Sputnik galvanized the American scientific community. Ambitious plans were quickly mounted in the late 1950s to prepare rocket flights to the planets.

Early this century, Soviet spaceflight theorist Konstantin Tsiolkovsky stated, "When the solar system is conquered, man will have three dimensions." Long before modern rocketry, Tsiolkovsky realized the importance of expanding the human presence away from the home planet Earth. His 1929 prophesy is now close to fulfillment. Earthbound humans have spread their presence throughout the solar system within the span of single generation. During the next generation, thanks to the pioneering exploration by American and Soviet unmanned probes, people of Earth will set foot on Mars and complete Tsiolkovsky's prophesy.

The First Target: Venus

The limited power of our booster rockets restricts the orbital path allowed toward other worlds. Because of this limited rocket power, celestial mechanics dictate that only certain periods of time, so called "launch windows" are available to send probes to other planets. The first of these launch windows toward Venus was available in June 1959.

Initial American plans called for the launch of two Venus orbiters using the Atlas-Able booster. However, after the Soviet *Luna 1* success, these plans were changed and the flights aimed toward the Moon in an effort to match Soviet lunar efforts.

Alternate plans were readied to launch a 95-pound (43-kilogram) payload on a Venus fly-by trajectory using a Thor-Able booster, but delays hampered the project.

When the craft was finally launched on March 11, 1960, *Pioneer 5* was simply aimed toward near-Venus space. The flight came eight months after the close of the Venus launch window and the planet simply would not be there when the craft approached. Data about previously uncharted space was received from the craft for three months as it looped toward the orbit of Venus. Daily press reports fascinated the public as each day the craft set new long distance communications records. Finally on June 26, 1960, at a distance of 22 million miles (36 million kilometers), *Pioneer 5*'s faint signal was lost.

The Soviet Union answered the Venus challenge on February 12, 1961, when the giant seven-ton *Sputnik 8* spacecraft entered orbit. Shortly after, an "escape stage" separated and boosted the 1,419-pound (644-kilogram) *Venera 1* on its way to Venus. However, on February 27, the craft's orientation system failed and contact with *Venera 1* was lost.

American planetary exploration efforts in the early 1960s were hinged to the development of the powerful Centaur upper stage to boost half ton payloads toward the planets. However, extended delays made it clear the Centaur was not going to be ready for years. In the interim, engineers at NASA's Jet Propulsion Laboratory determined that a modification of the Ranger lunar spacecraft could carry a scientific payload to Venus using the Atlas-Agena B booster. Thus the Mariner spacecraft series was born.

At the next Venus launch opportunity, *Mariner 2* lifted off from Cape Canaveral on August 25, 1962. The craft was the first of a new breed of sophisticated interplanetary explorer. Weighing only 447 pounds (203 kilograms), the craft was composed of 54,000 separate components, each of which must not fail during the 109 day trip to Venus.

On December 14, 1962, *Mariner 2* passed 21,500 miles (35,000 kilometers) from Venus and performed the first close-up studies of another planet. Mariner radioed data which seemed to confirm that Venus had a very high surface temperature. No difference was noted between the 900 degree Fahrenheit (450 degree Centigrade) readings on the day and night side. The craft also detected no appreciable magnetic field or evidence of meteoric particles near Venus.

Earth Takes Aim at Mars

The Soviets were also the first to mount a Mars probe program. Twin launch attempts in October 1960, successfully entered an Earth parking orbit, but the escape stages failed to ignite, stranding the probes. While the Mars shots were an Earth orbital success, they were Mars probe failures and the Soviets never publicly revealed their existence.

The next Mars launch window occurred in late 1962. This time, the Soviets readied three launch attempts. The second of the three, on November 1, 1962, succeeded in sending the *Mars 1* probe toward Mars to relay close-up photos of its surface. For nearly four months all seemed well with the 1,970 pound (893 kilogram) craft. *Mars 1* even exceeded *Mariner 2*'s long distance communications record of 55 million miles (88 million kilometers). However, three months before arriving at the planet, contact with *Mars 1* was lost.

Two years later, it was America's turn at Mars. As the Centaur booster was still not operational, two more of the small Ranger-based craft had been prepared, each carrying six experiments and a television system to relay close-up images of the Martian surface. Using an Atlas-Agena D, *Mariner 3* was successfully launched on November 11, 1964, but a malfunctioning payload shroud prevented the craft from deploying its solar panels, thus starving *Mariner 3* of its power.

After a 17-day, around-the-clock crash program to develop a modified payload shroud for a second American Mars attempt, *Mariner 4* rocketed away from Earth on November 28, 1964. Eight and one-half months later, it arrived at Mars on July 15, 1965. As the craft approached within 6,100 miles (9,800 kilometers), it snapped 21 overlapping images forming a swath covering one percent of the planet's surface.

Up to this time, planetologists held hope that Mars could support some form of primitive life, but the startling results beamed back from *Mariner 4* forever quashed such hopes. Initial results showed the planet's atmospheric pressure was only five millibars, much thinner than the anticipated 80 millibars. This was only one percent of the pressure of Earth's atmosphere at sea level.

As *Mariner 4* transmitted its photographs (see Figure 3-4), Mars revealed the biggest surprise: the planet was covered with craters. While they were invisible from Earth, Martian craters up to 75 miles (120 kilometers) in diameter were clearly imaged by the spacecraft.

Mariner 4 remained in contact with Earth for another two years, setting a long distance communication record of 192 million miles (309 million kilometers). As the craft reported from deep space, it encountered several meteoroid streams, some producing one strike per minute.

The Soviets did not ignore the 1964 Mars launch window and sent a probe called *Zond 2* toward the planet on November 30. Spacecraft reliability problems again plagued the Soviets and contact was lost in April 1965.

Eight months later, the Soviets launched again. As the launch occurred far beyond the close of the Mars planetary launch window, the planet was not there when the *Zond 3* spacecraft arrived at Mars' orbit. In light of the lack of reliability in previous Soviet planetary probes, it is speculated this flight was really a diagnostic mission to test a new spacecraft design. Improvements were successful as *Zond 2* was still functional in March 1966, at a distance of 96 million miles (154 million kilometers).

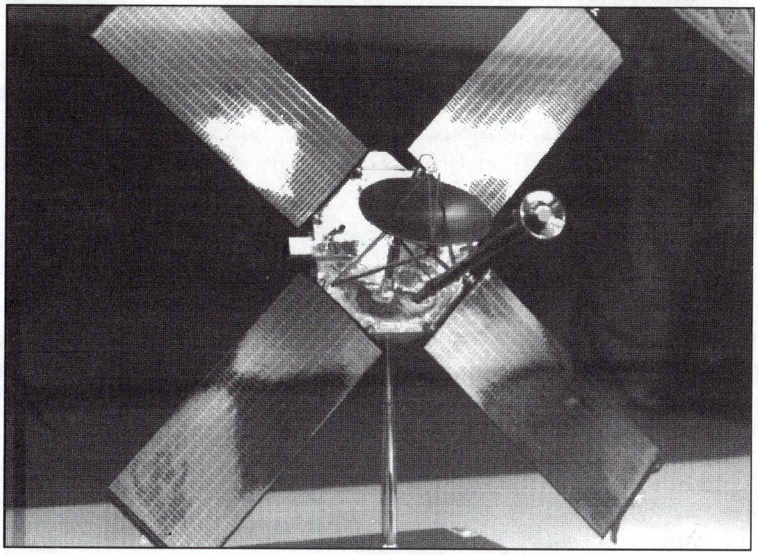

Figure 3-4. *Mariner 4* sent back photographs of Mars

Soviets Set Claim to Venus

Soviet persistence in trying to reach Venus began to pay off as two launches in November 1965 succeeded in placing twin one-ton probes on course toward that planet. *Venera 2* carried a photographic payload while *Venera 3* carried a 742-pound (337-kilogram) atmospheric entry capsule.

Venera 2 was targeted to pass 15,000 miles (24,000 kilometers) from the planet on February 27, 1966. *Venera 3* was placed on a path so accurate that it would strike the planet near the center of its visible disk. The twin missions seemed to be going well until shortly before planetary encounter, both craft fell silent. While *Venera 3* hit the planet on March 1, 1966, no data was returned from the mute spacecraft.

At the next Venus launch opportunity, the Soviet *Venera 4* left Earth on June 12, 1967, carrying with it a 843-pound (383-kilogram) atmospheric entry probe. Following a 210 million mile (338 million kilometer) trip around the Sun, *Venera 4* broke the Soviet string of planetary bad luck and successfully completed its assigned mission of dropping an atmospheric probe into the clouds of Venus. Ten Soviet Venus probe failures had preceded the *Venera 4* triumph on October 18, 1967.

After decelerating through the dense atmosphere of Venus, the capsule began its parachute descent. Initially, it was thought that data transmissions began at 16 miles (26 kilometers) altitude and continued to the surface where final readings indicated an atmosphere almost entirely of carbon dioxide at a pressure of 22 Earth atmospheres and a temperature of 260 degrees Centigrade.

Later comparison with *Mariner 5* data gathered the next day showed the *Venera 4* descent capsule had actually begun transmitting at 34 miles (55 kilometers) altitude and had ceased at 17 miles (27 kilometers) where the atmospheric pressure reached 22 Earth atmospheres, crushing the descent capsule. Extrapolations of the data showed actual surface conditions were closer to 75 atmospheres pressure and 500 degrees Centigrade.

The last American spacecraft to visit Venus for the next six years was *Mariner 5*, launched by an Atlas-Agena D on June 14, 1967. The 520-pound (245-kilogram) craft was in reality a modified backup spacecraft for the previous Martian *Mariner 4*.

Mariner 5 was directed much closer to Venus than its predecessor, *Mariner 2*, in order to search for any Venusian magnetic field. On October 19, 1967, it passed 2,555 miles (4,100 kilometers) from the surface and returned data which agreed with Soviet observations taken the day before by *Venera 4*. The craft found only a negligible magnetic field.

Venera 5 and *6*, launched on January 5 and 10, 1969, continued the persistent Soviet exploration of Venus. After *Venera 4* had encountered harsh atmospheric conditions at Venus, there was not enough time before the next launch window to redesign the atmospheric probe carried by both craft. For this reason, it was expected that both capsules would be crushed by the Venusian atmosphere before they reached the surface.

Both Veneras successfully parachuted their capsules into the planet's cloudy atmosphere on May 16 and 17, 1969. They confirmed previous temperature, pressure, and composition readings before they collapsed when the pressure reached 27 Earth atmospheres.

Planetary Attention Returns to Mars

In the mid-1960s, the American Centaur upper stage booster was finally perfected, allowing at last greater instrument loads to be carried to the planets. The first to take advantage of this were the *Mariner 6* and *7* probes to Mars. The craft were able to carry more powerful transmitters and computers, increasing the data transmission rate to 2,000 times that received from *Mariner 4*.

Launched on February 25 and March 27, 1969, both of the identical craft succeeded in reaching the planet. Both relayed photos of the cratered Martian terrain via TV scan platforms carrying wide angle and telephoto cameras and confirmed earlier atmospheric data.

While American plans concentrated on the preparation of a Mars orbiter mission the following year, the Soviets returned to Venus again in 1970. The lone *Venera 7* arrived at the planet on December 15. A decade of dogged Soviet perseverance paid off as the redesigned *Venera 7* descent capsule penetrated the atmosphere of Venus and survived 23 minutes on the hellish surface. Readings directly from the Venusian surface confirmed earlier extrapolations, showing the pressure was 90 Earth atmospheres with a temperature of 900 degrees Fahrenheit (475 degrees centigrade).

Finally having achieved success at Venus, the Soviets were hopeful their new class of Mars explorers would find equal success. Launched by the giant Proton rocket, the five-ton craft were to enter orbit around Mars after ejecting a 990-pound (450-kilogram) landing capsule toward the surface of the planet. The orbiters were to photographically map the planet while the lander analyzed the surface and photographed the Martian terrain.

Mars 2 and *3* departed Earth on May 19 and 28, 1971. The following November, both craft arrived at Mars. *Mars 2* then ejected its landing capsule and entered orbit around Mars. Unfortunately, as the probe arrived, high velocity winds enveloped Mars in a planetwide dust storm. During this storm, the *Mars 2* landing capsule crashed on November 27 and returned no data.

Five days later, *Mars 3* arrived and also entered orbit around the planet. Its landing apparatus fared better and touched down on the planet's surface in the Electris region. Unfortunately, the area was also blanketed by the huge dust storm and contact with the craft was lost 20 seconds after touchdown. Both Mars orbiters continued to study the planet and returned considerable data. Among the data returned was the first direct studies of large amounts of dust suspended in the atmosphere of a planet, leading to the refinement of "nuclear winter" theories about the global devastation which would be created on Earth in even a small scale nuclear war.

American Mars efforts continued with the launch of *Mariner 9* on May 30, 1971. The 2,198-pound (998-kilogram) craft was to orbit the planet and complete a remote survey and high resolution photographic map of Mars. While *Mariner 9* left Earth later than its Soviet *Mars 2* and *3* sister ships, it arrived in Martian orbit two weeks ahead of the Soviets.

Mariner 9 also encountered the same planetwide dust storm which plagued the Soviets; however, the craft was able to wait in orbit. On January 2, 1972, the mapping mission was able to begin. This time, the Mariner's camera revealed not only craters, but huge volcanoes, mysterious channels which appeared to have been caused by running water, and giant canyons cutting across the surface.

Into Interstellar Space

In the early 1970s, interplanetary space beyond Mars was still unexplored territory. No probe had yet been sent to the giant planet Jupiter. This was to be the target of *Pioneer 10*, launched by the U.S. on March 3, 1972, aboard an Atlas-Centaur booster.

At Jupiter, solar radiation is only four percent of that at Earth, far too weak to power solar cells. Because of this, *Pioneer 10* was nuclear powered.

The craft carried 11 instruments, including an imaging system, for studying Jupiter and interplanetary space. At launch, the 568-pound (258-kilogram) craft was accelerated to 32,106 miles (51,670 kilometers) per hour, the greatest speed attained by any spacecraft so far. Upon arrival in December 1973, *Pioneer 10* found the giant planet's radiation field was 10,000 times greater than Earth's. The probe also returned 300 images and data about near-Jupiter space.

After passing Jupiter, *Pioneer 10*, which is shown in Figure 3-5, continued outward away from the Sun. On June 13, 1983, the craft passed the orbit of the outermost planet at 2.8 billion miles (4.5 billion kilometers) from the Sun. *Pioneer 10*'s nuclear battery will keep the craft in contact with Earth until the next century as it continues to investigate the boundary between the Sun's influence and true interstellar space.

Pioneer 11 joined its twin in space on April 6, 1973. This time, the craft was targeted toward both Jupiter and Saturn. After a scientifically successful encounter with Jupiter on December 3, 1974, *Pioneer 11* used the planet's immense gravity to slingshot the craft toward Saturn. Jupiter flung the spacecraft almost 93 million miles (150 million kilometers) above the plane of the solar system before it looped back and intercepted Saturn on September 1, 1979.

Figure 3-5. *Pioneer 10*

The first visit to Saturn was rewarded with the discovery of a previously unknown 124-mile (200-kilometer) diameter moon and another ring around the planet. The craft also found the planet's magnetic field was more than 500 times stronger than Earth's.

Like its twin, *Pioneer 11* is also leaving the solar system. The craft passed Pluto's orbit in 1991 heading toward the center of the galaxy in a direction of the constellation Sagittarius. This is opposite the path taken *Pioneer 10*.

While America was taking an interest in the outer planets, Soviet sights remained on Venus and Mars. The one-ton *Venera 8* spacecraft left Earth on March 27, 1972, and carried an improved landing capsule. After a trip of nearly 310 million miles (one-half billion kilometers) around the Sun, *Venera 8* dropped its lander into the clouds of Venus on July 22, 1972.

Upon reaching the surface, the lander confirmed the conditions found by its predecessor. *Venera 8* also carried a gamma-ray spectrometer with which it could analyze the composition of the local Venusian surface. The craft determined the local rocks were of volcanic origin and closely resembled terrestrial granite.

Wishing to improve upon their unfortunate Martian luck from two years before, the Soviets prepared an impressive armada of four spacecraft for the 1973 Mars launch opportunity. In July and August of 1973, all four, *Mars 4* through *7*, were successfully launched by Proton booster rockets toward Mars.

However, Soviet Mars luck did not improve. *Mars 4*'s retro-rocket failed and the craft sped past the planet without entering orbit. *Mars 5* entered orbit, only to tumble out of control two weeks later. *Mars 6* ejected its lander, but the capsule crashed. *Mars 7* prematurely ejected its lander and completely missed the planet.

In 1973, America again launched toward the inner planets. The half-ton *Mariner 10* lifted off aboard an Atlas-Centaur on November 3, destined for Venus. The craft was to make a close approach to the "veiled planet," then use Venus' gravity to fling it toward the innermost planet, Mercury.

As *Mariner 10* passed Venus on February 5, 1974, it relayed the first close-up photographs of that planet. Using ultraviolet filters, the craft showed startling detail in the clouds of Venus. Sequential photographs showed that 185 mile (300 kilometer) per hour winds in the upper atmosphere circle the planet every four days.

Next, *Mariner 10* set sail for Mercury, arriving at that planet on March 29, 1974. As the craft approached, it revealed the planet was very lunar-like, being almost completely covered by craters and lava fields. Because of its carefully selected orbital path, *Mariner 10* circled the Sun and intercepted Mercury two more times, first on September 21, then again on March 16, 1975. On the later encounters, the Mercurian surface was imaged to landforms as small as 440 feet (130 meters) high.

Because of Mercury's small mass, only six times that of the Earth's Moon, one would not expect any atmosphere. However, because it is so close to the Sun, the planet is exposed to a great deal of alpha and proton radiation. This results in a thin but detectable helium and hydrogen atmosphere. Mercury's weak magnetic field, about one percent of Earth's, fails to protect the sunlit side of the planet from intense radiation, while the night side leaves a wake of charged particles similar to a comet.

Mariner 10 also determined that Mercury rotates on its axis three times for every two orbits around the sun. While there is no permanent hot or cold side, there are two hot spots where the Sun is directly overhead during perihelion (closest to the Sun). The average planetwide temperatures range between 660 degrees Fahrenheit (620 degrees Kelvin) and –270 degrees Fahrenheit (100 degrees Kelvin). The mysterious hot spots were estimated to reach 1,380 degrees Fahrenheit (840 degrees Kelvin).

On all three passes, *Mariner 10* always encountered the same side of Mercury, and always viewed the same 35 percent of its surface. This included what is probably the largest feature, the Caloris Basin, a huge impact crater that may have come close to shattering Mercury 3.5 billion years ago.

The Soviets continued their domination of Venus exploration by sending *Venera 9* and *10* toward the planet on June 8 and 14, 1975. Being a new generation Venera weighing five tons, each carried a totally redesigned 1,450-pound (660-kilogram) lander which had, among many other instruments, an imaging system to show the surface of Venus.

Both craft entered orbit around Venus on October 22 and 25, 1975, respectively. Each successfully dropped landers to the surface which revealed a dusty, rubble strewn terrain made primarily of volcanic origin basaltic rock. The orbiters continued to return data about the planet for many months and greatly enhanced the understanding of our inner planetary neighbor.

America's final Mars effort to date began with the dual launching of the *Viking 1* and *2* spacecraft on August 20 and September 9, 1975. Each five-ton spacecraft comprised an orbiter and a lander. The nuclear powered landers were designed to search for signs of biological activity in the soil and report conditions on the Martian surface for several years. The orbiters were also to observe and image the planet during their multi-year mission.

Viking 1, shown in Figure 3-6, arrived at Mars in June 1976 and released its lander on July 20. The successful Mars landing by *Viking 1* caused a holiday-like festive atmosphere in America, which was mesmerized by the first search for life on another planet. Although *Viking 1*, and its companion lander which arrived on the Martian surface on September 3, 1976, were completely successful, they did not find any signs of life on Mars.

A Trip to the Outermost Planets

The American *Voyager 1* and *2* spacecraft marked the most ambitious era of exploration in the Space Age. After both craft visited Jupiter and Saturn, *Voyager 2* would

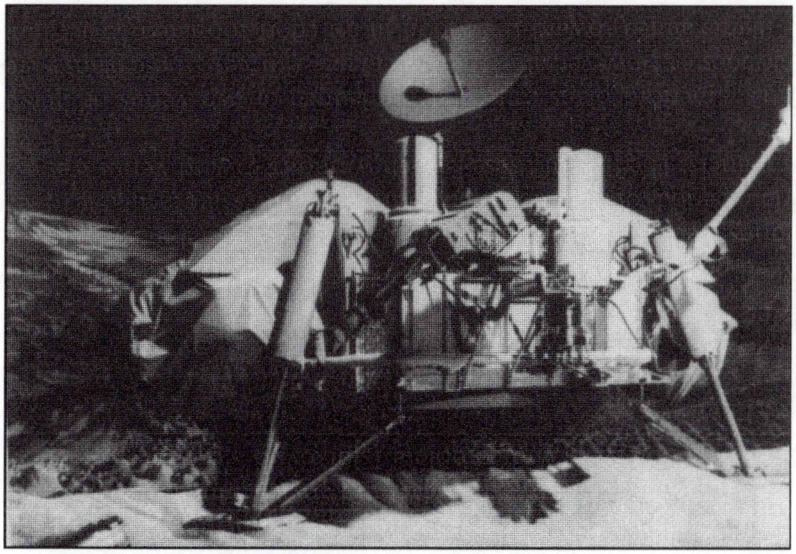

Figure 3-6. *Viking 1*, the first successful lander on Mars

continue on to the yet unexplored planets Uranus and Neptune. The twin 1,816-pound (825-kilogram) craft were launched in reverse order by Titan-Centaur boosters on August 20 and September 5, 1977. This reverse order was used because *Voyager 2*, launched first, would actually arrive at Jupiter after the later-launched *Voyager 1*.

Carrying vastly improved instruments over their *Pioneer 10* and *11* pathfinders, the Voyagers were to conduct in-depth studies of the outer planets and the miniature solar systems which swarmed around them in the form of dozens of moons. Carrying ten instruments, including high resolution color cameras, *Voyager 1* passed Jupiter on March 5, 1979, and continued on to Saturn for a November 12, 1980 rendezvous. *Voyager 2* followed, passing each planet on July 9, 1979 and August 26, 1981. As the Voyagers passed Saturn, a worldwide television audience was treated to live views of the distant planet.

After passing the "ringed planet," *Voyager 1* continued outward from the Sun. It will continue to transmit data about the far reaches of the solar system until its nuclear battery dies in the year 2012 at a distance of 11 billion miles (18 billion kilometers) from the Sun.

A grander fate awaited *Voyager 2*, shown in Figure 3-7. After passing Saturn, celestial mechanics allowed Saturn's gravity to push that craft toward the seventh

Figure 3-7. The Voyager satellite series

planet, Uranus. Completing a voyage of nearly 10 years, *Voyager 2* arrived at Uranus on January 24, 1986. The planet itself presented a featureless green disk to observers, but its large moon, Miranda, displayed a tangle of geological features which delighted planetary scientists.

Three and a half years later, *Voyager 2* paid its final planetary visit. On August 25, 1989, the craft swept past Neptune, providing the first data about many aspects of the outer "gas giant" planet.

Venus Again the Focus of Attention

The only American effort to land on Venus has been the dual *Pioneer Venus 1* and *2* missions launched May 20 and August 8, 1978. The two craft were launched by Atlas-Centaur boosters and carried a total of two dozen instruments between them. The 1,283-pound (582-kilogram) *Pioneer Venus 1* was a planetary orbiter which carried a radar mapping apparatus to image the surface of the planet through the obscuring cloud cover. *Pioneer Venus 2*, weighing in at 1,993 pounds (904 kilograms), carried four separate atmospheric entry probes, each targeted toward a different spot on the planet's globe.

The Venus orbiter arrived at the planet on December 4, 1978, and set about the task of radar mapping the planet. It revealed that Venus is much smoother than the Earth. The orbiter also studied the Venusian atmosphere and confirmed many of the previous Venera findings.

The *Pioneer Venus 2* atmospheric probe mission arrived at Venus on December 9, 1978, and peppered the planet with five separate spacecraft. While each encountered some form of difficulty because of the harsh Venusian environment, all were largely successful and painted a planetary picture similar to that received from the previous Soviet landers.

The Soviets used the 1978 Venus opportunity to land two more craft on the planet's surface. Unlike the previous pair, *Venera 11* and *12* were not orbiters, but instead flew past the planet as their landers descended. This allowed the craft to act as radio relays for the landers for a longer period of time.

After its September 9, 1978 launch, *Venera 11* arrived on Christmas day. *Venera 12* was launched five days after its twin, yet it arrived at Venus four days sooner. The two craft succeeded in landing probes on the surface. While the imaging system on both failed, other valuable data about atmospheric composition was gathered.

Not to be deterred, the Soviets pressed ahead with *Venera 13* and *14* two years later. This time, landings on March 1 and 5, 1981, were completely successful. Higher resolution color views of the landing site showed a lava-like surface composed of flat, cracked rock. The landers also carried drill mechanisms designed to dig into the surface and deliver soil samples to an X-ray fluorescence spectrometer, which then analyzed for 22 different chemical elements.

Veneras 15 and *16* followed in 1983, but carried no landers. Instead, both carried a large radar mapping apparatus for imaging the surface of Venus through the planet's cloud layers. Arriving on October 10 and 14, 1983, both orbiters carried out complimentary radar observations which extensively mapped the northern hemisphere of Venus.

International Study of Halley's Comet

The Soviets, taking note of *Voyager 2*'s success in using planetary gravity to propel the craft toward another target, proposed a daring mission for their 1984 Venus probe launches. The craft were to use Venus' gravity to boost them toward Halley's Comet in 1986. In addition to Venus landers, each craft carried a scan platform which housed specialized comet study instruments.

Called *VEGA 1* and *2*, the two five-ton craft arrived at Venus on June 11 and 15, 1985, and ejected landers onto the Venusian surface. During descent, each lander also released a balloon probe which drifted high in the Venusian atmosphere for 48 hours, relaying wind and cloud particle data to Earth. The landers also performed complex chemical analyses of the atmosphere and surface of Venus.

After looping past Venus, the two VEGA spacecraft continued toward a rendezvous with Halley's Comet. *VEGA 1* flew within 6,000 miles (9,000 kilometers) of the comet's nucleus on March 6, 1986, relaying close-up photographs. These were valuable for targeting the European Space Agency's *Giotto* spacecraft which would later pass even closer to the comet. *VEGA 2* approached within 5,000 miles (8,000 kilometers) three days later.

The first two Japanese deep space missions were designed to participate in the international Halley study. The first, launched on January 7, 1985, was called *Sakigake*, or "Pioneer." It was essentially a practice mission to gain experience in deep space tracking. The second, called *Suisei*, or "Comet," was launched on August 18, 1985, and carried the actual cometary study instruments.

Unlike the Soviet and European Halley spacecraft, the Japanese comet probes would not make a close approach to the comet. On March 8, 1986, *Suisei* came within 93,000 miles (150,000 kilometers) of Halley and studied the interaction of solar radiation with the comet by photographing it with an ultraviolet imaging system.

Likewise, Europe flew its first deep space mission with the July 2, 1985 launch of the 1,545-pound (960-kilogram) *Giotto* spacecraft toward the nucleus of Halley's Comet. *Giotto* was specially armored for a close approach through the dusty cometary coma (head) in order to get high resolution images of the nucleus of Halley's Comet.

On March 14, 1986, the craft passed 376 miles (605 kilometers) from Halley. Ten instruments, including a color camera, studied the comet extensively as the probe passed the comet at a relative velocity of 42 miles (68 kilometers) per second. At this speed, impact with any appreciable sized comet debris would destroy the spacecraft. Miraculously, the craft survived and relayed valuable data about activity near the comet.

The Soviet Union attempted to return to Mars in 1988 with twin launches of its new class of Mars explorer, the 13,715-pound (6,200-kilogram) Phobos spacecraft. Loaded with an international cargo of instruments and special landers which were to actually touch down on the largest of Mars' two moons, *Phobos 1* and *2* left Earth on July 7 and 12, 1988.

Worldwide optimism was high that Soviet probes would finally succeed in reaching their goal after 28 years of Martian failures. However, this was not to be. In early September, an accidental ground controller error irrevocably shut down the *Phobos 1* probe before it reached Mars. On January 29, 1989, *Phobos 2* arrived at Mars and entered orbit. After several months of maneuvering in Martian orbit, the craft was within days of depositing its lander on the moon Phobos when it suffered control failure and the mission was lost.

A total of 11 years was to pass after the twin flight of Pioneer Venus before the United States would launch another planetary probe. The *Magellan* Venus radar mapping spacecraft was launched from the Space Shuttle in May 1989, on a long, looping trajectory which did not reach Venus until August 10 of the followingyear.

After a radar calibration and trouble shooting period, the craft began to use its synthetic aperture radar on September 15, 1990, to penetrate the Venusian cloud cover and map the surface with a 400-foot (120-meter) resolution. In the first three days, it covered only 1.5 percent of the planet, yet returned more data than *Mariner 9* did during its entire mission to Mars. When the first mapping cycle of the whole planet was complete, the imagery created 3,070 gigabits of unprocessed data. This was more raw data than all previous planetary missions combined.

Magellan is completing mapping cycles of the entire Venusian globe every 243 days and has revealed an evolving planetary surface featuring active volcanoes and huge impact craters.

As of early 1992, another probe is en route to Jupiter. The NASA sponsored *Galileo* probe is scheduled to orbit the giant planet and return images with up to ten times the detail shown by the Voyagers. At the same time, the European Space Agency (ESA) is sponsoring the *Ulysses* probe. After looping around Jupiter in early 1992 and using the planet's immense gravity to boost the spacecraft above the plane of the solar system, *Ulysses* will then arc back over the top of the Sun and enter the never before studied solar polar regions.

From the flight of *Vostok 1* in April 1961 up to Apollo-Soyuz in July 1975, the U.S. and Soviets had 59 manned space missions. This is a short biographical list of some of the more prominent astronauts and cosmonauts who were involved in the space race.

An Abbreviated Who's Who of Astronauts and Cosmonauts

There are so many brave space travelers that deserve mention in the chronicles of space exploration. Only a brief few can be covered in this chapter. There are numerous books and publications that properly cover individual careers in deserved detail. For more information on astronauts and cosmonauts, the author recommends *Who's Who in Space* by Michael Cassutt.

Astronauts of the United States

During the space race, there were numerous groups of astronauts recruited for this daring endeavor, space exploration. Here is a list of a few of America's space explorers during the race to the Moon.

Buzz Aldrin: "Doctor Rendezvous"

Edwin (Buzz) Aldrin was born on January 20, 1930, to parents who both had military backgrounds. His father was an aid to the visionary leader of "air power," General Billy Mitchell during the 1920s and his mother was the daughter of an Army Methodist chaplain.

While in high school during World War II, Aldrin was very active in sports and dreamed of someday becoming a fighter pilot. Buzz received an appointment to the United States Military Academy and graduated from West Point in 1951. As the Army Air Corps now had become a separate branch of the military, Aldrin transferred into the Air Force. After 18 months of training, Buzz qualified as a F-86 fighter pilot. In December 1952, he was shipped to Korea and went head-to-head against the vaunted MIG-15. By the end of the war, he had flown 66 combat missions and was credited with two MIG kills.

A decade later, the space race between the superpowers quickly caught Aldrin's interest and attention. After Gagarin's flight and Kennedy's Moon speech, Buzz was determined to be an astronaut. While stationed in Germany, Aldrin learned of his friend Ed White's selection as an astronaut. White had many hours as a test pilot which helped in his selection. Lacking this qualification, Aldrin went to the Massachusetts Institute of Technology for 18 months and completed his doctorate in 1963, specializing in orbital rendezvous. At the time, Aldrin was considered the world's foremost expert on rendezvous and docking procedures.

NASA was obviously impressed with the results of Aldrin's thesis and selected him in late 1963 to be part of the third group of astronauts. Michael Collins, with whom he would eventually fly to the Moon, was also in that group.

As an astronaut candidate, Aldrin experienced some initial frustration in being an orbital rendezvous specialist yet not being slated for any of the early Gemini rendezvous missions. NASA policy on spaceflight crew scheduling dictated that the backup crew on Gemini missions became the prime crew three missions later. Aldrin and fellow astronaut James Lovell were the backup crew on the *Gemini 10* mission, which meant they would be the prime crew for the nonexistent *Gemini 13*. Tragedy as well as fortune then crossed Aldrin's life. On February 28, 1966, the *Gemini 9* prime crew, Charlie Bassett and Elliot See, fatally crashed during a routine jet flight to St. Louis to inspect their spacecraft at the factory. As result of this accident, the backup crew of Tom Stafford and Gene Cernan flew the *Gemini 9* mission and all the other crews moved up one slot in mission rotation. This put Aldrin and Lovell on *Gemini 12*.

Throughout the development of the Gemini program, Aldrin actively campaigned his theories on manually piloting orbital rendezvous with the help of simple reference charts instead of by computer. The majority of the NASA leadership ignored Aldrin's theories and felt that a computer was required for the complex task. During his *Gemini 12* flight, fortune crossed Aldrin's path again. Just before

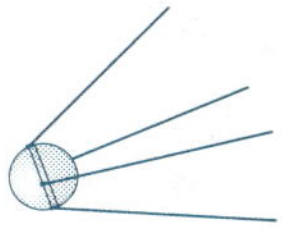

the rendezvous sequence was about to begin, there was a computer data loss. Rather than delay rendezvous for several orbits while the problem was corrected, Aldrin, shown in Figure 3-8, used his charts to manually plot the correct rendezvous with great success.

Aldrin trained extensively for EVA (Extra Vehicular Activity, or "space walk") and is shown training in Figure 3-9. In preparation for his *Gemini 12* mission, Aldrin came up with a novel method of training for his space walk. After he discovered that swimming underwater simulated weightlessness, he began training for his EVA in a water tank. As a result, his EVA on the final Gemini mission was the only EVA considered a total success. Previous spacewalkers had trouble controlling their motions in the weightlessness of space.

Three years after his Gemini flight, Aldrin participated in the first manned landing on the Moon. As *Apollo 11* descended, Aldrin called out the altitude and speed coordinates as his crewmate, Neil Armstrong, piloted the lunar module *Eagle* to a touchdown on the Moon. The landing spot is now known to the world as "Tranquility Base."

After the Moon mission and a world goodwill tour which lasted nearly a year, Aldrin resigned from NASA to serve briefly as the commandant of the Air Force Research Pilot School at Edwards Air Force Base, California. He is now retired and spends his time as a writer as well as spokesman and advocate for future space exploration.

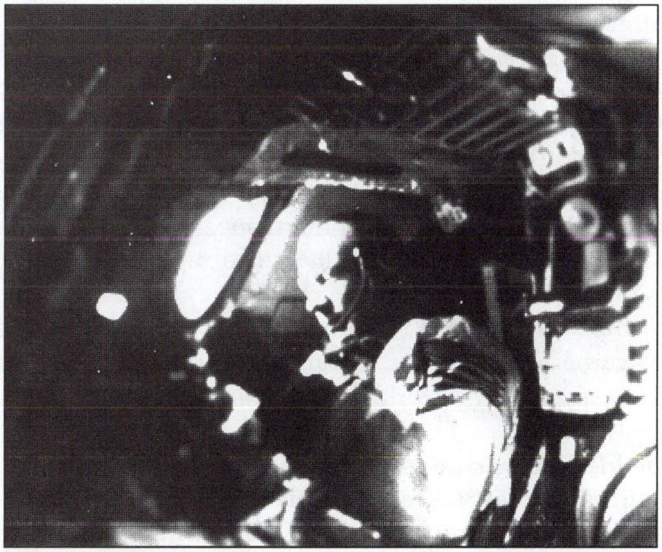

Figure 3-8. Buzz Aldrin aboard *Gemini 12*

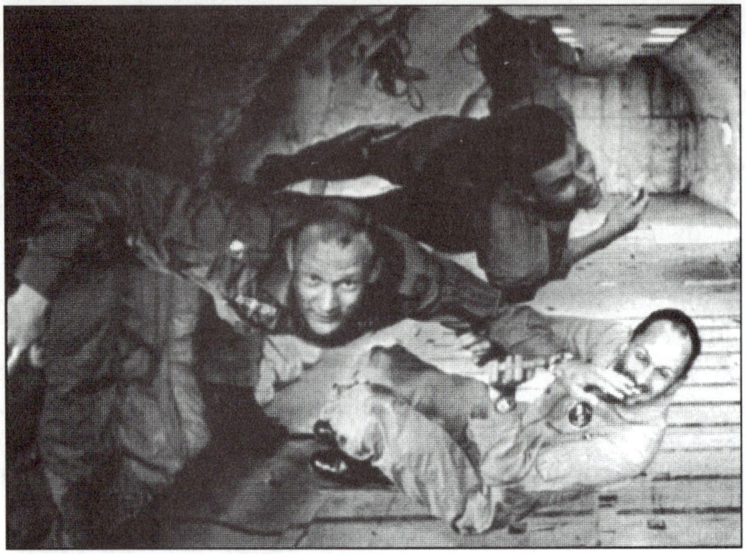

Figure 3-9. Buzz Aldrin training with astronauts Bassett and Freeman in the "vomit comet"

Aldrin received the Presidential Medal for Freedom, the NASA Exceptional Service Medal, the NASA Distinguished Service Medal, and was decorated by 11 different countries.

Neil Armstrong: First to Step on Another World!

On July 20, 1969, civilian astronaut Neil Armstrong, shown in Figure 3-10, was the first human to walk on the lunar surface.

As he stepped off the ladder of the lunar module, Armstrong spoke the now historic phrase, "That's one small step for man, one giant leap for mankind."

Testing the firmness of the lunar soil, Armstrong quickly proved it was safe enough to walk on as he gathered a small contingency sample. His crewmate, Buzz Aldrin, joined Armstrong and the two men setup various scientific experiments. They continued gathering soil and rock samples and then planted the American flag. President Nixon made a historic phone call (via NASA communications) to congratulate the two men.

In his youth, model plane building helped spark Armstrong's interest in flight. On his 16th birthday, he received his pilot's license. Under a U.S. Navy scholarship, he enrolled at Purdue University to study aeronautical engineering. In 1948, Armstrong was called to active duty and sent to flight school. Before

Figure 3-10. Neil Armstrong (inside the LM) was the first man on the Moon

he could return to Purdue to resume his studies, the Korean War erupted. Armstrong was assigned to Fighter Squadron 51 and flew numerous bridge-breaking and tank-busting sorties. On one mission, Armstrong had to eject after his F9F-2 jet fighter lost a wing.

Returning to the States, he graduated from Purdue with a degree in aeronautical engineering in 1955.

As a civilian, Armstrong went to work for NASA's forerunner, the National Advisory Committee on Aeronautics (NACA) at Lewis Flight Propulsion Laboratory in Cleveland, Ohio. Later, Armstrong transferred to Edwards AFB as a test pilot.

At Edwards, he gained valuable experience by flying the F-102A, the X-5, the Parsev, and six flights in the X-15 rocket plane.

With such credentials, Armstrong was accepted into the second group of NASA astronauts in 1962. After survival training, he assisted in the development of the Gemini spacecraft. Armstrong was assigned as backup commander for *Gemini 5* and *Gemini 11* and was capcom for *Gemini 5* and *Gemini 9*.

Armstrong had several close calls with death during his astronaut career. His first space mission was with David Scott on *Gemini 8*. All was going well on the flight as the spacecraft reached orbit and the astronauts performed the first docking to the unmanned Agena. Suddenly the Gemini capsule and the Agena Docking Module began to swing violently. The astronauts assumed that the Agena was the culprit and detached the capsule. Then the astronauts' worst fear was realized: the Gemini capsule tumbled at an even faster rate. It was actually an erratic thruster on the Gemini capsule spewing precious fuel which was the cause. Spinning one revolution per second, Armstrong and Scott were close to losing consciousness as they struggled to regain control. At this point, several NASA officials thought the crew would blackout and never return home. By using their re-entry thrusters the astronauts were able to stabilize the spacecraft. NASA quickly aborted the mission, sending *Gemini 8* on a splashdown 600 miles south of Japan. The unplanned re-entry left the two men bobbing in the sea for nearly four hours. The destroyer *Leonard F. Mason* picked up two seasick, exhausted astronauts. A planned three-day spaceflight turned into a tense, ten-hour terminated mission.

> *"That's one small step for man, one giant leap for mankind."*
>
> —Neil Armstrong, as he stepped down onto the lunar surface, July 20, 1969

On May 6, 1968, while practicing lunar landings in the free-flight trainer also known as the "Flying Bedstead," Armstrong "punched out" in the nick of time before the craft crashed and exploded. When the vehicle malfunctioned and became uncontrollable, the cool-headed pilot ejected and landed with the parachute finally deploying just as his feet just touched the ground.

After *Apollo 11*, and a brief stint as deputy associate administrator for aeronautics at NASA Headquarters in Washington, D.C., Armstrong resigned from NASA in 1971 and became a professor of engineering at the University of Cincinnati. In 1984, he served on the National Commission on Space. After the *Challenger* disaster, Armstrong was appointed vice chairman of the *Challenger* Committee to determine the causes of accident.

A private individual, Armstrong avoids public appearances and lives with his family in Lebanon, Ohio.

He received the Presidential Medal for Freedom, the NASA Distinguished Service Medal, the NASA Exceptional Service Medal, the Congressional Space Medal of Honor, and has been decorated by 17 different countries.

Frank Borman: First Around the Moon!

"In the beginning God created heaven and earth...." On Christmas Eve, 1968, *Apollo 8* orbited the Moon. Commander Frank Borman and crewmates Jim Lovell and Bill Anders read the stirring passage from the Book of Genesis in the Bible. As these words were spoken, the public also watched on television and saw a small fragile distant Earth rise above the lunar surface. The success of this mission put the United States in the lead position for landing men on the Moon.

Originally, *Apollo 8*'s mission plan was for an Earth-Orbital LM test, but the LM was far behind schedule and stories of an upcoming Soviet launch altered the mission. As the unmanned Zonds returned from the Moon, the White House, CIA, and the NASA leadership nervously awaited the next Soviet launch for the Moon. They knew it would be manned and the Soviet launch window was two weeks earlier than any American attempt.

The United States responded to the threat of a Soviet lunar Pass by announcing that *Apollo 8* would orbit the Moon. Frank Borman was offered the choice of later lunar landings missions or the first manned launch on a Saturn V rocket. Borman opted for the "gutsy" lunar orbital. These plans were all made even before the first test flight of *Apollo 7*.

It was only a year earlier that the crew of *Apollo 1* tragically died in a fire in the capsule. The cause of the fire had to be found and the spacecraft had to be redesigned from top to bottom. Borman's tough no-nonsense approach led to his appointment as the astronauts' representative to the review board in charge of the investigation. He also led the "tiger-team" which oversaw the new safety procedures and re-engineering of the Apollo spacecraft.

It was Borman's persuasive leadership that helped to get the Apollo spacecraft flying again. *Apollo 7*'s test flight was flawless which made it "go" for Borman's historic flight around the Moon. After completing ten lunar orbits, the Apollo capsule fired its engine for trans-earth-injection burn and sent the crew home.

Borman graduated from West Point in 1950 and received an M.S. in aeronautical engineering at Caltech in 1957. He transferred to the Air Force and became a pilot in 1951. After an overseas assignment in the 44th Fighter-Bomber Squadron in the Philippines, Borman moved up to flight instructor stateside several years before attending Test Pilot School at Edwards AFB.

He was selected in the second group of astronauts and was quickly groomed for assignment in the first Gemini flight, until his commander Alan Shepard was medically grounded. He later served as backup to *Gemini 4* before he became commander of *Gemini 7*. On December 4, 1965, Frank Borman (shown in Figure 3-11) and Jim Lovell soared into space on a Titan rocket. Returning in good health, the two men set a new space endurance record, just shy of 14 days. This long flight gave the medical clearance for lunar duration missions.

After *Apollo 8*, Borman became field director of NASA's Space Station Task Force. In 1970, he resigned from NASA and became chairman of Eastern Airlines. After 16 years, and a tumultuous relationship with the unions, he resigned from Eastern Airlines.

Borman received the NASA Exceptional Service Medal.

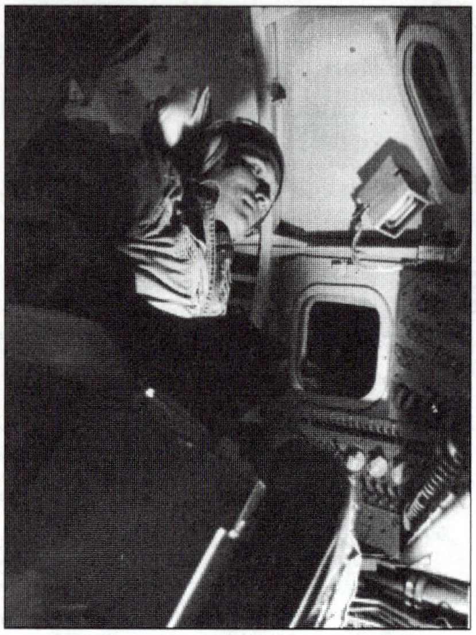

Figure 3-11. Frank Borman in training before *Gemini 7* flight

Mike Collins: First Manned Mission to the Moon

In the command ship *Columbia*, Mike Collins circled the Moon while Buzz and Neil stepped down on its surface. For nearly 24 hours he was alone, monitoring spacecraft systems and photographing the lunar surface before he was reunited with his crewmates.

He flew into space once before on *Gemini 10*, on July 18, 1966. The mission was a successful three-day flight, the craft joining the Agena Docking Module with Collins performing two separate space walks.

Following a family of military officers (his father was a major general, one uncle was brigadier general, and another became Army Chief of Staff), Collins graduated from West Point in 1952. He elected to transfer to the new Air Force and completed pilot school in 1953. In 1960, he completed USAF Test Pilot School at Edwards AFB and remained a test pilot until he was accepted as an astronaut by NASA.

He was part of NASA's Group Three astronauts, recruited in October 1963. His first assignment was to assist in the development of space suits for EVAs. Collins was the first Group Three astronaut assigned to a flight crew (as backup pilot on *Gemini 7*). After his flight on *Gemini 10*, he was assigned to *Apollo 8* as lunar pilot, but unfortunately was diagnosed with having a bone spur on his neck and was removed from flight status for surgery. After recovery he was available for assignment on *Apollo 11*. Deke Slayton offered Collins the opportunity to backup *Apollo 14* and then command *Apollo 17* (which would have allowed him to walk on the

Moon). Collins opted for the success of *Apollo 11* and wished to spend more time with his family.

He resigned from NASA in January 1970, and became Assistant Secretary of State for Public Affairs. The following year, he began first as Director of the National Air and Space Museum and later became Under Secretary for the Smithsonian.

Collins received the Presidential Medal for Freedom (the highest award for a citizen), the NASA Distinguished Service Medal, the NASA Exceptional Service Medal, and decorations from 11 different countries.

John Glenn: First American to Orbit the Earth

It was on February 20, 1962, that John Glenn and the Mercury capsule *Friendship 7* lifted off into space. Even though two Soviets had already orbited the Earth, it was Glenn's flight that captured the world's imagination. The launch and flight were broadcast live to millions around the world.

Upon reaching orbit, Glenn was given a "go" for seven orbits. At the beginning of the second orbit, flight controllers detected that the heat shield had torn loose. This would have been a catastrophic end for Glenn during re-entry. For several minutes, controllers debated procedure and kept Glenn in the dark. Finally, flight control instructed Glenn keep the capsule's retro-rocket pack attached to the capsule. Glenn demanded to be informed of the situation. Now fully aware of the potential danger he was in, Glenn began re-entry on his third orbit. Chunks of burning metal and flames flew past Glenn's window. There was no way to tell if it was the retro-rocket pack and heat shield, or indeed the entire capsule burning up.

Glenn splashed down in the Atlantic ocean, 800 miles southeast of Cape Canaveral. He was greeted with a thunderous welcome and ticker tape parade in New York City. Glenn's flight was reminiscent of Charles Lindbergh, making the astronaut a national hero overnight. He was greeted by President John Kennedy, who saw Glenn as future political ally and urged him to consider a seat in the Senate.

John Glenn was born July 18, 1921, in Cambridge, Ohio. He was attending Muskingum College when World War II began. Glenn entered the Naval Aviation Cadet Program and received his commission in the Marine Corps. After flight training, Glenn was assigned to the 155th Marine Fighter Squadron. He flew 59 combat missions in the South Pacific.

He returned to the States in 1948 as a flight instructor at Corpus Christi, Texas. Glenn was transferred to Fighter Squadron 311 and flew 63 combat missions in Korea. He flew 27 combat missions for the U.S. Air Force as an exchange pilot. During his last nine days, Glenn manage to shoot down three MIG fighters near the Yalu River.

In 1954, Glenn transferred to the U.S. Naval Test Pilot School at Patuxent River, Maryland, and also received his B.S. in engineering from Muskingum College. He became project officer and was assigned to the Navy Bureau of Aeronautics in

Washington, D.C. He proposed, planned, and piloted an F8U on a record setting transcontinental jet flight from Los Angeles to New York in three hours and 23 minutes on July 16, 1957. This feat garnered him guest appearances on television's "I've Got a Secret" and "Name that Tune" on which he won $25,000.

After the furor of Sputnik, Glenn volunteered as a test subject for studies on manned spaceflight with the NACA. He was often used as a test subject in the centrifuge. When the call for astronauts came, Glenn, shown in Figure 3-12, was selected in the first group in April 1959. He served as backup pilot for Alan Shepard and Gus Grissom before he piloted *Friendship 7*.

After his Mercury flight, Glenn was assigned to the Apollo program. It was not known by Glenn, but Kennedy secretly instructed NASA not to risk sending Glenn into space again. After sensing that he would be the oldest permanent astronaut in training, Glenn left the astronaut group. In January 1964, he announced his candidacy for the Senate. One month later, Glenn slipped and fell while taking a shower and suffered a serious head injury. Facing a long period of recuperation, Glenn was forced to withdraw from the race.

Glenn became a consultant to NASA and remained in Houston. He also worked as an executive for Royal Crown Cola and was very successful with investments in hotel developments. In 1970, he was defeated in the Senate primary against Metzenbaum. Badly organized and underfunded, Glenn learned from this defeat and won a seat in the Ohio Senate in 1974.

Figure 3-12. John Glenn has his space suit fitted before his historic flight on *Friendship 7*

Glenn served his first term with distinction and was considered as Carter's running mate in the 1976 presidential election. In 1983, he announced his candidacy for the office of President of the United States. The lack of national organization forced Glenn to withdraw during the democratic primaries.

As senator, he suffered from association with the Keating scandal and only narrowly won re-election in 1992.

Glenn received the NASA Distinguished Service Medal.

Virgil Ivan "Gus" Grissom: Second American in Space

On July 21, 1961, Gus Grissom became the second American in space. On the suborbital Mercury flight, *Liberty Bell 7*, Grissom quickly demonstrated manual control of the capsule and made visual observations of the Earth as navigational references. The flight was flawless up through splashdown. While the capsule bobbed in the Atlantic Ocean under strong crosswinds, a faulty circuit caused the explosive hatch bolt to initiate. With the hatch open, the capsule became flooded and began to sink. Grissom jumped clear into open water. He was saved in the nick of time by recovery divers before the weight of the space suit could drag Grissom underwater.

Grissom was born April 3, 1926, in Mitchell, Indiana. In 1944, after graduating high school, he enlisted in the U.S. Army Air Corps. Grissom had begun pilot training when World War II ended. He left the Army Air Corps and graduated in 1950 from Purdue University with a B.S. in mechanical engineering. Grissom re-enlisted after graduation and received his wings in 1951.

Grissom was assigned to the 75th Fighter Squadron and flew 100 combat missions in the Korean War.

In 1952, he became an instructor pilot and later a test pilot at Edwards AFB. Grissom, shown in Figure 3-13, had over 3500 hours in jets and was selected in April 1959, as one of NASA's Original Seven astronauts. Grissom, along with Shepard and Glenn were selected by NASA as the Gold Team, the first three astronauts to fly into space.

NASA assigned Grissom as commander with rookie astronaut John Young as the first crew on *Gemini 1*. With a sense of humor, Grissom named the two-man capsule "Molly Brown" after the Broadway play *The Unsinkable Molly Brown*. Grissom and Young went into space on March 23, 1965. On the test-flight of the *Gemini 1*, Grissom maneuvered the spacecraft and returned after three orbits. Grissom and Young were later assigned as backup crew for *Gemini 6*.

Grissom was then chosen to command the first flight of the Apollo spacecraft. Grissom, Ed White (America's first spacewalker), and rookie Roger Chaffee were scheduled to fly a 16-day shakedown mission of the Apollo spacecraft in February 1967. On January 27, 1967, the *Apollo 1* crew was rehearsing a simulated countdown inside the capsule atop the Saturn 1B rocket. Suddenly a wire shot out a spark that

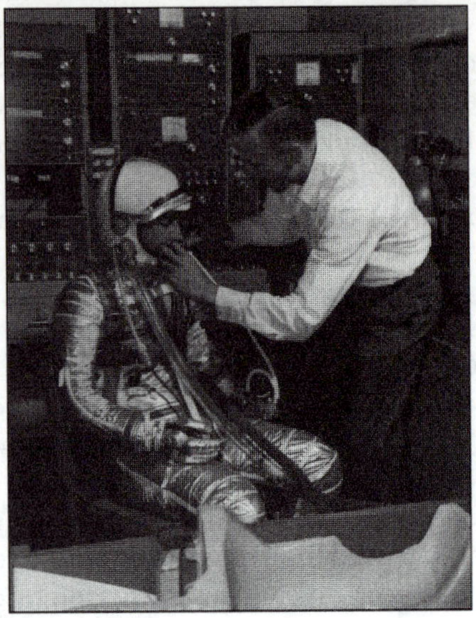

Figure 3-13. Gus Grissom during his Mercury pre-flight check

quickly ignited the pure oxygen inside the capsule into an inferno. Pressure and heat suddenly rose inside the sealed capsule. Helpless engineers stood by hearing the men plead for help via radio. Grissom, White, and Chaffee, trapped inside their capsule, died of asphyxiation. Grissom was buried with full military honors at Arlington Cemetery.

NASA was severely criticized for lack of safety standards. As a result, the Apollo program was grounded for over 18 months, and the capsule was completely redesigned.

According the reports, if the *Apollo 1* disaster had never occurred, Grissom would have been assigned to the first Moon landing. He might have been the first man to walk on the Moon. Grissom was awarded two NASA Distinguished Service Medals, the NASA Exceptional Service Medal, and posthumously, the Congressional Space Medal of Honor.

Alan Shepard: America's First Spaceman

On May 5, 1961, at 9:34 A.M., on a televised live broadcast to millions of Americans, Alan Shepard rode aboard the Mercury-Redstone into space. The entire suborbital flight took less than 16 minutes and Shepard experienced zero-G for only a few minutes. More importantly it established that the United States was back in the space race. Three weeks earlier, the Soviets had launched Yuri Gagarin into orbit as the first man in space. Originally, Shepard's flight was scheduled to launch in early April, several weeks before the Gagarin flight. But NASA decided to have one more test flight with a chimp occupant, which performed flawlessly. Shepard would have been the first man in space.

Shepard was born November 18, 1923, in East Derry, New Hampshire. He graduated from the U.S. Naval Academy in 1944. He served aboard the *Cogswell*, a destroyer in the Pacific. After World War II, Shepard entered flight school and received his wings in 1947. After several stateside base assignments, he served as operation officer of Fighter Squadron 193 for two tours, between 1953 to 1956, on the carrier USS *Oriskany*.

In 1956, he became a test pilot and instructor at the U.S. Navy Test Pilot School and also graduated from the Naval War College. Shepard was on his way to becoming commander of his own carrier squadron when he was selected by NASA to become an astronaut.

Shepard was highly rated by his fellow astronauts as a pilot and engineer. NASA and the other six astronauts both selected Shepard, who is shown in Figure 3-14, to be the first astronaut to fly into space. After his historic suborbital flight, he was awarded a Congressional Medal of Honor by President Kennedy.

Shepard was then assigned as capcom for the second (Grissom) Mercury flight and third (Glenn) Mercury flight. He was also backup pilot to Gordon Cooper on the last Mercury flight.

Figure 3-14. Alan Shepard gets suited up before his Mercury flight

After six successive manned Mercury missions, a seventh manned Mercury flight was lobbied heavily by several NASA engineers and Shepard. This three-day flight would have been scheduled for late 1963. NASA leadership, however, thought that the Gemini program should be given top priority. Possibly to replace the loss of the proposed Mercury mission, Shepard was assigned with Frank Borman to the first Gemini flight.

Unfortunately, an inner ear ailment grounded Shepard from flight status. He was reassigned Chief of the Astronaut Corps and was responsible for astronaut training, pilot evaluations of spacecraft designs, and mission planning.

In early 1969, Shepard underwent a successful ear operation and was restored to flight status. Deke Slayton assigned Shepard as Commander of *Apollo 14*, enabling Shepard to become the fifth man to walk on the surface on the Moon. Shepard and Edgar Mitchell explored the hills of the Frau Mauro region. They were on the lunar surface for 33 hours and performed over 9 hours of extra-vehicular activities. Shepard, shown in Figure 3-15, was the only Original Seven Mercury astronaut to walk on the Moon. He was also the first to ever play golf on the Moon.

Figure 3-15. Alan Shepard, the only Mercury astronaut to walk on the Moon

After returning from the Moon, he was promoted to the rank of Rear Admiral and resumed his duties as Chief of the Astronaut Office. In 1971, Shepard was a delegate to the United Nations General Assembly.

Shepard retired from NASA and the Navy in August 1974.

Shepard was awarded the Congressional Space Medal of Honor, the NASA Distinguished Service Medal (twice), and the NASA Exceptional Service Medal.

Deke Slayton: Test Pilot's Test Pilot

As one of the Original Seven Mercury astronauts, Slayton was designated for the first orbital manned mission. Originally, astronauts Shepard, Grissom, and Glenn were all scheduled for suborbital flights.

NASA's manned space agenda was approached with caution, and this gave the Soviets the lead in the space race with two cosmonauts already having flown separate orbital missions. With two successful suborbital flights (Shepard and Grissom) under their belt, NASA was probably feeling some political pressure to alter Glenn's flight to orbital status. This gave Slayton the second orbital mission.

After John Glenn's historic orbital flight, NASA announced that Glenn's backup pilot Scott Carpenter would replace Slayton as the next pilot of the second orbital mission. Slayton was grounded due to an erratic heart rate. Known as idiopathic atrial fibrillation, it was nothing more than an occasional irregularity of a muscle at the top of the heart. It was a minor occurrence that would not interfere with any of Slayton's abilities. Nonetheless, it would cost him his pilot seat the next space mission. NASA decided that since there were other astronauts available who were in "perfect health," Slayton should be replaced.

Slayton would have to wait 13 years before he would finally fly into space.

Donald "Deke" Slayton was born on a Wisconsin farm on March 1, 1924. In 1942, at the age of 18, he enlisted in the U.S. Army Air Corps and received his wings as a B-25 bomber pilot. He flew 56 combat missions in the European theater and seven combat missions over Japan. After the war, Slayton graduated from the University of Minnesota with B.S. in aeronautical engineering. At the outbreak of the Korean Conflict, he was recalled to active duty and assigned to the 12th Air Force Headquarters. In 1955, Slayton transferred to the Air Force Flight Test Pilot School at Edwards AFB. He was a test pilot at Edwards from 1956 to 1959. During his career, Slayton piloted over 7000 hours.

Slayton was selected as one of the Original Seven astronauts in 1959 and was considered the best pilot among the group. After his cancellation of flight status, Slayton became Director of Flight Crew Operations which oversaw the astronaut training and the selection of flight crews for all Gemini and Apollo missions. Slayton had complete control of astronaut crew selection. It was only when he assigned

Alan Shepard to command *Apollo 13* that NASA overruled his decision. NASA felt that Shepard needed more training and was reassigned to *Apollo 14*.

From 1965, Slayton, shown in Figure 3-16, persisted in his attempts to regain his flight status. He often joined his fellow astronauts in the training sessions and in 1966, he was nearly assigned to *Apollo 2*. There was a small chance that he might have the opportunity to fly the last Apollo flight to the Moon. In 1972, he finally regained flight status. Unfortunately, budget cuts had already curtailed any additional Moon flights. The Apollo program was coming to a close, and there would be no more flights into space until the Space Shuttle became operational. By then, Slayton would be too old for consideration.

In 1972, the Cold War began to thaw. Détente offered the opportunity of a joint U.S.-Soviet space mission.

In the mission profile, an Apollo would rendezvous and dock with a Soyuz spacecraft in orbit. After some campaigning with the NASA leadership, Slayton was assigned as docking-module pilot for *Apollo 18*, the Apollo-Soyuz Test Project (ASTP). It was the last Apollo flight, the last Saturn launch, and the beginning of the end of the Cold War.

Slayton and his crewmates trained for two and a half years for ASTP and were the first astronauts to train in the Soviet Star Town (cosmonaut training center). He also became close friends with his Soviet counterparts and former rivals, Alexei Leonov and Valery Kubasov. On July 17, 1975, *Apollo 18* docked with *Soyuz 19* and

Figure 3-16. Deke Slayton waited 13 years to fly in space

the crews exchanged handshakes and conducted joint experiments. After two days, the spacecraft separated and five days later, *Apollo 18* returned to Earth. At splashdown, there was a fuel leak that seared the astronauts lungs. The astronauts were treated and suffered no permanent ailments.

After ASTP, Slayton supervised the Approach and Landing Tests of the Space Shuttle and the first four orbital shuttle flights. In 1981, he resigned from NASA and started up a Texas space firm which launched the Conestoga rocket.

Slayton was awarded three NASA Distinguished Service Medals and the NASA Exceptional Service Medal.

Ed White: America's First Spacewalker!

On June 3, 1965, astronaut Ed White, in a specially designed space suit, stepped outside the Gemini spacecraft hatch and "walked" in space. Tethered with a life-supporting umbilical cord to *Gemini 4*, White propelled himself with a maneuvering gun. White remained outside the capsule for 22 minutes as Jim McDivitt photographed him (see Figure 3-17). White was the first American to walk in space. On March 18, 1965, cosmonaut Alexei Leonov was the first Soviet to space walk.

White was born November 14, 1930, in San Antonio, Texas. He graduated from the U.S. Military Academy in 1952 and later received an M.S. in aeronautical engineering at the University of Michigan.

Figure 3-17. Ed White, first American to walk in space

After joining the Air Force and receiving his wings in 1953, White was stationed for four years in Europe and flew F-86s and F-100s. In 1959, he entered the U.S. Air Force Test Pilot School at Edwards AFB.

As a test pilot, White applied to NASA and was chosen into the second group of astronauts on September 17, 1962. After *Gemini 4*, he was assigned as backup commander to *Gemini 7*. NASA was grooming White for the eventual lunar mission. He was assigned as senior pilot for *Apollo 1*. White and crewmates Grissom and Chaffee tragically died in a capsule fire during a simulated countdown on the pad. He was buried at West Point.

He received the NASA Exceptional Service Medal.

John Young: Six Times in Space!

John Young was the only astronaut to have flown four different spacecraft, Gemini, Apollo, Lunar Module, and the Space Shuttle. Young was selected in the second group of astronauts, September 17, 1962. He was teamed up with Gus Grissom for the first flight of the Gemini spacecraft. Young's first flight into space on *Gemini 1* was for three orbits. (In lieu of normal space cuisine he smuggled a pastrami sandwich on board!) He was assigned again with Gus Grissom as backup pilot for *Gemini 6*.

For nine years, Young was in constant training. In addition to the missions he flew into space, he acted as backup command module pilot for *Apollo 7* and was the backup commander on *Apollo 13* and *Apollo 17*.

Young was born September 24, 1930, in San Francisco, California. He graduated from the Georgia Institute of Technology with a B.S. in aeronautical engineering in 1952. Young was also in ROTC and after graduation served aboard the destroyer USS *Laws*. A year later, he entered pilot training and was eventually assigned to Fighter Squadron 103 from 1955 to 1959.

In 1959, Young entered U.S. Navy Test Pilot School and afterwards was assigned to the Naval Air Test Center. Young flew the F-4 Phantom for a world time-to-climb record. In 1962, NASA selected Young into the second group of astronauts. He was the first of his group to go into space.

After *Gemini 1*, Young next flew into space as commander on *Gemini 10*. With Mike Collins, the two astronauts maneuvered their spacecraft to rendezvous with two separate Agena docking modules. They also set an altitude space record of 477 miles. The three-day mission splashed down within 3.5 statue miles of the recovery carrier.

In May 1969, as command module pilot of *Apollo 10*, Young flew with Stafford and Cernan on the dress rehearsal flight for the lunar landing. Stafford and Cernan flew the LM within ten miles of the lunar surface while Young remained in the Apollo command module.

It was *Apollo 16* that took John Young, shown in Figure 3-18, to the lunar surface. It was also his fourth trip into space. With Charlie Duke, Young flew the

Figure 3-18. John Young salutes while on the lunar surface

LM to a touchdown in the highlands near the crater Descartes. The two men remained on the Moon for three days and collected over 200 pounds of Moon rocks. In the Lunar Rover, they traversed over 17 miles.

It was nearly a decade before Young would fly into space again, this time on the maiden flight of the STS-1.

After a lengthy production delay, the Space Shuttle lifted into space. On April 12, 1981, the Shuttle *Columbia* carried commander Young and rookie pilot Robert Crippen on a 54-hour test flight. At Edwards AFB, the shuttle performed the first controlled landing. It gently glided to a perfect landing on the dry lakebed.

In December 1983, Young commanded his second shuttle flight, making this his sixth time in space. On STS-9, with a record crew of six, the shuttle also carried the European-constructed spacelab. This mission lasted ten days and the crew completed all of the 70 planned experiments. The *Columbia* landed again at Edwards AFB.

Young was originally scheduled for a seventh space mission in September 1986, which would have carried the Hubble Space Telescope. This was postponed for several years after the *Challenger* disaster. Young appeared as a witness at the *Challenger* Investigation, during which he criticized NASA for the compromises in flight safety.

Young received two NASA Distinguished Service Medals, two NASA Exceptional Service Medals, and the Johnson Space Center Certificate of Commendation.

Cosmonauts of the USSR

The Soviets began their manned space program in 1958. Korolev recruited 20 fighter pilots as potential space travelers. After an exhaustive series of tests, Korolev narrowed the field down to seven for extensive training. These seven cosmonauts were the Soviet's "original seven." Here is a limited list of some of the cosmonauts who were involved in the beginning of space exploration.

Valentin Vasilyevich Bondarenko: A Forgotten Hero

Valentin Vasilyevich Bondarenko, a senior lieutenant in the Soviet Air Force, has the unfortunate distinction of being the first cosmonaut to die in a training accident. He was the youngest of the original cosmonaut group and was known as "Valentin Junior" or "Tinkerbell."

Born in the Ukrainian city of Kharkov in 1936, Bondarenko attended an air force school and was briefly a jet pilot before he was selected for spaceflight training with the first cosmonaut group in March 1960.

After Gagarin and five other cosmonauts were selected for advanced training in preparation for the first manned flight, Bondarenko and the other cosmonaut trainees continued support activities. Bondarenko participated in isolation chamber tests.

On March 21, 1961, just three weeks before Gagarin made the first manned flight, Bondarenko was completing a ten-day isolation experiment in a pressure chamber at a training facility near Moscow. The pressure in the chamber had been reduced during the experiment, requiring the oxygen content to be elevated.

After a medical examination, Bondarenko removed sensors and was cleaning himself with an alcohol wipe, then carelessly tossed the wipe aside. To his horror, it landed on an electrical hot plate and ignited.

Although he tried to extinguish the blaze, in the elevated oxygen level the fire quickly spread into an inferno. Before technicians could open the chamber, Bondarenko was burned over 90 percent of his body. He was rushed to Botkin Hospital, but he died eight hours later.

One account of the tragedy tells of Yuri Gagarin accompanying him to the hospital the day before Gagarin left for Baikonour in preparation for the first manned spaceflight.

For 25 years, the West knew little of Bondarenko's existence or of his death. The Soviets' propaganda machine covered up the accident and never even informed Bondarenko's relatives of the cause of his death or that he was a cosmonaut involved in the space program. It wasn't until the 25th anniversary of Gagarin's flight in 1986 that the Soviet Union divulged the full information and account of his accident. Valentin Bondarenko was survived by his wife, Anya, and son Alexandr.

Konstantin Petrovich Feoktistov: First Rocket Engineer in Space

Konstantin Petrovich Feoktistov was born in Voronezh on February 7, 1926. At the age of ten he was fascinated by astronomy and was already nurturing the dream of spaceflight. The young Feoktistov had even worked out detailed plans for exploring the Moon. He studied mathematics and physics in order to understand the formulae of spaceflight pioneer Tsiolkovsy.

Dreams of spaceflight were interrupted by the Nazi invasion of the USSR during World War II. When the Nazis occupied the Voronezh region, Feoktistov became a scout for partisan fighters. In 1942, when he was 16, Feoktistov was captured. Condemned as a spy by the Nazis, he was shot and left for dead. Fortunately, his wounds were not fatal and after nightfall, Feoktistov escaped. Years later, he joked that he survived because the Nazis had shot him with his own gun.

After recovering, Feoktistov entered Moscow's Bauman Higher Technical College in 1943. He graduated in 1949 with a master of science degree in engineering and went to work as an industrial engineer in Zlatonst.

After completing graduate work as a candidate of a technical sciences degree in 1955, Feoktistov joined the rocket development team of Sergei Korolev. He worked on the original Sputnik program and was involved in the development of manned spacecraft. He soon became the head of a group of young engineers known as "Korolev's Kindergarten."

Feoktistov had long argued that engineers would be the best choice for space travelers and was "filled with despair" when military pilots were chosen instead. Feoktistov was an academic lecturer to the first group of cosmonauts. He did not give up his spaceflight quest however. In late 1961, he arranged to have himself put through the cosmonaut medical tests and in early 1962, he was approved for spaceflight.

At the Korolev bureau, Feoktistov was second in command to Korolev in the designing of the Vostok spacecraft. When the multimanned Voskhod was designed, he persuaded Korolev to allow him to be on the flight crew. In June 1964, at age 38, he began spaceflight training.

On October 12, 1964, Feoktistov's boyhood dreams were realized when he and cosmonauts Komarov and Yegorov were launched aboard the first three-man spacecraft, *Voskhod 1*. During his one-day flight, Feoktistov studied the Earth and horizon, and tested spacecraft control systems.

After his spaceflight, Konstantin Feoktistov continued his education, earning the degree of Doctor of Technical Sciences. Since 1967, Feoktistov has been instrumental in the design of the Salyut and Mir space stations and the Soyuz T transport spacecraft. In 1975, he served as flight director for the Salyut missions. At the age of 55, he again trained for spaceflight aboard *Soyuz T-3* in 1980, but health problems grounded him before the flight.

Feoktistov has been twice awarded the Order of Labor Red Banner, and is a Hero of the Soviet Union with the Order of Lenin and the Gold Star.

Yuri Gagarin: The First Man in Space

The first man in space was Soviet Air Force Major Yuri Alexiyevich Gagarin. Gagarin was born on March 9, 1934, into a family of collective farmers. His father Alexei was a carpenter and his mother Anna was a milkmaid in the village of Klushino.

Childhood brought young Gagarin unusual hardships. In October of 1941, only a month after he started school, the invading Nazi Army occupied his village and burned the only school. The Nazis forced all families in Klushino from their homes and Gagarin and his family had to live in a cold dugout in their own back yard.

> *"Now let the other countries try to catch us!"*
>
> —Cosmonaut Yuri Gagarin, at a post-flight rally, April 1961

After liberation from the Nazis two years later, the Gagarins were assigned to help rebuild the town of Gzhatsk. Here, Yuri continued his education. After completing vocational school, Gagarin was sent to an industrial college in Saratov. In his fourth year at college in 1954, Gagarin joined the Saratov Flying Club and found his true love in life, flying!

Although Gagarin was trained as a foundryman and moulder, his love of flying led him to join the Red Air Force. Shortly after the birth of space travel with the launch of *Sputnik 1*, he was stationed in the far north with a fighter squadron.

Two years later, when the *Luna 3* spacecraft photographed the far side of the Moon, Yuri could not resist the call of space. Gagarin filed a formal application to be admitted to a cosmonaut team, "should one exist."

In 1960, Gagarin was called before a special medical review board. After passing, he was invited to Moscow for even more extensive tests. Finally, Gagarin was declared "healthy enough to go beyond the stratosphere."

The first group of 12 cosmonauts was assembled in March 1960. Gagarin and the others received extensive classroom instruction on theoretical subjects as well as training in spacecraft mockups, isolation chambers, centrifuges, zero-G aircraft, and parachute jumping. As tests of the Vostok spacecraft progressed in 1960, a group of six cosmonauts was selected for more intensive training in preparation of the first manned flight. Gagarin was among the six.

Upon completion of the final unmanned orbital test flights of the Vostok spacecraft, a decision was made on April 3, 1961, to proceed with the launching of the first manned Vostok. Yuri Gagarin and the five other cosmonauts left for the launch site on April 5.

On April 10, Gagarin was chosen to be the first to ride the Vostok into space. He wrote his wife Valentina, "Just imagine an ordinary man like me being entrusted with such a great task of national importance!"

On the morning of April 12, 1961, Gagarin was strapped into the passenger seat of the Vostok spacecraft. At 9:07 A.M. Moscow time, the 20 rocket nozzles of the Vostok booster rumbled to life and Gagarin was off into the history books as the first man in space.

Fate prevented Gagarin from flying in space again. From 1963 to 1968, he was Deputy Chief of the Cosmonaut Training Center and worked hard to prepare for his second spaceflight. He was even a member of the Soviet lunar landing team when it was first formed. However, after the April 1967 death of cosmonaut Vladimir Komarov during the maiden flight of the Soyuz spacecraft, Gagarin was removed from flight status. As a revered Soviet hero, he was regarded as the "First Citizen of the Universe." Gagarin was too valuable as a figurehead and propaganda showpiece to risk being lost in a flight accident.

Gagarin bitterly fought this decision and won reinstatement to flight status in 1968. On the morning of March 27, 1968, Gagarin and flight instructor Vladimir Seryogin took off for a routine training mission in a twin-seat jet trainer. At 10:31 A.M., Gagarin crashed not far from the airfield and was killed instantly.

Yuri Gagarin's death rocked the Soviet Union. The national shock was similar to that felt by Americans at the assassination of John Kennedy. Gagarin was buried in the place of highest Soviet honor, the Kremlin Wall.

Georgi Mikhailovich Grechko

The first person to repair a manned spacecraft in orbit already had a rewarding space career even before becoming a cosmonaut. Georgi Mikhailovich Grechko was born in Leningrad on May 25, 1931. After spending his teens in the Urals where his family was evacuated after the Nazi invasion, Grechko returned to school. He graduated from the Leningrad Institute of Mechanics with honors in 1955, specializing in computers. He then went to work for the Korolev rocket design bureau.

Grechko was part of the team which perfected the first Sputnik satellite. He calculated, often by hand, the trajectory of the spacecraft. By 1964, Grechko was head of his own spacecraft design team and was instrumental in the development of the early Luna Moon landers. For that work, he was awarded the Lenin Prize. From his experience with the Luna landers, Grechko earned a master of science degree in 1967.

The spaceflight of fellow engineer Feoktistov in 1964 spurred Grechko to try for the cosmonaut team. In 1966, he began spaceflight training. His career was nearly cut short, however, when he broke his back during a parachute jump. After recuperating a year, he rejoined the team and was included in the lunar landing team. He practiced simulated lunar landings and water recoveries at a secret base on the Black Sea. When the manned Moon landing was abandoned, Grechko joined the Salyut project.

Grechko's first flight assignments were as backup for the *Soyuz 7, 9,* and *12* missions. In 1975, Grechko, a member of the teams which perfected the first satellite and the first soft landing on the Moon, entered space for the first time. Serving as flight engineer on *Soyuz 17*, he spent 30 days aboard *Salyut 4*, setting a new Soviet space endurance record. Grechko performed medical and astronomical experiments, including realuminizing a telescope mirror in space. He found spaceflight unpleasant, however, and admitted he was sick most of his time in orbit.

Becoming heavily involved in the *Salyut 6* project, Grechko was to command a future short duration Soyuz flight to the station. However, when *Soyuz 25* accidentally rammed the main Salyut docking port, Grechko, an engineer with spaceflight experience, was quickly assigned to *Soyuz 26* so he could inspect the station for damage. Launched December 10, 1977, Grechko spent 90 days aboard *Salyut 6*, setting another record. During a December 19 space walk, Grechko determined the Salyut docking mechanism was undamaged and the station was declared operational again. Grechko also performed a terrifying space rescue. During the space walk, commander Romanenko failed to secure his safety line and accidentally drifted away from the station. Only Grechko's quick action in grabbing the safety line prevented death in space.

Grechko's final spaceflight came in 1985 as flight engineer aboard the eight-day *Soyuz T-14* mission to the *Salyut 7* space station.

During a career which spanned from the birth of the space age to permanent manned space stations, Georgi Grechko has spent 134 days in space. He has been awarded the title of Hero of the Soviet Union with the Order of Lenin and the Gold Star.

Yevgeny Vasilevich Khrunov

Born of a peasant family in the village of Purdy on September 10, 1933, Yevgeny Khrunov was one of eight children. His father died in 1948 and his mother raised all the children herself. After completing seven years of secondary school, he attended the Agricultural Polytechnic.

Khrunov became interested in flying, however, and never turned to farming. In 1952, he joined a pilots school, then went on to the Air Force School and Flying College in 1953. After graduating in 1956, he served as a pilot in the Air Force until his selection as a cosmonaut in 1960. As a pilot, Yevgeny was recognized for his flying abilities and his "remarkable ability" with machines.

The first spaceflight assignment for Khrunov was as backup for Alexei Leonov on the EVA mission of *Voskhod 2* in 1965. At the same time, he was also attending the Zhukovsky Air Force Engineering Academy and graduated with honors in 1968.

In 1967, Khrunov's EVA training was extended when he was chosen as research engineer for the planned flight of *Soyuz 2*, which was to dock with *Soyuz 1*. Mission plans called for Khrunov and cosmonaut Yeliseyev to "space walk" from *Soyuz 2* to *Soyuz 1*. This plan was cancelled after fatal difficulties cut short the flight of *Soyuz 1*.

Nine years after selection as a cosmonaut and at age 35, Khrunov, now a lieutenant colonel, finally reached space aboard the January 15, 1969 launch of *Soyuz 5*. After linking with *Soyuz 4*, he and Yeliseyev completed the space walk planned for two years before. After two days in space, he returned to Earth in the *Soyuz 4* craft, completing the first crew transfer in space.

Following his spaceflight, Khrunov earned a master of science degree and became an "aviator scientist."

Khrunov's next flight assignment came 11 years later when he was backup research engineer for the *Soyuz 38* flight to the *Salyut 6* space station in 1980. After that, he was offered command of *Soyuz 40*, but turned it down.

Yevgeny Khrunov holds the title of Hero of the Soviet Union with the Order of Lenin and the Gold Star.

Pyotr Ilyich Klimuk

The hardship of war pressed Pyotr Klimuk at an early age as he was born during the Nazi occupation of the Belorussian village of Komarovka on July 10, 1942. Two years later, his father was killed by the Nazis.

Regarded as a child genius, Klimuk entered Chernigov Higher Air Force School in the Ukraine when he was 18. Upon graduation at age 22, he served as a fighter pilot. Only a year later, he was invited to take the cosmonaut medical examination. After passing, he was told to report to Star Town in October of 1965 for spaceflight training. Thus, Pyotr Klimuk became a cosmonaut at age 23.

Klimuk's group, the third team of cosmonauts, was a pivotal force in the course of Soviet manned spaceflight. While previous teams had been chosen from older, more experienced jet pilots, the third group of 20 pilots was specifically chosen from pilots and engineers whose average age was only 24. The intent was to mold this team into a cadre of career cosmonauts who were trained specifically for spaceflight.

During his two-year basic cosmonaut training, Klimuk's progress was so outstanding that his entire group was known as the "little Klimukis," and as "Klimuk's Group." Following basic training in 1967, Klimuk was assigned to the secret lunar team in preparation for manned Zond flights around the Moon. He was also destined for an eventual lunar landing mission. When that program was abandoned, he was transferred to the Salyut program.

Over the next nine years, Klimuk would make three spaceflights. The wisdom of selecting and intensively training very young cosmonauts in Klimuk's group became apparent. He made his third flight before he was 36, the age when many cosmonauts make their first flight, then soon have to retire due to age.

Klimuk's first flight was as the 31-year-old commander of the *Soyuz 13* astronomical mission in December 1973. On this week-long mission, he and Valentin Lebedev tested equipment to be mounted on future Salyut space stations.

Between May and July 1975, Klimuk and flight engineer Sevastyanov manned the *Salyut 4* space station. His 63-day *Soyuz 18* flight remained in space during the historic Apollo/Soyuz rendezvous that July.

In 1977, Pyotr Klimuk graduated from the Gagarin Red Banner Air Force Academy. A year later, Klimuk commanded the eight-day flight of *Soyuz 31* during a week-long visit to the *Salyut 6* space station. After the completion of his third mission, Klimuk was elevated to the rank of general at the age of 36.

In 1978, Klimuk was promoted to a deputy director of the Gagarin Cosmonaut Training Center. Between 1980 and 1982, he was a member of the Supreme Soviet. In 1984, Klimuk also graduated from the Lenin Military Political Academy.

In 1991, General Pyotr Klimuk became the head of the Cosmonaut Corps. He has twice been awarded the title of Hero of the Soviet Union with the Order of Lenin and the Gold Star.

Vladimir Mikhailovich Komarov: A Pilot's Pilot

The son of an intellectual, Vladimir Mikhailovich Komarov was inclined toward academic studies. He was born in Moscow on March 16, 1927. As World War II raged, Komarov entered prep school at age 15, attending the Moscow Air Force School. He graduated in 1945, too late to see action during the war. Komarov continued his education at the Third Sassov Air Force School, the Chkalov School in Borisoglebsk, and then graduated from the Serov Flying School in Bataisk in 1949. He served as a pilot and instructor pilot until 1954 when he entered the Zhukovsky Air Force Engineering School in Moscow. Graduating in 1959, Komarov became a tester of aircraft equipment. In 1960, he was invited to try out for the cosmonaut team.

After selection as a cosmonaut, he became known for his intelligence and flying ability. However, Komarov's luck soon changed. A hernia operation on May 15, 1960, nearly ended his space career.

After recuperating for six months, he rejoined the cosmonaut team. His first flight assignment was as backup for Pavel Popovich on *Vostok 4*. However, during centrifuge tests, doctors discovered he had an "extra systole," or irregular heartbeat, and grounded him (similar to Astronaut Deke Slayton's condition).

Komarov went to leading cardiovascular specialists in Moscow and gathered endorsements stating that his irregular heartbeat would cause no problems in space. The head of the manned space program, Sergei Korolev, was also impressed with Komarov's abilities. Regarding Komarov, he said, "I see in him a competent tester of manned craft." Komarov's persistence paid off; he was reassigned to flight status.

True to Korolev's vision, Komarov was named commander of the first multimanned spacecraft, *Voskhod 1*. Launched October 12, 1964, Komarov and two crewmates spent a day in space testing the Voskhod and performing medical experiments.

As the chief cosmonaut in the Soyuz design team, Komarov was once again assigned in September 1966, to pilot the maiden flight of a new spacecraft. Launched alone aboard *Soyuz 1* on April 23, 1967, Komarov was to command the first docking of a two-manned spacecraft. However, serious control and electrical problems plagued the Soyuz in orbit and the launch of *Soyuz 2* was cancelled.

Fighting increasing control problems, Komarov was forced to abort his second spaceflight after one day. Succeeding after three attempts, he brought the troubled craft back to Earth. However, the first cosmonaut to make two spaceflights was also the first human to die during a spaceflight. Komarov was killed when the Soyuz parachute failed to open and the craft crashed near Orenburg.

Had he lived, Komarov probably would have been closely involved in the manned lunar program. A pilot with such highly regarded skills would have been surely slated for a lunar landing mission. The crew of *Apollo 15* left a small statue on the Moon as tribute to American and Soviet spacemen who died in the race to the Moon. Komarov and 13 other American and Soviet space pioneers were honored.

Vladimir Komarov was buried in the Kremlin wall. He was twice awarded the title of Hero of the Soviet Union with the Order of Lenin and the Gold Star.

Alexei Leonov: First to Walk Among the Stars

On March 18, 1965, while several hundred kilometers above the Soviet Motherland, Lt. Col. Alexei Arkhipovich Leonov eased out of the airlock of the *Voskhod 2* spacecraft and gently pushed himself into the vacuum of space. As Leonov somersaulted at the end of a 17-foot (five-meter) tether, he entered the history books as the first man to leave his spacecraft while in orbit.

Leonov later said of his pioneering 12 minutes outside Voskhod, "I saw the universe in all its grandeur. The view of untwinkling stars...was followed by views of the Earth...as though I were swimming over a vast colorful map..."

Leonov was born on May 30, 1934, in the village of Listvayanka, in the Tisyl district of the Kemerovo region. In 1953, at the age of 19, he was encouraged by the Young Communist League to attend pilot training school. Later, he entered the Air Force School at Chuguchev from which he graduated with distinction.

Alexei served during the next three years as a Third Class Military Pilot, and also became an expert parachutist with over 100 jumps to his credit. He also became an Air Force Paratrooper Instructor in 1960.

On March 5, 1960, Leonov was chosen, along with 19 other young Air Force pilots, to be part of the first cosmonaut team. The short, moon-faced Leonov was called "sun-blessed" by his fellow cosmonauts because of his sandy hair, freckles, and sunny disposition. Leonov was known as cheerful, witty, and happy-go-lucky, but was also volatile, poetic, and ready to take a risk.

Leonov is an excellent artist. His paintings often illustrate Soviet space exploits. Alexei serves as the editor of *Neptune*, the cosmonaut newsletter, and is a member of the Union of Artists of Russia.

Alexei began preparing for his space walk mission in 1963, and joined the *Voskhod 2* training group in April 1964. Once, he showed quick wits and calm nerves during a near fatal parachute accident. A strap on his suit caught on the ejection seat and was forcing him to land on his head at high speed. He kicked the seat loose just 150 feet (45 meters) above the ground.

Fate again almost deprived Leonov of his place in history in late 1963. As he and his wife were being driven back to Star Town by his driver, their car slid off the icy road and crashed through a frozen lake. Leonov struggled free and rescued his wife and driver from under the water.

Such steel nerves were to serve Leonov well in space. During his 1965 space walk, he discovered his space suit was too rigid to allow him to reenter the air lock and reach the safety of the Voskhod cabin. Only after lowering the pressure in his suit to a dangerously low level could he regain the mobility to enter the craft.

Following his space walk experience, Leonov was promoted to second in command of the Cosmonaut Corps in 1965. Two years later, he was placed in charge of the manned lunar landing team. Leonov and 17 other cosmonauts were assembled and shipped to the Black Sea to practice water landings and fly helicopters in preparation for lunar missions.

Various reports have related that in 1968, Leonov was slated along with Oleg Makarov to be the crew for the first manned circumlunar flight aboard a Zond spacecraft. This launch was scrubbed in December 1968, because of technical problems. Leonov and Makarov escaped certain death when the Zond was launched unmanned a month later. The booster exploded at the fringes of space and the crew escape rocket failed to operate.

Further unofficial reports state that in 1969, Leonov was chosen to be part of the lunar landing crew. He would have been the first Soviet to step on the Moon. It was the flawed N-1 Moon rocket that prevented Leonov from ever making his historic second "walk."

When the Soviet manned lunar effort was abandoned, Leonov transferred to the Salyut space station effort. He was to have been the commander of the ill-fated *Soyuz 11* mission, but the illness of his co-pilot, Kubasov, forced the last minute replacement of the entire *Soyuz 11* crew.

When plans were formulated for the joint U.S./USSR Apollo-Soyuz Test Project, Leonov, shown to the left of fellow cosmonaut Valery Kubasov in Figure 3-19, was appointed commander of the Soviet flight crew. The success of the flight was crucial to the restoration of Soviet space prestige after the loss of the Moon race and the deaths of the *Soyuz 11* crew. Leonov was a natural ambassador for the Soviet space effort, being an excellent pilot and commander as well as quick to smile and friendly to all. Leonov is also shown in Figure 3-20, with American astronaut Tom Stafford training for ASTP.

Today, Leonov remains one of Russia's most visible space heroes. Alexei took over Yuri Gagarin's place as the most popular of the cosmonauts after Gagarin's death in 1968.

Figure 3-19. Alexei Leonov and Valery Kubasov walking to their launch of ASTP *Soyuz 19*

Figure 3-20. Leonov and astronaut Tom Stafford in training for ASTP

Currently, Alexei Leonov is retired with the rank of major general after serving as the Deputy Commander of the Yuri Gagarin Cosmonaut Training Center. He is married with two children. He has been twice honored with the title of Hero of the Soviet Union and is a holder of the Tsiolkovsky Gold Medal.

Oleg Grigoryevich Makarov

A keen chess player and hockey fan, Oleg Makarov came to the Cosmonaut Corps as a civilian via the Korolev spacecraft design bureau. Born on January 6, 1933, in the Kalinin region near Moscow, Makarov comes from an Army family. In 1951, he entered the Bauman Higher Technical School in Moscow. Following his graduation in 1957, he went to work as an engineer for the Korolev design bureau in the office responsible for the construction of *Sputnik 1*.

As an engineer, Makarov developed the control board for the Vostok spacecraft and designed systems for the Voskhod and Soyuz spacecraft. His selection as a cosmonaut came in November 1966.

Makarov's original assignment was as part of the Soviet manned lunar landing team. Some reports identified Makarov as the person selected to make the first Soviet manned landing on the Moon. If history had been slightly altered, Makarov would have made the first footprint on a new world. Instead of Armstrong and Aldrin, school children would have been taught the name Makarov. Following the cancellation of the manned lunar program, Makarov like his fellow cosmonauts, were quietly transferred to the Salyut space station project.

Makarov's first flight assignment was backup flight engineer for the *Soyuz 9* long duration medical flight in 1970. Following the deaths of the three *Salyut 1* space station crewmen aboard *Soyuz 11* in 1971, the spacecraft was extensively redesigned as a two-man craft. Makarov, with his engineering and design background, was assigned to the next flight to help evaluate the new spacecraft. When the replacement for the *Salyut 1* space station failed in orbit, *Soyuz 12* was launched in 1973 (see Figure 3-21) as a two-day solo test flight, commanded by Vasily Lazarev with Makarov aboard as flight engineer.

Makarov's next spaceflight was nearly his last. Again paired with Lazarev, he was launched in April 1975, aboard what was to be a *Soyuz 18* visit to the *Salyut 4* space station. At over 100 kilometers altitude, the Soyuz booster malfunctioned and the crew escape system pulled Lazarev and Makarov free from the stricken rocket for the first in-flight abort of a manned space launch.

Three years later, Makarov reached space again aboard the *Soyuz 27* flight when he and commander Dzhanibekov docked with the *Salyut 6* space station during a six-day resupply mission.

Oleg Makarov flew in space for the last time aboard the *Soyuz T-3* spacecraft in 1980. Makarov spent 13 days in space as the *Soyuz T-3* crew reactivated the mothballed *Salyut 6* space station and performed a number of repairs on the craft. Following the *Soyuz T-3* mission, Makarov, then 47, made no further spaceflights.

Figure 3-21. Soyuz rocket booster

Pavel Romanovich Popovich

When the Cosmonaut Corps was established in 1960, Popovich was the first chosen. He thus served as unofficial greeter and quartermaster for the later arrivals. Possessing a fine tenor voice, Popovich was popular as a singer. During cosmonaut training, he was among the six finalists for the first manned flight. He served as the cosmonauts' Communist Party secretary and was the capcom during Gagarin's pioneering flight.

Pavel Popovich, shown in Figure 3-22, first rode into space aboard the single seat *Vostok 4* spacecraft on August 12, 1962. At the young age of 31, he held the rank of colonel. In space, he joined the already orbiting *Vostok 3* for the first dual flight of manned spacecraft. After a three-day flight, he safely returned to Earth.

A language barrier and occupying Nazi forces during World War II made early life difficult for Pavel Romanovich Popovich. Born in the Ukrainian town of Uzin, near Kiev, on October 5, 1930, he spoke no Russian, only the native Ukrainian language. Popovich was first a shepherd, then he gained entry to a trade school in Belaya Tserkov. Later, he joined the Magnitogork Industrial Polytechnic where he learned the Russian language and graduated in 1951 as a building technician.

Popovich joined the aero club at school and his love of flying led him to join the Air Force. From 1951 to 1954, he attended the Kacha Air Force College, then

Figure 3-22. Cosmonaut Pavel Popovich demonstrates weightlessness during his flight on *Vostok 4*

served in the Arctic with a fighter squadron. Here, he earned the prestigious Order of the Red Star for an undisclosed secret assignment.

Popovich is married to Air Force Colonel Marina Lavrentyerna, a famous test pilot who has more flying hours than Popovich. His wife was at one time considered for selection as a cosmonaut for the mission to launch the first woman into orbit.

After his spaceflight, Popovich entered the Zhukovsky Air Force Engineering Academy, graduating in 1968. At this time, he was member of the cosmonaut group training for the Soviet manned lunar landing program. He probably would have been assigned to the second or third lunar landing mission.

Twelve years after his first flight, he completed his second space mission as commander of the *Soyuz 14* launch toward the *Salyut 3* space station. After liftoff on July 3, 1974, Popovich and crewmate Yuri Artyukhin spent nearly two weeks performing Earth observations and medical experiments.

For his two spaceflights, Pavel Popovich has been twice awarded the title of Hero of the Soviet Union with the Order of Lenin and the Gold Star. He was also a deputy of the Supreme Soviet of the Ukraine and is a member of the Ukrainian Union of Writers. He has also published an autobiographical account called "Takeoff In The Morning."

Valentina Vladimirovna Tereshkova: First Woman in Space

In the heady days of the first Vostok flights, Soviet premier Nikita Khrushchev sought further propaganda gains with manned spaceflight. Thus, the idea of orbiting a woman was developed. After Titov's one-day flight in 1961, hundreds of Soviet citizens volunteered for future spaceflight. One of them was Valentina Vladimirovna Tereshkova, a young sport parachutist who worked at the Krasny Perekop textile mill in Yaroslav.

Tereshkova was born on March 6, 1937, in the village of Maslenikovo. When she was six, her father was killed fighting the Nazis. At the age of 17, Valentina went to work in a tire assembly factory, but a year later switched to the textile mill where her mother and sister worked.

Tereshkova completed industrial school and joined the factory aero club in 1958, becoming an amateur parachutist. Over the next few years, she completed nearly 150 jumps. In 1961, her letter to Yuri Gagarin volunteering for spaceflight was answered by an interview summons to Moscow. Valentina and three women parachutists were given the rank of junior lieutenant in the Soviet Air Force and assigned to begin cosmonaut training in March 1962.

Rumors persist that Tereshkova was the backup, taking the flight assignment at the last minute when the other woman failed her preflight physical. When *Vostok 6* was launched on June 16, 1963, Valentina, at age 26, became the first woman to fly in space. Using the call sign "Seagull," she orbited the Earth 48 times over the next three days during a dual flight with *Vostok 5*, launched two days before. Soviet Flight control became alarmed during the flight when communication could not be re-established with Tereshkova after a rest period. She apparently overslept and rather soundly, for it was several hours before ground control could awake her. Chief Designer Korolev was disappointed in her failure to complete all of her assigned experiments.

Although women cosmonauts were chosen to show the equality of Soviet women, the existing cosmonauts were not pleased with the arrival of the female recruits. Years later, Tereshkova admitted that the cosmonauts avoided her "because I have invaded their little playground and because I am a woman." Tereshkova was paired for training with bachelor cosmonaut Andrian Nikolayev. Valentina was qualified for flight aboard MIG-15 trainers and received centrifuge and zero-G training.

Tereshkova and Nikolayev were married the November following her flight. The wedding of two space farers was a political heyday for the flamboyant Khrushchev, who personally gave away the bride. The following June, she bore a healthy daughter, Yelena.

After a series of propaganda tours, Tereshkova entered the Zhukovsky Air Force Engineering Academy and graduated in 1969. She later earned a candidate of technical sciences degree in 1976 and attained the rank of engineer colonel in the

Soviet Air Force. Tereshkova also entered politics, becoming a Deputy of the Supreme Soviet in 1966 and member of the Central Committee in 1972.

Tereshkova briefly trained for a second spaceflight to make an EVA in the Voskhod program, but her flight was cancelled. There were also rumors of a planned flight with Yuri Gagarin on the first lunar pass on the eve of the 50th anniversary of the Russian Revolution in 1967. More than likely, this was probably drummed up by various Kremlin departments with an eye more towards public relations instead of on technical merit and mission qualifications. The required space hardware for such a mission was also far behind in production.

Valentina Tereshkova holds the title of Hero of the Soviet Union with the Order of Lenin and the Gold Star.

Gherman Stepanovich Titov: The Youngest

At age 25, Gherman Stepanovich Titov was the youngest person to orbit the Earth. Born in the Siberian village of Verkhneye Zhilino on September 11, 1935, he decided he wanted to become a pilot after his uncle, an aviation war hero, visited the village. Well-read and fond of poetry, Titov is a sophisticated man.

> *"In these days, you can't even fly an airplane all by yourself. How much more enormously difficult it would be for one person to try to solve the problems of conquering outer space all by himself...there is room in outer space for everyone."*
>
> —Cosmonaut Gherman Titov, April 30, 1962

As a youngster, Titov was involved in a bicycle accident which broke his wrist. Rather than seek treatment, he endured the pain and rehabilitated himself with "secret exercises." In 1953, he entered the Kacha Higher Air Force School in Stalingrad (Volgograd). Titov feared his old wrist injury would ruin his chances of becoming a pilot so he got up an hour early every morning to work out on parallel bars. In 1957, Titov graduated from pilot training with distinction.

In 1960, Titov, shown in Figure 3-23, was selected for the first cosmonaut team. His weak wrist went undetected, but the hot-tempered and outspoken Titov came close to dismissal by rebelling against "silly questions" during psychological testing.

While studying at the Zhukovsky Air Force Engineering Academy, he met and married Tamara Vasilyevna. Shortly after the birth of their son Igor, the child became ill and died. Titov was then temporarily relieved from cosmonaut training.

After rejoining the cosmonaut team, Titov was one of six finalists for the first manned orbital flight of the Vostok spacecraft, serving as backup for Yuri Gagarin. After Gagarin's flight, Titov was chosen for the second manned flight, scheduled for three orbits. This was later lengthened to a full day in space.

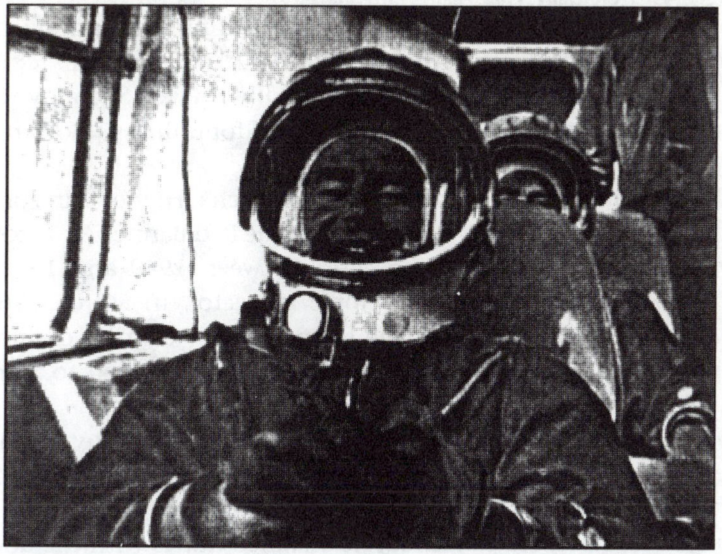

Figure 3-23. Gherman Titov, second Soviet in space

Titov was launched aboard *Vostok 2* on August 6, 1961. During his 17 orbits, he carried out many Earth observations and experiments. Titov also became the first person to sleep in space. His flight is most noted, however, for the discovery of "space sickness," or the disturbance of the inner ear caused by zero-G. After his spaceflight, doctors discovered his old wrist injury and said if it had been known, he would not have been allowed to make the flight.

After his only spaceflight, Titov attended the Zhukovsky Academy and graduated in 1968. In 1964, he was also made commander of the cosmonaut corps. In addition, Titov was on the editorial board of the magazine *Aviatsiya i kosmonavtika* (Aviation and Spaceflight).

From 1970 to 1972, Titov temporarily left the cosmonaut program and trained as a test pilot. After serving as flight controller of the *Salyut 3* mission, he returned to aircraft testing until 1978. He was also a member of the planning committee for the 1980 Moscow Olympics. Today, Titov holds the rank of general and remains active in the administration of the Russian manned spaceflight program.

Gherman Titov holds the title of Hero of the Soviet Union with the Order of Lenin and the Gold Star.

Vladimir Alexandrovich Shatalov: Four Flights into Space

The first cosmonaut to make three spaceflights, Vladimir Alexandrovich Shatalov, was born in Petropavlovsk on December 8, 1927. He spent his childhood in Leningrad. His father's experiences during World War II led Shatalov to enter the military where he attended the Lipetsk prep school for military pilots in Leningrad until 1946.

Shatalov continued his schooling at the Kacha Higher Air Force School, graduating in 1949. In 1953, he was admitted to the Red Banner Air Force Academy in Moscow and graduated three years later. Between 1949 and 1962, Shatalov advanced from squadron leader to air force inspector. In March 1962, he was training to become a test pilot when he learned a new team of cosmonauts was being chosen. Shatalov applied and was accepted. He reported to Star Town on January 11, 1963.

Being the leader of the new group of cosmonauts, Shatalov was assigned to support the group flight of *Vostok 5* and *6* in 1963. His first flight assignment was as backup for the two-week *Voskhod 3* flight.

When that mission was cancelled, he assumed duties with the development of the Soyuz spacecraft. Shatalov served as capcom to the doomed Vladimir Komarov aboard *Soyuz 1*. After the program recovered from the crash of the maiden flight, Shatalov served as backup for *Soyuz 3*.

Vladimir Shatalov entered space for the first time on January 14, 1969, as the solo pilot aboard *Soyuz 4*. The next day, *Soyuz 5* was launched with a crew of three. Shatalov successfully rendezvoused and docked with the *Soyuz 5*, a first for two manned spacecraft. Two of the *Soyuz 5* crew, Yeliseyev and Khrunov, transferred to *Soyuz 4* via a space walk and returned to Earth with Shatalov. This mission was an important step for crew transfers to and from a lunar module on flights to the Moon. If the N-1 Moon rocket had properly performed, the next missions would have been lunar bound.

Ten months later, instead of going to the Moon, the manned *Soyuz 6* and *7* were launched in quick succession. On October 13, *Soyuz 8* carried Shatalov to join the other two craft in space. (A Soyuz spacecraft in earth orbit is shown in Figure 3-24.) As the "space troika" orbited the Earth, Shatalov commanded the flagship of the first "space squadron." For the Shatalov and the other lunar trainees, this circus stunt must have been a pale substitute to the American lunar landings.

In 1971, Shatalov's spaceflight experience cast him as the commander of the planned 30-day *Soyuz 10* mission to man the newly launched *Salyut 1* space station. After a successful launch on April 23, the craft docked with the Salyut the next day. *Soyuz 10* then had a brush with death as the hatch between the Soyuz and Salyut refused to open, locking the crew out of the station. Worse, the stuck hatch also

Figure 3-24. Soyuz capsule in orbit

refused to release the craft from the station so it could return to Earth. With the limited life support of the Soyuz expiring, Shatalov was able to disengage the two craft and return to Earth just as the air in the cabin turned poisonous. Indeed, crewmate Rukavishnikov lost consciousness during the return.

In October 1971, the cool-thinking Shatalov was promoted to chief of cosmonaut training, a post he held for the next 20 years. In 1972, he earned a candidate of technical sciences degree and by 1975, he was promoted to lieutenant general. Active in politics, Shatalov was deputy to the Supreme Soviet between 1977 and 1980.

Vladimir Shatalov has been twice awarded the title of Hero of the Soviet Union with the Order of Lenin and the Gold Star.

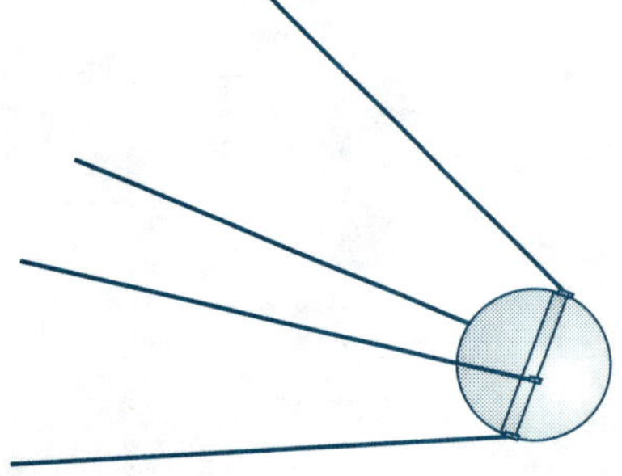

CHAPTER 4

Disasters and Near Catastrophes

During man's quest to explore space, the specter of death has always been nearby. Although both NASA and Russian engineers take precautions for every conceivable spaceflight emergency, the element of the unknown still shadows the young art of space travel. With each achievement in space, there were those who dared to pay the ultimate sacrifice, and some rendered payment in full! Setbacks and even death chose no political allegiance. Both the United States (Figure 4-1) and the former Soviet Union have had their share of space disasters and near catastrophes. Safety features were redesigned and incorporated after these catastrophes so others may move forward.

Hidden Soviet Disasters

In their zeal to display Soviet "scientific advances," Soviet authorities rarely admitted trouble or delays in the early days of their space program. Problems were admitted only after it was clear to Western observers that something had gone wrong with a Soviet space mission. Additional details of failures had to be sifted from evidence, often years after the event.

The first of these Soviet failures was also the worst so far. On the night of October 23, 1960, at the Baikonour Cosmodrome in the Soviet Union, there was a

Figure 4-1. America's *Apollo 1* funeral

tremendous fiery rocket explosion on a launchpad that later became known as the "Nedelin Disaster." Nearly 30 years after the incident, Soviet authorities revealed details about the horror unleashed that night.

Soviet Premier Nikita Khrushchev, who was in New York attending the 1960 United Nations Summit Meeting, thirsted for a major new rocket development to boast about before the world gathering. Field Marshall Mitrofan Nedelin, who had been appointed by Khrushchev as Commander of Strategic Rocket Forces, was obviously under pressure to produce results in the perfection of a new generation of Soviet intercontinental missile.

On the launchpad, the new rocket known as the R-14 reached the end of its countdown, then nothing happened! Ignition did not occur. Instead of draining all of the missile's explosive fuel before a close inspection

IF AT FIRST YOU DON'T SUCCEED...

"From Pre-launch to after the mission is complete, telemetry transmits data from the orbiting payload back to the ground for review. If there is a failure or anomaly, the data is analyzed and attempts are made to duplicate the readings on an engineering model. In the early days, the information was transferred from radio to a chart recorder using a pen and a strip of paper. A failed missile test could involve the review of literally miles of paper."

of the rocket, Nedelin ordered technicians to the pad with hopes of finding a quick solution. While scores of workers were on the launchpad, along with Nedelin himself, an electrical fault ignited the missile's second stage engine and an explosion engulfed everyone. Casualties were reported to be nearly a hundred dead, including Nedelin. The deadly inferno was so intense that some workers were never found; only personal effects such as keys and other metal objects remained.

> *"Soviet space research serves peaceful purposes exclusively! I can tell you confidently that Vostok was not built to carry bombs."*
>
> —Cosmonaut Gherman Titov
> August 1961

On the twenty-fifth anniversary of man in space, the Soviets released previously unknown information regarding the first death of a cosmonaut. Only twenty days before colleague Yuri Gagarin became the first man in space, Valentin Bondarenko, 24, died after being engulfed in a flash fire. He was just completing ten days of tests in the oxygen-rich atmosphere of a simulated space cabin when the accident occurred. Bondarenko was removing cotton padded medical sensors that were dampened with alcohol. One of the pads accidentally landed on a hot plate causing a flash fire reminiscent to the later *Apollo 1* tragedy. He survived the conflagration, only to die later. When Bondarenko was carried out of the cabin, he said, "It was my fault, don't blame anyone."

During the Vostok spacecraft hardware testing prior to the flight of Yuri Gagarin, the Soviets discovered a grisly flaw in their Vostok manned spacecraft. In February 1961, a manned parachute recovery test was conducted. A Vostok reentry capsule carrying veteran parachutist Pyotr Dolgov was dropped from a transport aircraft. As Dolgov was ejected from the descending cabin, disaster befell him. The cabin's hatch was too small and his helmet struck the hatch frame, snapping his neck and killing him instantly. A hasty emergency redesign enlarged the hatch, and orders were given that the pilot chosen for the first manned flight should be small.

Close Calls for America

For America, the excitement of manned spaceflight took on new dimension in the summer of 1961. The second American suborbital manned spaceflight was a success, with Gus Grissom landing safely in the ocean aboard his *Liberty Bell 7* spacecraft. Then without warning, as the craft bobbed in the water awaiting recovery, the explosive bolts securing the spacecraft's hatch detonated and ejected the hatch! Seawater quickly filled the cabin and Grissom had no choice but to quickly abandon his rapidly sinking craft. The recovery helicopter, shown in Figure 4-2, attached to the swamped capsule was nearly pulled into the sea before it released *Liberty Bell 7*. While all of the attention was on the sinking spacecraft, Grissom was also sinking as his space suit filled with water. Navy divers were able to rescue Grissom, but *Liberty Bell 7* was lost in 18,000 feet (5,500 meters) of water.

While the world thought John Glenn's historic orbital flight seemed to be going well, behind the scenes, a critical nail-biting situation had developed in space. During Glenn's second orbit around the Earth, an instrument reported that the heat shield was not in position on the rear of Glenn's *Friendship 7* spacecraft. If this indication was correct, the heat shield would detach during reentry and the craft would burn up. Such a catastrophe on America's first manned orbital flight would have a profound negative impact on the future of spaceflight! For that reason, the public was not informed of this crises until after the mission.

Ground controllers quickly realized the only way to keep the heat shield firmly in place in case it had indeed come loose was to leave the craft's retro-rockets strapped over it after they had been fired. Normally, they are jettisoned before reentry, but now in an untried emergency procedure, they were called on to keep the heat shield in position until the pressure of the atmosphere took over and held the shield in place. The decision proved to be a good call, and Glenn and *Friendship 7* came home safely. A NASA illustration of a Mercury capsule is shown in Figure 4-3.

Figure 4-2. Grissom's capsule, *Liberty Bell 7*, sinks into the ocean

Figure 4-3. This cutaway view depicts various spacecraft systems, the crew compartment, heat shield, and retro-rockets

As fate would have it, Glenn's crisis was not caused by a real failure, but by a faulty indicator, and Glenn had not been in real danger. The quick response by flight controllers to the perceived crisis set the future NASA pattern of using daring innovation to overcome spaceflight emergencies.

In May 1962, Walter Cronkite reported to a national television audience that an astronaut may have been lost during the flight of *Aurora 7*, America's second manned orbital flight. Scott Carpenter had expended nearly all of his spacecraft's maneuvering fuel and was late initiating his reentry, causing the craft to land off-target. For half an hour, recovery teams searched for *Aurora 7*, causing considerable alarm in NASA and the news media. Eventually, Carpenter was found safe and dry in a flotation raft beside his floating spacecraft, but he was 250 miles (400 kilometers) off course.

Faith 7, America's last Mercury flight, was proceeding smoothly in May 1963, until the nineteenth orbit. Instrument failures then forced astronaut Gordon Cooper to take manual control of the craft. Stabilization and control systems short circuited and nearly all instrument lights shut down. Fortunately communications were still working and Cooper calmly performed a successful manual reentry at the conclusion of the mission. In the process, Cooper, shown in Figure 4-4, landed the closest of all the Mercury flights to the recovery ship.

Figure 4-4. Gordon Cooper, prior to entering the Mercury spacecraft

In 1964, Theodore Freeman became the first NASA astronaut to lose his life. While flying a T-38 jet aircraft, he collided with a flying goose. Freeman died instantly in the resulting crash.

Greater Risks, Severe Consequences

By this time, the glow of apparent space success was beginning to fade in the Soviet Union. Under extreme pressure from Premier Khrushchev to fly the first multi-man spacecraft before the Americans, the first Voskhod mission disregarded all safety measures in order to squeeze three cosmonauts into the redesigned one-man Vostok cabin. Due to the cramped conditions and extra weight, there were no backup systems and the cosmonauts couldn't wear space suits. Soviet officials today acknowledge that the October 1964 mission had no scientific value and was nothing more than a very risky stunt. Fortunately, they were very lucky. Cosmonauts Vladimir Komarov, Boris Yegorov, and Konstantin Feoktistov safely returned after one day in space.

The second Voskhod flight was launched March 18, 1965, just five days before the first American Gemini flight. It was a flight that brushed with disaster twice in twenty-four hours. The mission was to showcase the first extra-vehicular activity (EVA) with Alexei Leonov's "spacewalk." Leonov and mission commander Pavel Belyayev were on their second orbit when Leonov crawled into a specially designed airlock attached to *Voskhod 2*. After depressurizing, the airlock door was opened and Leonov floated out to be the first man to leave his spacecraft while in orbit. The EVA went as planned until Leonov found his primitive space suit lacked mobility to allow him to reach safety inside Voskhod's airlock. In the vacuum of space, his suit had ballooned in size, making movement difficult. Leonov could not fit back into the airlock! Even after lowering his suit pressure, he still could not squeeze through. Only when Leonov lowered his pressure to .25 atmospheres and therefore risking the "bends" could he squeeze inside the airlock. The EVA lasted twice as long as planned and Leonov's suit oxygen had run dangerously low. When he reentered the cabin, he was greeted by a very relieved crewmate. A NASA diagram of Voskhod is shown in Figure 4-5.

Leonov and Belyayev faced further danger the next day while preparing to return to Earth. When it was time for reentry, the automatic retrorocket failed to ignite! After three attempts by ground control, Balyayev had to manually reorient the craft in order to fire the backup retrorocket. The craft did not maneuver well and was not properly aligned for reentry. *Voskhod 2* was way off course, landing a thousand miles (1,600 kilometers) from the recovery zone in the snowy mountain-

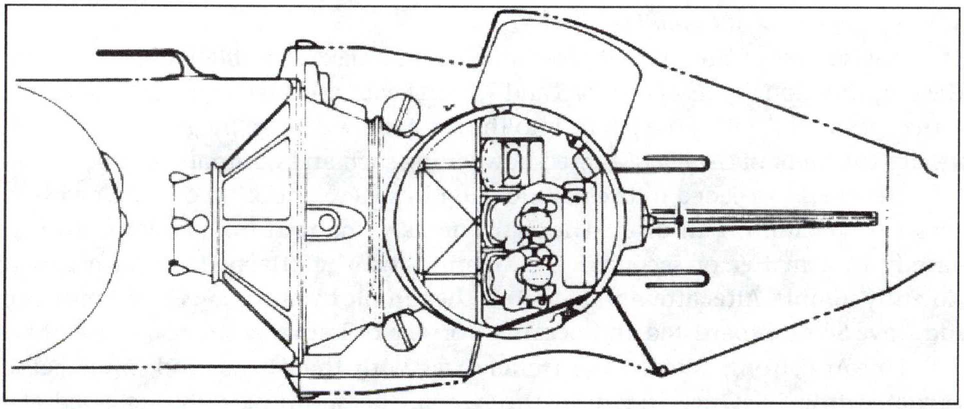

Figure 4-5. NASA diagram of Voskhod

ous terrain of the Urals. Alone and surrounded by wolves, the crew had to spend the night waiting for the recovery team.

The Loss of a Leader

In January 1966, Soviet spacecraft Chief Designer Sergei Korolov suffered a decline in health and was admitted to a hospital for routine surgery. Various reports state it was either for blockage of the bowels or hemorrhoids. The Soviet Minister of Health Boris Petrovsky, decided to personally perform the operation. During surgery, a tumor was discovered. Though Petrovsky was now a bureaucrat and had not operated for several years, he decided to continue the operation and remove the tumor. He thoroughly botched the procedure and Korolov suffered severe hemorrhaging. The stress was too much and Korolov's heart collapsed. Had there been a more skilled surgeon available, Korolov might have survived the operation and lived a few more years. This was a catastrophic disaster for the Soviet space program. Korolov was the visionary leader and designer of all the major Soviet civilian space programs. He was irreplaceable.

Setbacks and Tragedy for the Americans

For the Americans, another tragic T-38 jet crash occurred on February 28, 1966, and took the lives of astronauts Charles Bassett and Elliot See. Both were scheduled to be the prime crew of *Gemini 9*.

The success of the Gemini program began to make spaceflight appear routine. The long duration flights of *Gemini 5* and *7*, along with the dual flight and rendezvous of *Gemini 6* and *7*, lulled the public into thinking the American program could do no wrong. The flight of *Gemini 8* showed how wrong such an assumption can be.

Gemini 8 succeeded in the first docking in space, smoothly coupling with an Agena rocket waiting in orbit. The flight quickly changed from triumph to near tragedy in a matter of seconds. The Gemini and the attached Agena began to violently tumble. After attempts to correct the tumble by astronauts Neil Armstrong and Dave Scott, it persisted and became worse. Not knowing the real cause of the problem, Armstrong decided the trouble was with the Agena, and immediately separated from it. Once free from the Agena, the spinning only increased at a sickening rate. Rotating one revolution per second, the crew was in danger of blacking out and the Gemini spacecraft might become permanently damaged. Communication with ground control was also hampered by the rapid tumbling.

Now realizing his own craft's attitude control system had failed, Armstrong cut off all thruster systems and initiated the separate reentry control system. This gave Armstrong some control, allowing him to stabilize the craft. NASA quickly elected to order the astronauts home. It was determined that thruster number eight had become stuck in the 'on' position and had nearly killed the astronauts. The crew made a safe emergency landing in the Pacific Ocean after only seven orbits.

Apollo 1, planned to be the first mission of the American spacecraft that would eventually take men to the Moon, instead became a chariot of death. Originally slated to fly as joint mission with *Gemini 12* in November 1966, the craft was hampered by continuous delays common of new space systems. The cabin and engines had nearly twenty thousand test failures logged by the end of January 1966. Gus Grissom only a couple of days before jokingly hung a lemon on the Apollo cabin to make a point of his frustration. The crew of *Apollo 1* is shown in Figure 4-6.

Grissom, along with Roger Chaffee and Ed White, were inside the cabin rehearsing a simulated countdown on the evening of January 27, 1967. Then, the unthinkable happened. An electrical spark apparently started a fire under Grissom's seat and it spread quickly in the pure oxygen atmosphere. "Fire! I smell fire!" were the first words heard from the crew to indicate trouble. Then Grissom yelled, "Fire! We've got a fire in the spacecraft." The last words were from Chaffee: "We've got a bad fire...let's get out...we're burning up!" At this point control room instruments

Figure 4-6. The crew of *Apollo 1:* Grissom, Chaffee, and White

displayed that the cabin pressure had risen to 17 atmospheres and smoke was boiling out around the still sealed hatch. At 6:31 P.M., January 27, 1967, the crew of *Apollo 1* died of asphyxiation by toxic gases while they were trapped in the cabin of their burning spacecraft.

The nation was shocked by the tragedy and went into mourning. A thorough investigation later revealed the fire was started by an electrical short circuit. Perhaps one of the countless wires had been pulled loose by the continual opening and closing of the cabin's equipment during tests and troubleshooting. Other factors contributing to the crew's death were inadequate provisions for escape, rescue, and medical assistance. The sealed cabin hatch had no explosive bolts for quick release and took nearly five minutes to open. There were no fire extinguishers on board and none of the cabin's insulation was tested for fire potential in the pure oxygen environment inside the cabin. The wiring, plumbing, and overall workmanship on the spacecraft were ruled unacceptable. Also, there was no adequate fire fighting equipment for rescue workers.

A final grim note was reported from the autopsy that one or more astronauts were possibly still alive, although unconscious and badly burned, at four and a half minutes after the fire began. The cabin door was finally opened 30 seconds later.

NASA was criticized by the Review Board for complacency, but the investigation stopped short of asking if safety corners were cut. Nor was there any discussion of the mounting pressure of the space race with the Soviets. Congress was interested only in seeing NASA quickly back on the horse and in business again. The accident delayed the first manned flight of Apollo by nearly two years while the craft was extensively redesigned in the wake of the tragedy.

Soyuz 1

With the Americans searching for answers and rebuilding the Apollo program, the Soviets were confident now that they could move ahead. A very ambitious schedule was laid out in the spring of 1967. *Soyuz 1* was to be launched first with cosmonaut Vladimir Komarov controlling the maiden flight of the new generation Soviet spacecraft. Twenty-four hours later *Soyuz 2* carrying three crewmen, Valery Bykovsky, Yevgeny Khrunov, and Alexei Yeliseyev, would launch for a rendezvous and docking with *Soyuz 1*. After docking, Khrunov and Yeliseyev would have transferred via EVA from their cabin to join Komarov in *Soyuz 1*. Komarov with his two new crewmates would return to Earth, followed shortly by Bykovsky, now alone in *Soyuz 2*.

This would have been a very impressive demonstration. After watching the Americans move ahead with ten consecutive Gemini missions, one joint mission of this magnitude would have equaled all of the American achievements and put the Soviets back in the Moon race.

The exciting *Soyuz 1* and *2* joint mission was prematurely pushed into action by pressure from the highest political circles in the Kremlin. Unmanned test flights had revealed serious flaws in the Soyuz spacecraft. After the *Apollo 1* tragedy, one would have thought that the Soviets might have been a little more cautious. If Korolov were still alive, he might have prevailed in insisting on more thorough spacecraft testing. But the Soviet leadership saw that the window of opportunity was open for advancement and the Soviets went for it, and lost. A photograph of a Soyuz rocket is shown in Figure 4-7.

The launch of *Soyuz 1* on April 23, 1967, was routine during ascent. All systems were apparently performing well. Once in orbit, however, trouble occurred immediately and the *Soyuz 2* launch was quickly canceled. The Soviets have never fully explained the cause, but speculation and rumors center on the failure of one of the solar panels to deploy, leaving *Soyuz 1* with a critical power shortage. Even worse, Komarov's spacecraft was experiencing attitude control problems, often causing loss of communications as the spacecraft was tumbling out of control. The failed solar panel, still folded against the side of the Soyuz service module, blocked part of the radiators critical to cooling the craft's electronics. This likely affected the craft's guidance computer and attitude control system.

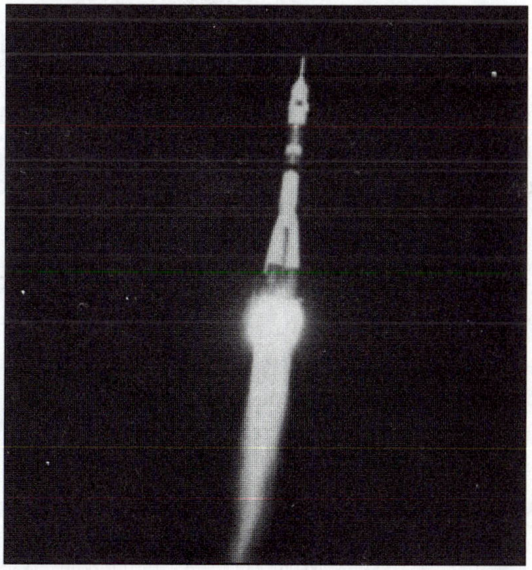

Figure 4-7. A Soyuz rocket at launch

Komarov and ground controllers put all their efforts on getting the spacecraft into proper alignment for an emergency reentry. On the sixteenth orbit, Komarov was to initiate reentry, but the craft started to tumble again. Reentry would be attempted again on the seventeenth orbit, but once again the Soyuz was out of control. On the eighteenth orbit, the retrorocket was finally fired.

By now, Komarov was nearly out of attitude control fuel. The only way left for him to stabilize the reentry cabin during its fiery plunge through the atmosphere was to place it in a roll around its vertical axis, similar to a spinning bullet. After reentry, *Soyuz 1* was still rolling as it fell to Earth. Komarov, with no fuel left for his control thrusters, was powerless to stop it. High in the stratosphere, Komarov desperately tried deploying the drogue chute and then the main parachute. Both became uselessly twisted and wrapped in their own shroud lines by the cabin's roll. Komarov made his last ditch attempt by deploying the reserve chute, and that too became hopelessly snarled with the first parachute. What thoughts could have flashed through his mind during the last moments of his fatal plunge? Duty? Honor? Country? Death? The spacecraft hit the ground at nearly 400 miles (650 kilometers) per hour, killing Komarov instantly.

The Soviets also conducted a thorough investigation of the accident, and like the Americans, never admitted skipping steps or bypassing safety in the race into space.

A Race to the Finish Line

In a matter of months, both space-faring nations had suffered horrible tragedies with their most advance spacecraft. Now, both the Soviets and the Americans would have to reassess their goals, find the causes, and rebuild their Moon-bound ships.

The Soviets faced death again only nine months later. The first man in space, Yuri Gagarin, was killed in an aircraft crash. This new loss affected the Soviet citizenry with the same kind of impact that the Kennedy assassination had on the American public.

As the United States rebuilt its Apollo program, astronaut training continued. To practice Moon landings, American astronauts flew a free-flight trainer called the "Flying Bedstead," shown in Figure 4-8. The name fit the description of an odd-looking contraption whose purpose was to practice vertical landings using the thrust of a jet engine. Flying it was very difficult and it became virtually uncontrollable if the craft leaned too much while balanced on its column of jet thrust. On May 6, 1968, Neil Armstrong was flying the craft for the twenty-second time. At approximately 200 feet (60 meters) altitude, the vehicle began to malfunction and carry Armstrong on a careening, looping trajectory resembling the swings of a clock pendulum. With all control of the trainer lost, Armstrong wisely bailed out via a

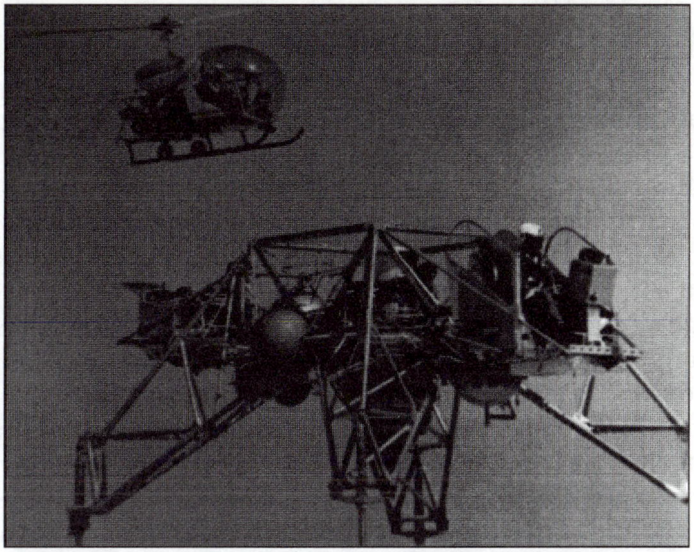

Figure 4-8. "Flying Bedstead" gave astronauts the chance to simulate lunar landings

rocket-powered ejection seat. His parachute deployed only just before touching ground. Armstrong was unhurt, landing a stone's throw from the burning rubble and potential death in the crashed trainer.

The cover of the December 6, 1968 issue of *Time* magazine, featured a foot race between an American astronaut and a Soviet cosmonaut running in space toward the Moon. The subtitle was "Race to the Moon," and it was indeed neck and neck. No one knew which country would send the first men around the Moon. The Soviets had the built-in advantage, thanks to orbital mechanics, of an earlier launch window. But mysteriously, the launch from Baikonour never came. Reports indicate that the *Zond 7* spacecraft, launched by the Proton rocket, was poised to carry a crew on a circumlunar flight. Three flight teams had been secretly assembled and trained for the mission; they were the prime crew of Alexei Leonov and Oleg Makarov, and backup crews Valery Bykovsky and Nikolai Rukavishnikov as well as Pavel Popovich and Georgi Grechko. Apparently as the countdown progressed, technical problems postponed the launch. The frustrated cosmonauts could only watch as *Apollo 8* circled the Moon instead. In January 1969, the Zond was launched unmanned and its booster exploded. The Soviets had wisely waited this time and could only hope the American program would hit a snag and be delayed.

The N-1: A Flawed Moon Rocket

The American momentum continued with success of *Apollo 9* and *10*. It was now time for the event dreamed of for centuries, the Moon landing. But only two weeks before the *Apollo 11* launch, the Soviets had two more hands to play. A last ditch manned lunar attempt and an unmanned robotic sample retriever were readied to usurp victory from the imminent American lunar landing. The secret N-1 Soviet Moon rocket was on the pad, perhaps ready to linkup with cosmonauts in space and then land on the Moon. No one knows for sure; the Soviets still haven't shed any light on the exact missions attempted by the N-1 launches. What is known is that at least one N-1 exploded on the pad on July 3, 1969. It has been reported that there were two N-1's ready on launchpads about half a mile (1000 meters) apart when the explosion occurred. A wall of shrapnel pierced the second N-1 and destroyed it also. The blazing fireball, almost nuclear in proportions, destroyed forever any Soviet hopes of beating the Americans to the Moon. A drawing of the possible payload configuration of the N-1 rocket is shown in Figure 4-9.

Were there casualties around the launchpads again? No clues, nor information regarding this have ever been announced by the Soviet government. However, private

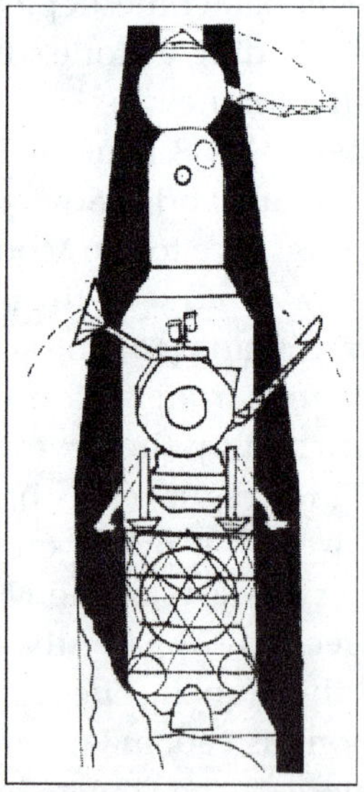

Figure 4-9. An illustration of the N-1 payload configuration

reports from Soviet citizens have referred to massive casualties on the second N-1 launchpad among observers gathered to cheer the liftoff of the gigantic Moon rocket.

Continued delays with the N-1 booster eventually led to the cancellation of the Soviet manned lunar landing project. Ultimately, after a 14-year effort, the N-1 booster project was abandoned in 1976.

The Moon Landings

After a seemingly perfect launch and relatively quiet flight from Earth, *Apollo 11* prepared on July 20, 1969, to make the first manned landing on the Moon. As Neil Armstrong and Buzz Aldrin made their final approach to the lunar surface, the lunar module's computer developed serious problems. Warning alarms alerted the crew and mission control that the system used for calculating altitude and descent rate information with the radar system had overloaded. A decision had to be made whether to abort the mission. A confident engineer named Steve Bales acknowledged, "Go Flight," and eventually the computer cleared itself. Another tense moment occurred at low altitude when the lunar module *Eagle* approached the planned landing site and found it was littered with boulders and craters. Armstrong took over manual control and flew in search of an alternative site while Aldrin monitored the altitude and read off the remaining fuel. After a tense minute, a suitable site was found and the *Eagle* landed. Engineers calculated there was only between ten to seventeen seconds of fuel remaining before the landing would have been aborted. The controls of the lunar module are shown in Figure 4-10.

After the huge success of the first Apollo manned lunar landing, the second landing by *Apollo 12* would seem to have been anticlimactic. But the beginning of *Apollo 12*'s voyage was anything but routine. The weather at the Kennedy Space Center was dark and gloomy at the time of liftoff on November 14, 1969. The decision to launch in such marginal weather was likely influenced by the presence at the launch site of President Richard Nixon.

As the lunar craft and its Saturn 5 booster climbed through rain clouds, they were struck by lightning 36 seconds after liftoff! The strike tripped the circuit breakers in the Apollo command module *Yankee Clipper* and temporarily shut down nearly all electrical components in the spacecraft. By the time the Saturn's second stage ignited, Pete Conrad and his crew had restored order in the spacecraft and had begun to realize the potential for disaster they had just flown through. Conrad radioed, "We had our problems here...I'm not sure that we didn't get struck by lightning." Conrad's suspicion was confirmed when he returned from the Moon aboard the lunar module *Intrepid*. When he met the *Yankee Clipper* in lunar orbit, scorch marks were visible on the side of the craft! Figure 4-11 shows the damage to the *Apollo 12* craft.

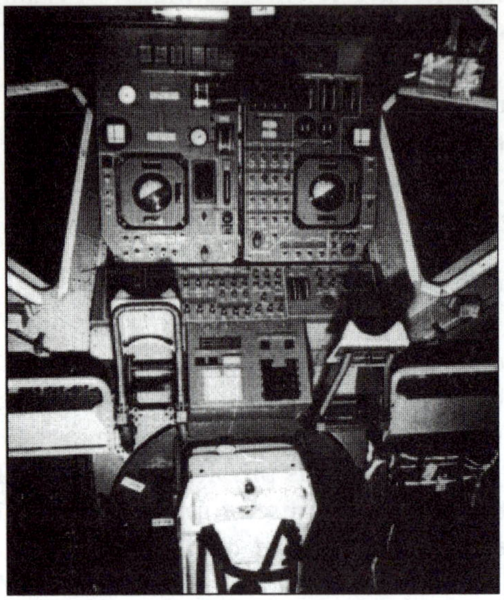

Figure 4-10. Internal view of the American lunar module

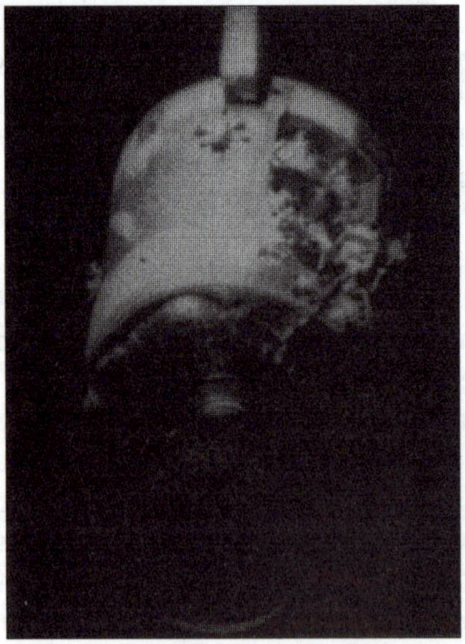

Figure 4-11. The extensive damage can be seen on the side of *Apollo 12*'s service module

Apollo 13: A Close Brush with Death

Disaster and death's hand nearly touched American astronauts again during *Apollo 13*, the third manned landing mission to the Moon. Two and one half days after their April 11, 1970 launch, James Lovell, Jack Swigert, and Fred Haise, were 204,000 miles (328,000 kilometers) from Earth. Deep inside the Apollo's service module, a liquid oxygen tank heater became stuck in the "on" position and raised the tank pressure to the bursting point. With no warning, it exploded like a bomb and blew the skin off the service module, destroying the life support and power generation systems.

Apollo 13 was in desperate condition. The Apollo craft itself was a powerless hulk, soon to run out of oxygen. The safety of Earth lay three days distant. Instead of going to the Moon, it became a mission of survival, and saving the crew of *Apollo 13* was one of the most concentrated endeavors in NASA's history. An all-out engineering effort by ground controllers devised ways of overcoming the maze of potentially fatal problems that afflicted *Apollo 13*. Using the lunar landing module as a "lifeboat," the crew was able to survive after the Apollo command module "died."

Conditions in the lunar module were not hospitable for the astronauts. Constantly damp and near freezing, the crew had to conserve precious power for the quarter million mile (400,000 kilometer) trip back home. The astronauts shivered for three days, averaging less than three hours sleep a day while eating semi-frozen food with an extremely limited water supply. After safely returning on April 17, 1970, all three astronauts were severely dehydrated, underweight, and weak. Haise also suffered a kidney infection for several weeks after the mission.

NASA evaluated *Apollo 13*'s brush with death and was able to quickly correct the cause of the explosion before the next lunar flight. But the real heroes of the *Apollo 13* flight were the hundreds of ground controllers and engineers who shepherded the crippled Moonship back to Earth. Without their brilliant seat-of-the-pants engineering, *Apollo 13* may well have been a frozen tomb for Lovell, Swigert, and Haise.

Death and Frustration for the Soviets

After 1969, Soviet manned space efforts shifted toward space stations in Earth orbit. In April 1971, Vladimir Shatalov docked *Soyuz 10* with the *Salyut 1* space station. However, Shatalov and his crew made a hasty emergency return to Earth before entering the *Salyut*. Shatalov later revealed that he and his crew were lucky to return alive, but he shed no light on the nature of their emergency.

A month later, *Soyuz 11* succeeded in manning the *Salyut* for a three-week mission. While the orbital part of the flight was largely successful, it was marred by panic when an electrical cable caught fire aboard the station. During the return to Earth, disaster struck again. The flight crew of Georgi Dobrovolsky, Vladislav Volkov, and Viktor Patsayev were following the now standard Soviet procedure of not wearing space suits. This cost them their lives when the craft accidentally depressurized during the return to Earth, causing the crew to suffocate in the vacuum of space. A photograph of the interior of the Soyuz capsule is shown in Figure 4-12.

On April 5, 1975, the flight that was to have been the *Soyuz 18* visit to the *Salyut 4* space station lifted off with crewmen Vasily Lazarev and Oleg Makarov. The ascent continued without incident until the third stage ignited at 90 miles (144 kilometers) altitude. As the upper stage fired, the second stage failed to disengage and drop away! The combined stages entered into a violent roll that threatened to destroy the craft until the Soyuz escape mechanism pulled the spacecraft free. The cabin parachuted into the Altai Mountains near Chinese border. After touchdown, the cabin slid down a mountain, stopping at the edge of a chasm. The parachute had fortunately become tangled on some trees and shrubs and averted total disaster. Lazarev and Makarov had successfully ridden out the only manned abort from a rocket in flight.

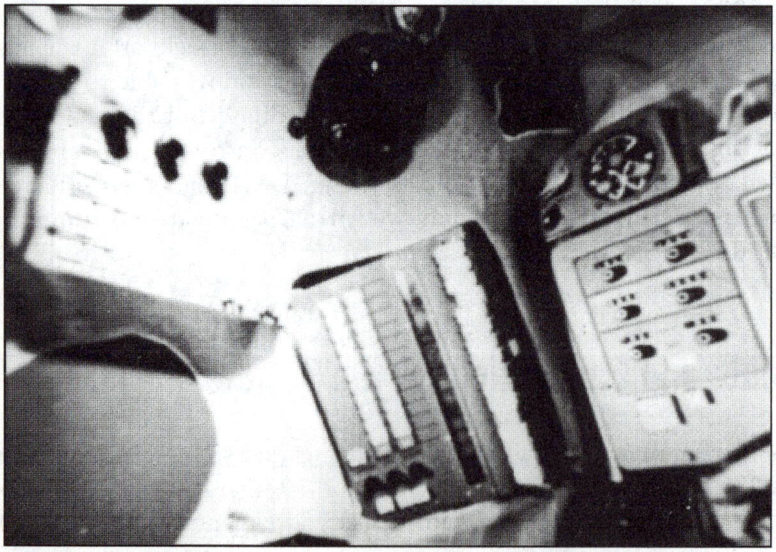

Figure 4-12. Soyuz instrument panel

Apollo-Soyuz

Two months later on July 17, 1975, all the world watched the historic American/Soviet space handshake of Apollo-Soyuz. For several weeks after the highly successful mission, it was hidden from the public that the American crew nearly died from toxic gases upon landing. During descent through the atmosphere, deadly nitrogen tetroxide gas seeped inside the *Apollo 18* cabin. Apollo command module pilot Vance Brand lost consciousness and all three crew members had their lungs blistered. A quick response by mission commander Tom Stafford in getting his crew into oxygen masks saved their lives. An artist's conceptualization of Apollo-Soyuz is shown in Figure 4-13.

Nearly Stranded and a Failed Launch

No major space mishap occurred over the next four years until the flight of *Soyuz 33* carrying Nikolai Rukavishnikov and Bulgarian cosmonaut Georgi Ivanov. Upon final approach to the *Salyut 6* space station, the Soyuz maneuvering engine failed and shut down, leaving the crew in peril of being stranded in space. The next day,

Figure 4-13. An artist's depiction of Apollo-Soyuz

an emergency return to Earth had its anxious moments. Retrofire, using the backup engine, encountered trouble as the engine would not shut off after the prescribed period! Rukavishnikov manually shut off the runaway engine and *Soyuz 33* safely landed, although far short of its target.

Four years later, the September 26, 1983 launch attempt of *Soyuz T-10* was punctuated by disaster. All proceeded well until 90 seconds before liftoff. Then, a violent fire erupted on the launchpad and engulfed the SL-4 booster and Soyuz. To make matters worse, the fire destroyed the cable connections needed by ground control to fire the Soyuz escape rocket. With only seconds to spare, radio commands successfully fired the escape rocket and carried cosmonauts Vladimir Titov and Gennady Strekalov away before the booster exploded on the launchpad.

The Space Shuttle Challenger

Early in its development, the United States Space Shuttle was plagued with persistent loose tiles on re-entry, and during one landing had a small fire in an engine. A more serious incident occured in mid-1985 when the shuttle had to abort during liftoff due to a premature main engine shutoff.

On January 28, 1986, spaceflight fatality became a reality for the United States as the crew of the Space Shuttle *Challenger* met their fate shortly after liftoff. It was America's worst space accident. Seven crew members, Commander Dick Scobee, Mike Smith, Ron McNair, Judith Resnik, Ellison Onizuka, Gregory Jarvis, and Christa McAuliffe, all perished with the destruction of the Space Shuttle. An inherent design flaw in the Shuttle's solid-fuel rocket booster (SRB) allowed hot gasses inside the left side rocket case to burst through the skin of the booster. Seventy-three seconds after liftoff, the burn-through severed the lower struts holding the SRB to the Shuttle's external fuel tank. In the structural collapse that followed, the external tank ruptured and exploded. The *Challenger* was torn free and disintegrated, as shown in Figure 4-14. Its crew cabin tumbled intact toward the ocean ten miles below. Commander Dick Scobee and his six crewmates were killed when the cabin slammed into the ocean.

A Presidential Commission was setup (similar to the *Apollo 1* Commission) which determined several causes for the accident. The main component that failed was the Joint O-rings in the Solid Fuel Strap-on Boosters. This was aggravated by extremly low temperatures at launch. The contracter Morton Thiokol, which designed the Strap-ons, overruled their own engineers who raised objections to the launch because of the low temperature. Besides being the coldest launch temperature for the shuttle, there were very heavy crosswinds which caused further stress in the joints. The commission also criticized NASA management. Several years

Figure 4-14. After 73 seconds, a failed O-ring causes the destruction of the *Challenger*

before, there had been warnings of joint flaws. Nine previous shuttle flights had experienced Joint O-ring erosion. NASA management ignored these warning flags and never informed the astronaut corps of these problems. It has been suggested that NASA was overconfident. In 1983, NASA analysis predicted a booster failure at 1 in 100,000. (Ironically, in 1985, the game designer of "LIFTOFF" determined a 1-in-25 failure rate for shuttle launches.)

Soyuz TM-6

The flight of *Soyuz TM-6* in August 1988 quickened the heartbeat of cosmonauts Vladimir Lyakhov and Afghan cosmonaut Abdullah Mohmand. After separating from the *Mir* space station for its return home, *Soyuz TM-6* was plagued by computer glitches that left it stranded after its retro-rocket failed to complete three firing attempts. Another attempt a day later was successful and the crew was recovered safely.

In the four years that have passed since the drama of *Soyuz TM-6*, thankfully there have been no further "cliff-hangers" in space. But it must be remembered that space boosters are among the most powerful machines on Earth, and also among the most fragile. Although manned space flight has occurred for over 30 years, each launch is fraught with the potential for disaster. Space travel is, and will be for a long time to come, an experimental and dangerous activity. The men and women of all nationalities who fly in space are well aware that disaster can be only a heartbeat away!

CHAPTER 5

More Secret and Forgotten Space Proposals

Due to the breakup of the Soviet Union and glastnost's release of information, historians have had to revise a great deal of history. Thanks to the persistence of such noted space authors as Philip Clark, Nicholas Johnson, Mike Cassutt, and experts such as Robert Reeves and Art Bozlee, a wealth of new Soviet space data has been brought to the surface. Each of these individuals are specialists on various aspects of the Soviet space program. Philip Clark, a rocket engineer, has written several books and articles on the Soviet manned lunar program. Nicholas Johnson, a science engineer, has written numerous books on Soviet spacecraft and an upcoming book on the N-1 entitled, *The Soviet Reach For the Moon.* Mike Cassutt wrote *Who's Who in Space* on all astronauts and cosmonauts and is about to release a revision. Author Robert Reeves has been researching the Soviet unmanned planetary programs for the past 30 years and is publishing two books on this subject. Art Bozlee, another dedicated researcher and writer, has gathered numerous Soviet secrets from various sources. Bozlee has numerous articles about space on GEnie, a national computer service.

I am indebted to these authors for the information and materials that they have provided. I strongly recommend the many fine books and numerous articles that they have written. There are many other fine sources on space listed in the back of this book in the bibliography.

German Spacemen?

While developing the V-2 ballistic missile for Germany's arsenal of destruction during World War II, Wernher von Braun and some of his coworkers often dreamed of spaceflight. In spite of the war, they even outlined plans for a larger version of the V-2 capable of orbital flight. The first versions, the A-5 to A-8, were enhanced V-2s, but the A-9 and the A-10 were to be combined as a two-stage rocket that would have a range of 3,200 miles (5,120 kilometers). Hitler was intrigued by the A-9/10 since it could reach New York or Washington. Von Braun continued with even more ambitious plans. The A-11 would have a third stage and could put a pilot into orbit. The last rocket, the A-12, would have the capability of sending a 60,000-pound (27,300-kilogram) payload into Earth orbit, or a 10,000-pound (4,500-kilogram) payload around the Moon. Because of his spaceflight dreams, Von Braun was charged by the German Gestapo with treason for his alleged lack of loyalty, and for not contributing to the military desires of the Fatherland. Only the intervention of Hitler himself saved von Braun from execution.

The X-15: America's First Spaceplane

On June 24 1952, the United States Air Force and NACA (the National Advisory Committee for Aeronautics) issued plans for a research aircraft that would explore flight characteristics at speeds in excess of Mach 4 and in the 50-mile altitude range. A development team was assembled to examine the problems of hypersonic speeds and space flight, which included Walter Dornberger, former commander of the Nazi's V-2 flight-research facility at Peenemunde.

During development, there was some internal friction and discussion on using existing experimental aircraft such as the Bell X-2. In 1954 the Air Force initiated some new policy guidelines for a more aggressive approach in manned aircraft research and approved the X-15 instead of modifying the X-2.

A very thorough test period began in late 1954 on structural design and stability. Various airframe shapes were proposed to withstand reentry temperatures exceeding 4000° F. A design by C.H. McCleean proved to be the answer to all of these hypersonic difficulties. He replaced the traditional thin airfoil shape to a 10° wedge design. A further refinement was to vary the wedge angle which increased stability and achieved a greater temperature tolerance on reentry.

After these designs were completed the project was presented to 12 airframe contractors for bidding.

In June 1955, the airframe contract was awarded to North American. Ironically, North American had just begun to withdraw their bid due to more lucrative contracts on other projects. Finally, after some consultation, North American agreed to build the most technologically complex experimental aircraft of that time.

Test pilot Scott Crossfield was assigned for the initial flight testing. His experience on previous experimental aircraft that included the Bell X-1, the Douglas D-558-II, and the Convair XF-92A was monumental in refinement of the X-15 development.

Due to the performance and maneuvering characteristics and the requirements of extended emergency landing sites, the Air Force constructed an array of tracking, telemetry, and communications sites between Edwards AFB and Wendover, Utah. This area became known as X-15 High Range.

> *"...the same week the Russians launched the first Earth satellite, we launched a magnificent selection of automobile models, including the uniquely elegant new Edsel."*
>
> — John K. Galbraith, former ambassador to India

In June 1956, North American was given the approval to begin construction of three airframes. On October 15, 1958, the first X-15 rolled out for static testing. Several B-52s underwent modification in order to carry the X-15 to launch altitude. After five months of static tests, Scott Crossfield flew the first captive flight of the X-15 attached to the B-52. In June 1959, the first glide flight was successfully conducted. On September 17, 1959, Crossfield piloted the first powered flight and reached an altitude of 52,341 feet.

It was on the fourth flight that a near mishap occurred when the flight engine exploded and required an emergency landing. Landing heavier than normal weight, the aircraft hit the dry lake hard, damaging the fuselage. While this model X-15 was being repaired, the second model rolled out for flight testing.

On January 23, 1960, Scott Crossfield set a new speed record of 1,669 mph and an altitude of 66,844 feet.

Now the testing program went into high gear. Six pilots were assigned to the X-15 program, NASA pilots Joe Walker, Neil Armstrong, John McKay, Air Force pilots Robert Rushworth and Robert White, and Navy pilot Forest Peterson. Later on additional pilots joined the project. NASA assigned Milton Thompson and William Dana; the Air Force sent William Knight and Michael Adams. Some of these X-15 test pilots are shown in Figure 5-1. On March 25, 1960, Joe Walker piloted the first official government X-15 flight. The Air Force had the next opportunity with Robert White flying on April 13, 1960.

Each successive flight of the X-15 broke new records and expanded the knowledge of hypersonic flight, and on April 30, 1961, Joe Walker flew the X-15 to the mission goal altitude 246,700 feet.

The X-15 program had a total of 199 flights and its last high altitude flight was in October 1968. Its 200th flight was scrubbed due to bad weather. Seven of the military pilots, White, Walker, Rushworth, Engle, McCay, Adams, Knight, and Dana earned their astronaut wings after qualifying flights over 50 miles in altitude. Two flights by Walker attained the Federation Astronautique Internationale guidelines of spaceflight by reaching an altitude of 62 miles.

Figure 5-1. America's X-15 test pilots

Variants of the X-15

Several variants of the X-15 were proposed during the life of the program. One was actually built. When one of the X-15s crashed, it was decided that, during the repair process, it would be retrofitted into the X-15A2 model. This model incorporated a fuselage stretch, increased fuel tankage, and the ability to carry expendable fuel tanks.

The proposed X-15A3 would have included delta wings for additional fuel. Other variants would have included an internally mounted scramjet for sustained high-speed flight.

During the panicky years surrounding the formation of the Mercury program, other crash programs were studied. One of the more interesting was the X-15B. This specially-built X-15 variant would be rocketed into orbit atop a rocket derived from the North American Navaho, but similar in capability to an Atlas. The X-15 could maneuver in space under the control of its single pilot, and initiate a re-entry burn. The X-15 would be coated with an ablative heat shield, but the heat of re-entry, and the melted ablative, would cause too many problems for a normal landing. This would force the pilot to eject from the X-15B when it had slowed to subsonic speed. The X-15B was thus a proposal for an expendable minishuttle—truly an oxymoronic concept.

While the X-15B proposal had merit (the X-15 was existing hardware at the time), it had several difficulties. First, the modifications to the X-15B configuration would be extensive. Also, testing of the original X-15 had yet to begin. Further, the Navaho-derived booster did not yet exist, nor was it likely to in time. Worst of all, however, the pilot was required to eject. This proved the deathblow to the concept, since the Federation Aeronautique Internationale guidelines recognized only those flights where the pilot took off and landed in control of his craft. Since the X-15B pilot did not return with his craft, the orbital flight technically did not count. Later, the Soviets would run into this problem with their Vostok flights. They got around it simply by keeping the design of the Vostok, and the pilot's recovery method, a secret for several years.

Theoretically, an X-15B could even have been converted into craft for a manned lunar pass, with a propulsion and power module, and additional consumables. While such an idea seems crazy today, it was actually under consideration by during the panic of the early '60s. In times of panic, and with the continuing momentum of early Soviet space spectaculars, there was no shortage of half-baked ideas.

> *"If the newspapers printed a dispatch that the Soviet Union planned sending the first man to Hell, our federal agencies would appear the next day crying, 'We can't let them beat us to it.'"*
>
> — Admiral Hyman Rickover, 1959

TOP SECRET

The X-20: America's First Minishuttle

The concept of rocket-launched minishuttles did not die with the demise of the X-15B. Far from it, in fact; designs for a reusable orbital spaceplane trace their origin to the German engineer Eugen Sanger, who proposed his spaceplane design, the Silbervogel, or Silverbird, in the 1930s. During World War II, that project was seriously studied as an intercontinental bomber, nicknamed the "Amerika Bomber." The idea was scrapped, as it was considered too advanced for the day.

Sanger's work interested Josef Stalin greatly, however, and he set Soviet space scientists to work on the Sanger proposal shortly after the war, although at a low priority. Eventually they began to make progress, however, toward the design of a small, reusable minishuttle, the Lapot, to be launched atop a conventional rocket. Ironically, it was U.S. intelligence's discovery of this project which provided the impetus to the X-20 project.

Sanger's proposal also inspired work in the U.S. within the Bell corporation, which refined the design of a rocket-launched minishuttle called the BOMI (for Bomber Missile) starting in 1955. Gradually, other engineers saw the benefits of Bell's hypersonic glider design, and the competition for the X-20 contract was held. Ironically, however, the contract was won, not by Bell, which had invested much in the design, but by Boeing, because Bell's older design had several technical flaws.

The X-20 was nicknamed the "Dyna-Soar," for "Dynamic Soaring." It was a manned, reusable hypersonic glider, or minishuttle, to be launched atop a conventional rocket. The actual rocket to be used varied with the changing weight budget of the Dyna-Soar, but was finally determined to be a Titan III (a stretched Titan II with solid-fuel strap-ons) for orbital flights (shown in Figure 5-2). The glider was flat-bottomed, for aerodynamic reasons, and had a sharply swept delta wing with wingtip rudders. An illustration of the X-20 is shown in Figure 5-3. Rather than using the ceramic tile system of today's space shuttle, the X-20's thermal protection consisted of exotic Rene-Columbium alloys attached in chevron-shaped sheets to allow for heated expansion. Due to the high heat load imposed on the undercarriage, conventional rubber tires would not survive re-entry. Instead, landing gear consisted of skis, with steel wire brushes beneath.

Since the X-20 was a military program, it came under the jurisdiction of the then new Secretary of Defense, Robert McNamara. McNamara felt that the X-20 did not adequately fill any particular military mission, and eventually cancelled the program, although not before selecting the pilots, beginning the construction of the first shuttle, and spending half the projected cost of the project.

Thus ended what would have been America's first reusable space shuttle. With the X-20 as a testbed, larger and better shuttles could have been built later, with

Figure 5-2. An illustration of an X-20 atop a Titan with strap-on boosters

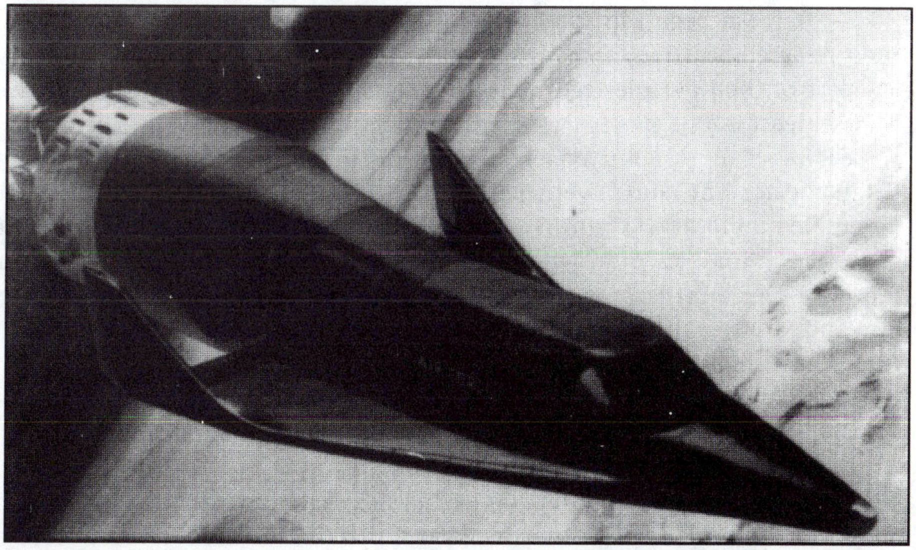

Figure 5-3. The Air Force-proposed reusable X-20 spaceplane

less risk. Instead, the Space Shuttle of today was designed without sufficient knowledge of the environments it would experience. The X-20, stretched and widened, could also have served as an excellent multimanned minishuttle, had history taken that turn.

Even cancelled, however, the X-20 program served the U.S. space program well. Several advanced systems, especially avionics components, were transferred from the cancelled X-20 program into the X-15 program, which was then actively in flight test. Thus, knowledge of hypersonic flight and spacecraft re-entry was aided through the use of X-20 hardware. Re-entry analysis data and techniques were also useful in the design of the Apollo, and later, were considerably more useful in the design of the space shuttle.

NOTE: For a brief time, the X-20 bore the designation XMS-1, for "experimental Manned Spacecraft, One." Within the game, it is assumed that the three-person minishuttle is the expanded, more capable follow-on, and is thus given the (completely fictional) designation XMS-2.

MOL: America's First Space Station

While NASA chalked up a string of spectacular space firsts, the U.S. Air Force chalked up a string of programs cancelled while nearly complete, like the X-20. Another of these was the MOL, or Manned Orbiting Laboratory.

The MOL was begun in 1964 as the X-20 was cancelled. It was intended as a manned research and reconnaissance platform (another of its designations was KH-10, part of the KeyHole reconnaissance satellite series). A total of eight were planned, at least initially.

MOL was a tall cylinder, ten feet in diameter, making it fit exactly the diameter of a Titan booster. The launch vehicle was to be an upgraded Titan III, and the crew would employ a modified Gemini capsule, called Gemini-B. Since orbital rendezvous was at the time an unproven technique, it was considered somewhat risky. Therefore, the Gemini-B would be on top of the MOL and launched. Once the Gemini and MOL reached orbit, the crew would transfer to the station via a hatch cut in the heat shield, shown in Figure 5-4.

Normal mission duration was to be 30 days, extended to 60 days for the later missions. To achieve this endurance, the Gemini had to be shut down in orbit to preserve its fuel cells and other consumables. This ability to shut down, plus the heat-shield hatch, plus a larger retro-rocket system, plus an oxygen-helium atmosphere mix (to match the station) were what differentiated the Gemini-B from the NASA version. At the end of the mission, the MOL, at least initially, would be abandoned, and would re-enter Earth's atmosphere.

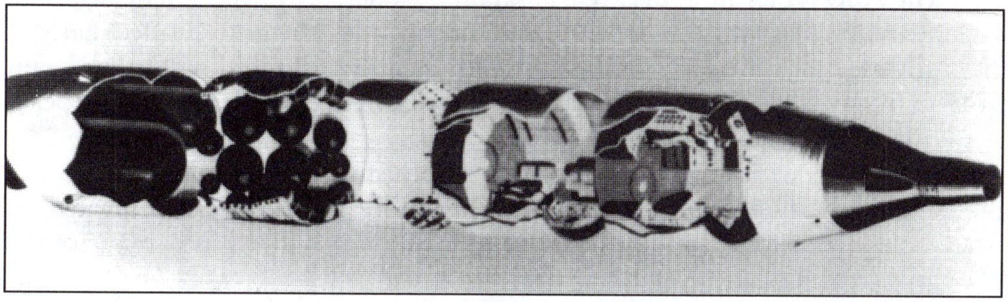

Figure 5-4. The Air Force-proposed Manned Orbiting Laboratory

Later plans called for docking at least two MOLs together to make larger stations. Two MOLs were to be owned and operated by the U.S. Navy, which would configure them for ocean surveillance.

One MOL test mission was flown in 1966. In that mission, a Titan III launched a long empty tank section (MOL mockup, to demonstrate that the structural and vibration margins were sufficient), atop which was a Gemini capsule with the heat-shield hatch. This capsule had previously served as the Gemini-2 unmanned test vehicle, and had been refurbished and modified for this test flight. Thus, it was actually the world's first known reusable spacecraft. The Gemini was recovered intact, and was judged to be "theoretically reusable" yet again! The heat-shield hatch, which many had considered a "dangerous joke," had performed perfectly. It was designed so that the heat of re-entry would melt it closed, rather than burn it open.

Contained within the MOL mockup were a number of experiments to take advantage of the several minutes of suborbital flight provided by the test flight. This would be the only scientific data provided from a MOL, however, as this program also was cancelled in 1969. At the time, it was judged to be less than two years from a flight, although this date had slipped from earlier predictions due to budget cuts in the latter years.

Aside from the MOL test flight, there are other legacies from the MOL program. The 7-segment Titan III solid strap-ons, which were improved over the 5-segment solids of the day, were designed and test-fired to support the considerable weight of the MOL. While the 7-segment design was shelved after the MOL cancellation, it was resurrected in 1985 to create the improved Titan III known and used today as the Titan IV.

The other legacy from the MOL program is sadder: the MOL launchpad, SLC-6 at Vandenberg AFB, was nearly completed for the MOL, then mothballed. Later, it was converted for use by the space shuttle for polar orbits, then mothballed again in 1986. Today this nearly 30-year-old pad still sits, in anticipation of a large rocket and a manned launch.

In the Soviet Union, of course, the highly successful Salyut space station program began as a direct response to the MOL program, and was its equivalent in nearly every way. Today, a Salyut-derived space station is in use as the core of the Mir space station.

Space Scooters: The Minimum Lander

When the Lunar Orbit Rendezvous (LOR) concept was still new, the design of the "bug," or Lunar Module, or LM, varied greatly. One NASA-Langley study attempted to define what was the minimum possible size for such an LM. They found it.

The "space scooter" weighed just 3200 pounds, as compared to the actual LM, which weighed over 30,000 pounds. The space scooter was simply an open-topped platform on which a single space-suited astronaut would stand. The pilot would land the vehicle using the waist-high control stand; there was no automatic control at all. Beneath the platform was a rocket engine, fuel tanks, and landing legs. On undocking from the Command Module, the scooter pilot would thrust against his orbit for a predetermined length of time to kill all orbital velocity. After a precalculated wait, he would rotate tail-down and thrust for another predetermined length of time, at which point he would be at low altitude and slow speed, and would land manually. Return launch was again precalculated, and relied on the Command Module to accomplish the rendezvous.

The space scooter was, of course, an excessively risky design, and was abandoned for that reason. There were other points against it, however; its limited payload, including only a single astronaut, allowed very little scientific exploration of the Moon. The limited stay-time of only about four hours was also a drawback. Additionally, the fuel-wasting, but easy-to-accomplish "right-angle" trajectory drastically limited the choice of landing sites. In the end, the more conventional enclosed lander was selected as the preferred approach. However, vehicles resembling the space scooter remained under consideration for years afterwards—not for lunar landing, but for mobility on the Moon, similar to (but faster than) the lunar rover.

Lunar Surface Rendezvous: The "Other" Option

Most space historians will be able to describe the great debate between the proponents of EOR, Earth Orbital Rendezvous (which included von Braun himself), Direct Ascent (which included NASA headquarters and several rocket contractors), and LOR, Lunar Orbital Rendezvous (which began with an engineer at Johnson Space Center in Houston and became more widely accepted as its merits were discussed). There was, however, another option that historians frequently overlook: Lunar Surface Rendezvous, or LSR.

LSR was never a serious contender to become the one-and-only method of going to the Moon, but it had several advantages. It required no unproven space-rendezvous techniques at all, nor did it require a giant rocket. In fact, use could be made of the Surveyor lunar probe, which could be converted into a lightweight, inexpensive supply transport, prelanded on the lunar surface. Its disadvantages included the fact that lunar landings in such close proximity were not yet proven, and were considered to be beyond the state-of-the-art in space guidance. Additionally, extensive Rube Goldberg devices would be needed to transfer equipment and fuel between various landers once on the lunar surface, unless the astronauts were to carry everything.

LSR was never a single, concrete plan either, but the general idea involved a manned lander that would proceed directly to the Moon and land, Direct Ascent style, but the vehicle would land completely empty, without equipment or return fuel. The lander was thus much lighter than the Direct Ascent vehicle. The return vehicle could be either the craft the astronauts arrived in, or another one, already landed nearby, but the procedure was the same. Preceding the astronauts to the Moon would be a series of unmanned payloads sent to the same spot. These would include consumables, scientific equipment, rovers, and other mobility aids, and—most importantly—fuel for the return trip. This would be transferred to the return vehicle, which would then have enough to return to Earth.

Thus, the mission proceeds much like a Direct Ascent, but instead of a giant rocket, it requires a smaller one, preceded by a long series of smaller supply launches. These would be on Atlas rockets with converted-Surveyor payloads, or larger payloads on larger rockets.

Lunar Surface Rendezvous, is a *technique,* however, not a specific plan. Any future lunar base, by definition, will use LSR (shown in Figure 5-5) and have several launches from Earth which will all join up on the Moon. Such future missions may also take advantage of lunar fueling as well, since liquid oxygen processed from lunar sources could be used to refuel any landing spacecraft, thus dramatically reducing the required initial mass.

Figure 5-5. A second LM joins an LM already on the Moon

The likely Soviet lunar landing included an LSR element. It now appears likely that the plan included an unmanned LM test which would be landed on the lunar surface. The manned mission would land nearby, and the cosmonaut would then have his choice of two landers for the lunar liftoff, in case his own developed problems.

The historical lunar missions provide mixed reviews for the proponents of LSR. *Apollo 11* missed its landing site by over four miles due to a computer failure. This might have been a fatal miss had *Apollo 11* required a lunar-surface refueling to get home.

On the other hand, subsequent missions, with improved computers, came much closer. And, ironically, LSR is a technique which has already been tried. The very next mission, *Apollo 12*, landed less than 300 feet from its target, the *Surveyor III* probe. In fact, human intervention in the landing sequence was required, as the computer might well have landed the LM on top of the Surveyor probe! The *Apollo 12* crew examined the inactive Surveyor in its landing site, and removed several parts from the probe and returned them to Earth to study the effect of three years of lunar space exposure on the craft.

Apollo Applications: Programs Beyond the Landing

Many people are aware that the Skylab space station sprang from the Apollo Applications Program (AAP). But some confuse the two, believing that the AAP and Skylab programs were the same thing. In fact, they were not.

AAP was a wide-ranging program to adapt Apollo hardware for new scientific or engineering uses. The Skylab program was but one example of this. In it, a Saturn V was used to launch a large space station (fashioned out of a Saturn S-IVB fuel tank, with several Apollo and LM components), which was serviced by Apollo capsules.

Another was the SIMM bay, which was a modular bay of scientific instruments carried within the Apollo Service Module on the later (*Apollo 15* and beyond) missions. These included lunar survey instruments, a cosmic dust collector, and even, on one occasion, a subsatellite, making the first time a manned spacecraft launched a satellite. Retrieval of the dust collector necessitated a lunar EVA.

The design of the Saturn and Apollo allowed additional missions to be undertaken with different hardware. Space within the Service Module allowed some modifications to be made there, but, more importantly, the LM could be modified to perform any of a number of missions, or be replaced altogether with a specialized module. Excess payload capacity and cavernous space within the Saturn-Apollo adapter allowed any of a number of payloads to be carried. The Apollo-Soyuz mission took advantage of this capability when the Apollo was launched, with the docking module stowed behind it. The Apollo rotated and docked with the module as if it were an LM, then chased down the Soyuz for the world's first international docking.

Other uses for this space could have included specially-designed, single-purpose laboratories, or an Apollo Telescope Mount, or ATM, a manned space observatory. The ATM was in fact mounted to the Skylab space station, but the unit was also capable of operating in a free-flying mode, carried into orbit, or serviced by, Apollo missions. The ATM controls were monitored by astronauts within the ATM. The shirt-sleeve environment within the ATM, and the control panel, existed because the ATM was in fact a heavily converted LM!

Other LM conversions were possible. The large velocity change the LM could execute meant that, in Earth orbit, it could change orbits to a great degree. This created a brief study to examine the LM's suitability for use as an anti-satellite weapon. It also gave birth to the concept of the "Rescue LM," shown in Figure 5-6, which would be based at a space station and could be dispatched to rescue stranded astronauts in any of several orbits. Shorn of its landing legs, the Rescue LM incorporated a "robot arm" for the grappling and movement of "stretchers" and

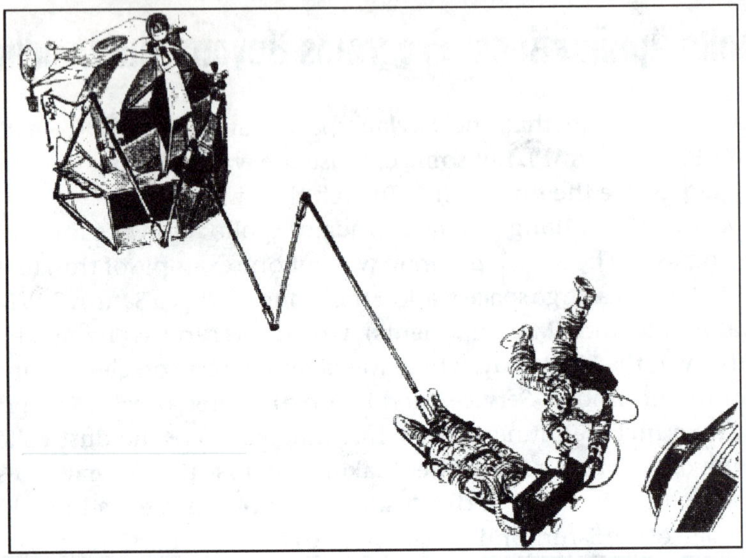

Figure 5-6. A "Rescue" LM retrieves a stricken astronaut

other apparatus, and any wounded astronauts within them. The Rescue LM could then return the astronauts to the station. The Rescue LM could also be used to retrieve or repair satellites in orbit.

An LM could be useful for more survey missions than just the ATM. It provided a versatile, flexible platform for any of a number of Earth-orbiting observatories, including X-ray observatories, or remote sensing platforms, or bioscience laboratories.

An LM could also be converted for use as a lunar-orbit survey module. An Apollo with a survey module would park in lunar orbit for 30 days, surveying the lunar surface in many frequency bands, using spectroscopes to determine surface composition, performing radar mapping, and other missions. The converted LM would carry not only the instruments, but additional consumables required for the longer mission. An LM lab is show in Figure 5-7.

It was calculated that if an LM were converted entirely to carry consumable supplies, an Apollo capsule could be supported for an independent mission of five or six months! This would be enough to plane-change out of the Earth's orbit about the Sun, and return on the other side of the orbit six months later, thus making the first brief foray into interplanetary space.

Of course, the best contribution the LM could make would be on the lunar surface itself. While a single LM could be improved or upgraded, more could be gained by sending a second LM, unmanned, ahead of time. This LM could carry

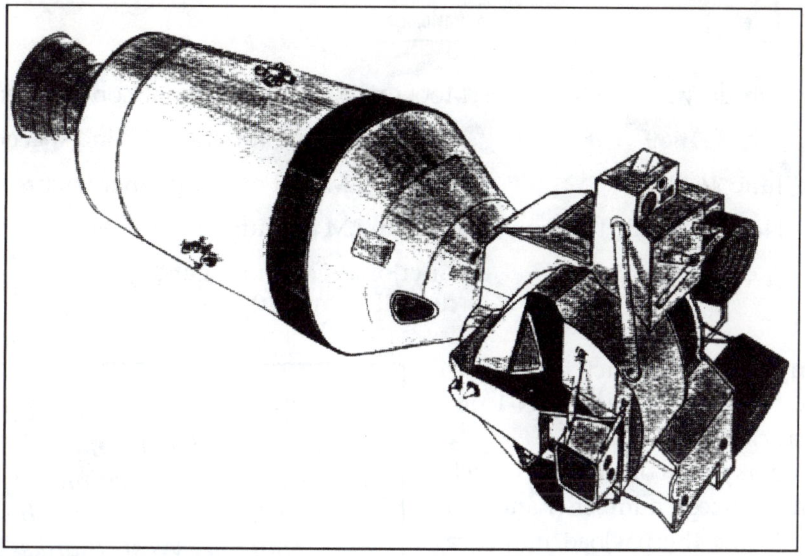

Figure 5-7. The LM lab

supplies for an extended stay, additional equipment, a larger, more powerful rover, or even a complete small station for stays of a month or more. This concept, called the "LM truck" (shown in Figure 5-8), was designed to be versatile enough to carry any of a number of such payloads without redesign. This would allow the gradual extension of man's stay on the lunar surface, and allow eventual expansion into a permanent lunar base.

Alas, it was not to be. LM conversions, LM trucks, and extended lunar stays were cancelled with the Apollo lunar landings and the Saturn V. The legacies of the Apollo Applications Program were Skylab, the Apollo-Soyuz mission, and the modular, multimission design of the present space shuttle payload bay. What was given up was a range of capabilities, from Earth orbit to interplanetary space, and including a lunar base.

The N-1: New Information

Since the publication of the historical manual provided in the "Buzz Aldrin's Race Into Space" computer game package, more information has surfaced regarding the N-1 flights. In particular, their payloads are known:

#	Date	Payload
1	February 21, 1969	Mockup LM and functional Zond spacecraft
2	July 3, 1969	Mockup LM and functional Zond spacecraft
3	June 26, 1971	Mockup LM and mockup lunar Soyuz
4	November 23, 1971	Mockup LM and functional lunar Soyuz
5	August 1974	Functional LM and lunar Soyuz (cancelled launch)

An additional launch in late 1974 was also scheduled when the N-1 program was cancelled.

Rumors persist of an actual launch, scheduled between numbers 1 and 2 on this list. While the payload numerical designators show a gap between those two launches, Soviet literature nowhere mentions such a launch.

From this, it appears that the Soviet program could have made a 1970

> *"We Russians, a backward people ourselves less than a lifetime ago, can now do even more spectacular things than the rich and pompous West, thanks to Communism."*
>
> —Editorial from the *London Economist*, October 1957

Figure 5-8. The LM truck

target date had they not suffered their N-1 launch failures. No functional LM was ever launched, even unsuccessfully, on an N-1. The only LM tests were the two unmanned tests launched by a pair of Proton rockets.

Voshkod to the Moon?

Among the ambitious schedule for the Voshkod two-man spacecraft, it appears that there was at least some serious discussion regarding lunar flights. The program was a rushed design, pressured by the political whims of Khrushchev. The three-man stunt was a perilous flight. Leonov's space walk on *Voshkod 2* (a diagram is shown in Figure 5-9) with its close calls, proved at least that man could survive outside of the protective womb of the spacecraft. The reliability of the Soviet space suit design (far more complex than the American Gemini EVA suit) proved that a cosmonaut's lifesupport could be self-contained. This was an ideal precursor of the Lunar EVA suit. Leonov's space suit is shown in Figure 5-10.

The planned 15-day duration flight of *Voshkod 3* would have given the medical clearance for eventual lunar missions. The all-female crew of *Voshkod 4* would have had a more ambitious space walk. On this flight, the female spacewalker would probably have rehearsed the necessary crew transfer stage for future lunar landing missions. *Voshkod 5* mission goal was an MMU (Manned Maneuvering Unit) test

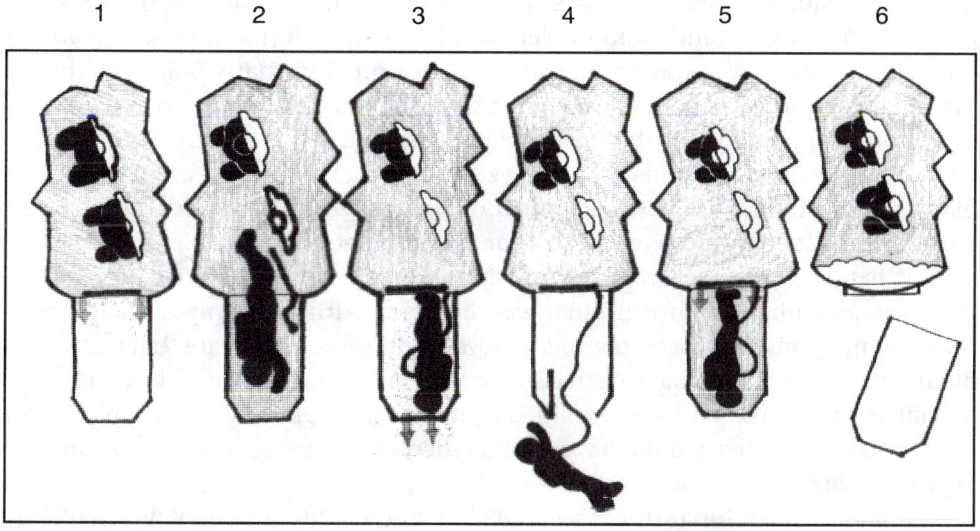

Figure 5-9. A diagram of various stages of Leonov's EVA with use of the attachable airlock

Figure 5-10. Leonov's space suit and the Voshkod airlock

identical to the scrubbed Gemini mission. A cosmonaut would have spacewalked and attached himself to a backpack that had various attitude thrusters for maneuvering. The sixth Voshkod flight would have been a biomedical mission to determine man's adaptability to long-term exposure in space. With all of the space proposals floating around, studies were being made for future potential manned Mars flights. *Voshkod 7* was an all-military space mission that would have tested surveillance equipment on nuclear launch sites, submarines, and troop movements.

Korolev probably felt that fate had been pushed twice already, and that any additional Voshkod missions might have disastrous consequences. The goals and plans for the follow-up Voshkod missions were shelved permanently. The real scientific missions would have to wait for the completion of Soyuz.

When these "cancelled" space missions were originally planned, apparently there was an additional mission that was discussed. Although only five additional Voshkod missions have been revealed, sources and rumors indicate that a manned lunar pass option might have been considered. This seems plausible, but only with several assumptions. Voshkod would continue as a program with all the mission goals successful. This would have further depleted resources from the already delayed Soyuz.

If an American lunar threat seemed near, a Soviet lunar pass via Voshkod was a possible option. This would have also required the development and man-rating (testing for human safety) of the Proton rocket, which was having its own production difficulties at the time. To this day, the Proton rocket has never been officially man-rated.

The initial Voshkod never had any docking capabilities and would have required a costly redesign to achieve this feat. So we could safely rule out the likelihood of Voskhod-Lunar Module combination.

This would have been in direct competition with other approved manned lunar landing spacecraft hardware and mission profiles. But a manned lunar pass in a Voskhod would have been technically possible.

We can only speculate that the lunar Voshkod flight would have had additional consumables and modifications for the safety of one or possibly two cosmonauts on the five to six day journey. This lunar Voshkod would have required major upgrades to backup systems in lifesupport, propulsion, and communications. It would have needed to be fitted with a TLI Booster (Kicker) that would have had the capabilities of additional mid-course correction burns.

A possible date for this conjecture is merely hypothetical. In the mid-1960s there were rumors of sending the first man and woman in space, Gagarin and Tereshkova around the Moon. At the time, no one ever said what type of spacecraft would have performed this task. The Soviets also had the habit of coinciding their space launches with a historical anniversary. One ideal date for such a historic spaceflight was the 50th anniversary of the Soviet Revolution in October 1967. Maybe this was a possible timetable for the Lunar Voshkod mission.

> *"To be the first man to fly to outer space to enter into a single combat with nature, can one dream of anything greater?"*
>
> — Yuri Gagarin, cosmonaut, April 1961

Of course none of this ever went passed the pencil-sketch phase. Korolev knew that a continued Voshkod program would delay any chance of Soyuz and any eventual lunar landing via Soyuz and Lunar Module.

LAPOT: The Soviet Minishuttle

As early as 1961, American military intelligence had a classified report that mentions Soviet testing of a reusable glider-type spacecraft. New information reveals that in the early and middle 1960s it was the Mikoyan Design Bureau that worked on the design and development of lifting bodies. The design goals were to test and develop a reusable spaceplane capable of reaching orbit and returning to Earth for a runway landing. Lapot, or Project Spiral, as it also became known, was modeled after the then 30-year-old German Sanger proposal. It also followed the development of the U.S. Air Force X-20 Dyna Soar.

One MIG Spiral test plane was constructed and flown for low-speed atmospheric tests. This plane was called Product 105 and had an upturn boat nose with retractable wheels that folded around the wings.

Several subscaled prototypes were built as small unmanned vehicles designated BOR-3, BOR-4, and BOR-5. From 1967 to 1970, the Soviets flight tested the scale model BOR-3 in suborbit and possibly in orbit and recovered the models at sea. The BOR-4 was a 1/3 scale orbital version and the BOR-5 was a suborbital variant. It is likely that hundreds of drop-tests were also conducted with these unmanned vehicles. The full-size minishuttle Kosmolyot, shown in Figure 5-11, would have a crew of three. A manned version has yet to have been flown.

Project Spiral continued up until December 1984. Mysteriously, no tests have since that time been conducted.

A rival of Spiral might have been the cause of discontinuing those tests. Buran, the Soviet version of the American Space Shuttle, began development in the early 1980s along with the construction of the super rocket Energia.

Flight Tests

The following list of test flights for Project Spiral/LAPOT is unconfirmed, but it is assumed that numerous drop-tests were made before orbital launches.

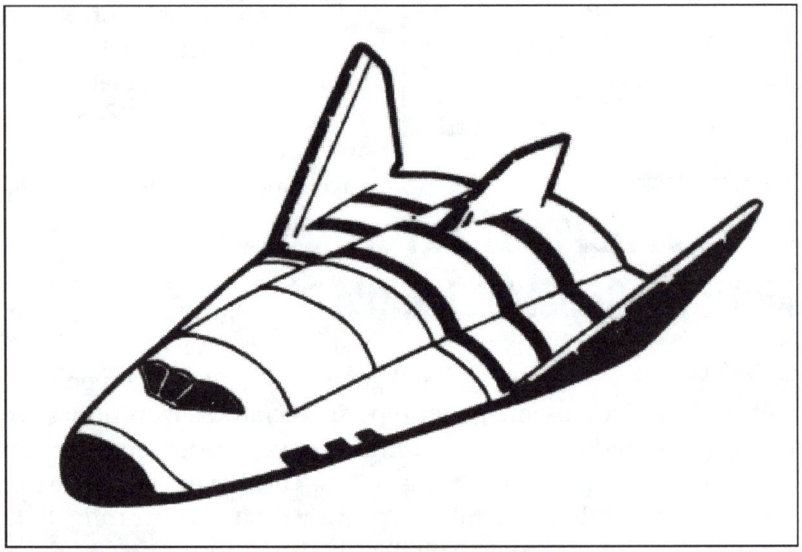

Figure 5-11. The Soviet minishuttle

Cosmos 1374 Unmanned test Delta-winged Minishuttle 1/6 scale, launched into orbit June 3, 1982 and successfully recovered in the Indian Ocean

Cosmos 1445 Launched on March 15, 1983; test similar to above

Cosmos 1517 Launched in December 1983; successful parachute-aided splashdown in the Black Sea after a controlled descent

Cosmos 1614 A 1/3-scale version launched December 19, 1984; successful parachute splashdown in the Black Sea

Soviet Parallel Programs

Unlike the U.S. space program, the Soviet "program" did not consist of a central agency with contractors competing for projects. Instead, design bureaus with space experience applied political pressure to get their programs started, then built their hardware, and finally handed them over to the launch complex for their launch. The difference is extremely important. While the U.S. had only one centralized program for a lunar capsule, or lander, etc., each Soviet bureau had its own projects.

The Korolev OKB

While Sergei Korolev was alive, his bureau, Korolev OKB, was undoubtedly preeminent. It was the Korolev bureau that contracted the A-series rocket, gave the USSR its first ICBM, and boosted the Sputniks and Gagarin into space. Korolev built the Vostok space capsule, converted it into the Voshkod multimanned capsule, and had the Soyuz capsule under development for a manned lunar landing. It was the Korolev bureau that began the development of the N-1 superbooster, although after Korolev's death, management of the program was changed to a disastrous system of management by committee.

> *"If the rocket does fly, and the second and third stages are substituted by iron mockups, when I leave the observation bunker, what have I gained?"*
>
> — Sergei Korolev, on the merits of complete testing, 1965

Korolev had the Kremlin's ear. It was his drive, his charisma, and his unassailable technical expertise that got him the political support he needed to carry out the programs that he designated as important. What

Sergei Korolev wanted, Khrushchev gave. Even when Khrushchev was ousted, Korolev's clout suffered little.

Korolev OKB's most serious competitor in the space business was not McDonnell Douglas or North American or NASA, but in fact the Chelomei OKB. The reason was that, unlike in the U.S. program, designer Vladimir Nikolayevich Chelomei competed with the Korolev bureau for the same funds and resources (including personnel), and even the same factories.

The Chelomei OKB Bureau and its Lunar Program

Chelomei's space experience encompassed much of the same rocket experience as Korolev. While it was Korolev who grabbed the glory of Sputnik and Vostok, it was Chelomei who was working hardest to put humans in the vicinity of the Moon, and later, to land them there. In the early and mid-1960s, it appeared that, in the Moon race, it was Chelomei beating Korolev.

The Chelomei lunar program aimed for both a manned lunar pass and a lunar landing. Both projects were begun in the early 1960s, about the same time as Korolev began work on the N-1.

The Chelomei manned lunar pass made use of a launch vehicle that was originally intended as a very large ICBM, but was adapted (in the design stage) for use as a space launcher. This was known by the Chelomei project number of UR-500. Today it is called the Proton, and it is shown in Figure 5-12. The Proton carried aloft the Proton experimental satellites, as well as the Luna, Lunokhod, Mars, and Venus missions. It was also intended to carry the Chelomei LK-1 spacecraft on a manned lunar pass.

The Chelomei LK-1 Space Capsule

The Chelomei LK-1 was an advanced, two-person space capsule very similar in design to the Gemini capsule. It was tested several times during the 1960s and early 1970s, and was intended from the start for reusability. With easy changeout of the expended heat shield, consumables, and short-lived components, a specific LK-1 spacecraft could make several flights. It consisted of three modules: the reusable descent module, the expendable propulsion system, and the instrument module, which incorporated solar panels.

The Proton launcher would launch the LK-1 on a free-return lunar trajectory. On its return to Earth, it would, like Apollo, use a lifting trajectory to keep re-entry loads under control.

The more powerful Chelomei involved a new, larger booster called the UR-700. Details on this program are still unclear in the West. However, it is known that this

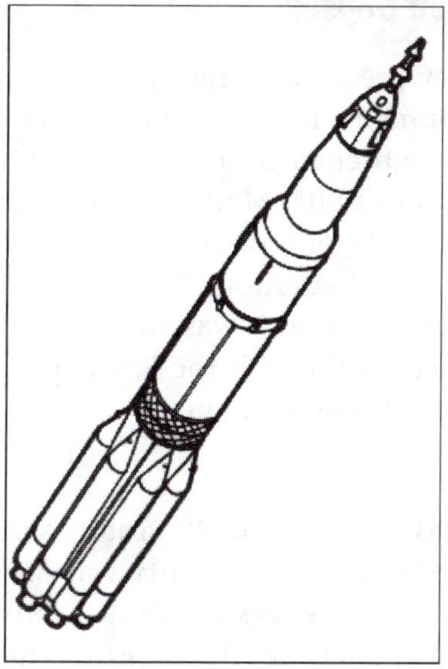

Figure 5-12. A diagram of the Proton rocket

was a modular booster, with the first stage consisting of eight clustered propellant tanks, each with its own engine (to ease transportation to and assembly at the launch site). The total payload of the UR-700 was somewhat more than either the Saturn V or the N-1.

The payload of the UR-700 was the LK-700. This was a conversion or adaptation of the LK-1 lunar capsule. For the conversion, a shortened Proton second stage was added as a TLI stage. A descent stage with landing legs was also added. The propulsion module was apparently enough to send the capsule on a Direct Ascent return to Earth. The LK-700 performed a Direct Ascent mission, like the Nova program, with no requirement for rendezvous. The LK-700 was thus a two-person capsule/module much like the Kvartet design within the game, and only slightly smaller. The game design of Kvartet is interestingly similar, especially since it was devised for gameplay eight years ago. The general appearance and the mission potential of the Kvartet and Chelomei ship are nearly identical. It wasn't until last year that the mere existence of the Chelomei manned lunar lander was released by the Soviets. Six months ago, Philip Clark discovered the true lunar goals of this spacecraft.

By 1964 the design of the LK-1 was finalized, the notorious problems of the Proton had not yet shown themselves, and it appeared that Chelomei's lunar landing plan was by far the most favored. However, fortune was about to frown on Chelomei OKB.

Korolev's Proposal: A New Capsule

In December 1965, Korolev presented a new proposal: he could modify his lunar Soyuz craft, under development, into a lighter vehicle for an early circumlunar mission. The vehicle was lightened by removing redundant systems and, more importantly, by removing the Orbital Module. Since the Orbital Module contained most of the consumables, only one or two cosmonauts could be carried, for a mission of six days. In this configuration, the circumlunar Soyuz conversion was called Zond. Its advantages were that it was already well into development, could be ready earlier, and was common with the Soyuz program. The Zond was light enough to be launched by a Proton launcher.

The Compromise

The compromise, then, was that Chelomei's Proton would launch Korolev's Zond on the Soviet circumlunar flight. Neither party was pleased with idea of working with their rival. The heated exchanges continued, bureau against bureau, head designer against designer. Though development of the UR-700/LK-700 program continued within Chelomei, the compromise made Korolev's Soyuz program the preferred one for the lunar landing, as well.

Chelomei's TKS Space Station Design

Although it would not be used for the lunar landing, Chelomei's manned spacecraft was not finished. As in the U.S. with their MOL program, a Soviet military space station program was under development, and the Chelomei bureau set out to build it. Since they would not be distracted by the requirements of a lunar landing, they could get the station project done quickly. The station was a military reconnaissance platform, designed to be launched by a Proton. Chelomei decided to modify the LK-1 spacecraft into a station-servicing vehicle, which would be called the TKS. Like the U.S. Gemini-B and MOL, the station would be launched with the TKS atop it. In orbit, the TKS crew would transfer into the station through a hatch cut into the heat shield, just as with the MOL design.

Korolev Joins the Space Station Design Program

Alas, this was not to be either. With the slowdown and then the cancellation of the lunar program, the Korolev bureau, with its greater resources, rapidly erased the head start on station design that Chelomei had accumulated. In fact, the first successful station design was a Korolev design; the second was a Chelomei design. The small station program was known as the Salyut station series, but in fact there were two separate designs. Korolev's Salyut design became the "civilian" station, and was used for *Salyut 1, 4, 6,* and *7*. Chelomei's "military" Salyut, which was better configured for reconnaissance, became *Salyut 2, 3,* and *5*, and later became the semi-independent "Star" photography module.

Results of the Salyut Space Station Design Program

One consequence of Korolev's entry into the Salyut program, however, was that the Soyuz capsule would be used for all crew transfers, with crews docking after the station was already in orbit. The Chelomei TKS capsule was used on the military Salyuts only, and only as a method of returning reconnaissance photos to the Soviet Union during the mission.

Altogether, the Chelomei capsule made several successful film-return flights under the TKS program, as well as several test flights with seats and dummy cosmonauts aboard. As far as is known, all were successful, at least two were reused, and at least one made at least three flights. Even with this excellent flight record, no human has ever ridden a Chelomei space capsule.

Yangel: Another Player in the Soviet Space Game?

The N-1 and the UR-700 were not the only proposed lunar rocket programs within the USSR. The Yangel OKB, builder of several successful (and otherwise) ICBMs, was also known to have proposed a heavy-lift lunar rocket, the R-56.

Details of the R-56 design are almost totally unknown in the West today. Some things can be discerned from the other Yangel vehicles which were actually built, however. All Yangel ICBMs used storable hypergolic propellants, and consisted of simple cylindrical shapes. It is known that the Yangel proposal would have used Glushko engines, which implies that a few large engines would be used, much like the U.S. Saturn.

It is believed that the Yangel bureau proposed an entire lunar program, including manned capsules and landers, but no details are known at this time. Yangel did, however, build the propulsion modules for Korolev's Soyuz capsules.

The Yangel program was doubtless scotched both by Korolev's strong political and technical abilities, and by Chelomei's strong competition therewith; there was little room in the space race for a third Soviet program. Doubtless Yangel's fortunes were not improved with the Nedelin disaster, in which a Yangel ICBM exploded on the pad after a misfire caused the entire maintenance crew to be recalled to the fueled booster. The explosion and burning of the toxic, volatile hypergolic fuel caused the burning death of over 200 people, including the visiting Marshal Nedelin, an important and knowledgeable space advocate within the Soviet military. This unfortunate disaster set back Soviet space to a degree, and Yangel's space program in particular.

When comparing the Soviet and American space programs, it is ironic to note that it was the Americans who used strong central planning: there was only one program, NASA's program, with contractors competing only to construct projects.

In the USSR, however, several bureaus, acting like companies, were in cut-throat competition, a feature generally associated with free markets and capitalism.

On closer inspection, it is easy to see the reason for this irony. Where in the U.S., competition existed and served to make the best product, in the USSR competition degenerated into political infighting. Rather than competing to make the best hardware, or even the best program, they competed on the basis of who could gain the most favor with the Kremlin. This destructive infighting was a contributor, although not the only one, in the Soviets' defeat in the race to the Moon.

A special thanks to Philip S. Clark, who provided invaluable information on the latest known details of the Chelomei, Yangel, and N-1 space programs. Phil Clark is a noted British author and expert on spacecraft systems and the CIS, the Russian (formerly Soviet) space program.

CHAPTER 6

An Interview with Buzz Aldrin

In a conversation with the author of this book, former astronaut Buzz Aldrin offers insight into his experiences in NASA's Astronaut Corps and makes observations about the space race and the *Apollo 11* Moon landing.

Competition Among the Right Stuff

Fritz: What was the competition like when you were first chosen for the astronaut program? Was it similar to the "Original Seven?" Lots of tests and confusion?

Buzz: That is a good observation. We didn't really know where we stood. We didn't realize it at the time, but a lot of weight was given to the "pure rating" query of our group of 14 (see Figure 6-1) after the first year of academics.

Fritz: I heard there was a lot of evaluating of fellow astronauts, is that correct?

Buzz: We were asked to rate each other. We were asked things like who we would fly first and in what order.

Figure 6-1. NASA's Group III astronauts (Buzz Aldrin, lower left)

Fritz: What did you think of those evaluations?

Buzz: I don't know if it is the most efficient way, but it is a way of coming up with some kind of a system. I am sure it had a certain amount of weight in the order of flying people within our group and bringing them into the Gemini Program. Of course, the system got upset by the jet crash which killed the first crew of *Gemini 9*, Charlie Bassett and Elliot See.

I did feel a little bit miffed at not being associated with an early rendezvous mission. Being an author and a well-qualified observer on the subject, I was aware of what it took to conduct a rendezvous. As it turned out, had not the Bassett and See accident occurred, I would have backed up *Gemini 9* and been eligible for the prime crew of *Gemini 13*, which of course never existed. So Lovell and I would not have flown at all but for that unfortunate event.

Fritz: Were there any other levels of competition among the astronauts?

Buzz: There was kind of a quiet undercurrent of competitiveness that frequently reared its evidence in light-hearted joking, but there was usually more behind it. There was a competitiveness in the form of one-upmanship over the other guys. Of course, there was the silent group versus group, and Air Force versus

Navy. Among the astronauts, there was a lot less structure than in purely military life.

I think the severest observation I could make is that I was rather dumbfounded and amazed to find out that once being selected as an astronaut, there was no yardstick, no qualitative measurement of a person's performance. There were no tests given that had any correcting grade. There was no evidence to anybody, including yourself, as to how well you were doing relative to other people.

Fritz: So that you could make improvements by evaluation?

Buzz: That's right, so you could gauge your own progress. Far more important, but not as evident at the time, was that the training people were in the dark as to how much was being absorbed. (See Figure 6-2.)

Fritz: What about physical conditioning?

Buzz: I think most lay people are amazed to find out that the physical conditioning was totally up to the individual. All one had to do was be healthy, pass a physical exam, and not have a major discrepancy.

Figure 6-2. Astronauts train under zero-G conditions on a C-131. The ride was also known as the "vomit comet!"

Doctor Rendezvous

Fritz: Before you were in the space program, you did a doctoral thesis on orbital rendezvous. Could you briefly describe it?

Buzz: The thesis that I did at MIT was a seat-of-the-pants sort of manned orbital rendezvous that was the basis for establishing simplified backup techniques for the co-elliptic or concentric orbits that the Gemini Program used. That is, departing one orbit for another orbit. The basis of what I wrote my thesis on was how you depart from one orbit to the other using very simplified guidance techniques. The guidance correction used crude angle measurements by the pilot to understand what was happening so that if something *did* go wrong, he had a solution as to what the mid-course maneuvers ought to be. This was a great confidence builder.

Fritz: It helped prove your point when the computer failed on your *Gemini 12* mission, didn't it? (See Figure 6-3 for a diagram of a type of Gemini spacecraft that Buzz flew.)

Buzz: Yes, but that was kind of a fluke. Actually, there was a dropout in data that resulted in the computer not being able to get a solution at the right time. We wanted to go for the rendezvous at the right time, but it (the computer) didn't have enough data points to come up with a solution. So we went anyway using my backup techniques.

Fritz: It must have been very satisfying for you to prove it to these people!

Buzz: Yes, I think so!

Approaches to the Moon

Fritz: What were some of the early alternatives that were discussed to get men to the Moon?

Buzz: There was a guy who was pretty famous, Bob Farquaer. He proposed using the gravitational libration point "L1" as a staging point for the Apollo missions. He pointed out the savings in fuel and the wider coverage of landing sights

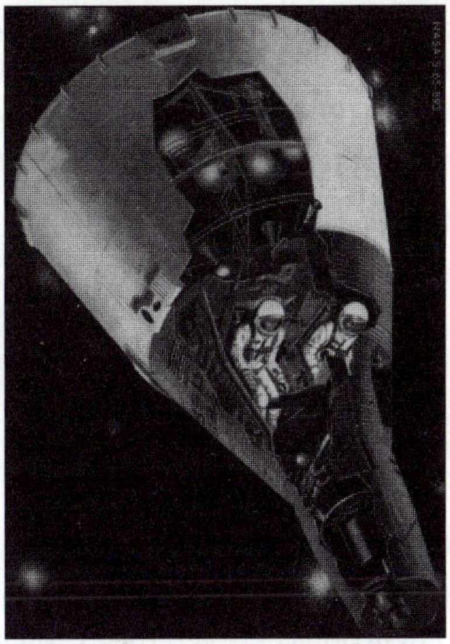

Figure 6-3. Cutaway view of Gemini spacecraft

on the Moon which you could get to by staging from the L1 point. The disadvantage was it took a longer time to get there. Also, once you were at the libration point, it took a longer time to get to the surface. Then to leave the surface was a lot more complex than neophytes could grasp at that time.

Fritz: How about some of those unusual long-shot plans?

Buzz: There was one earlier plan of getting somebody to the surface, then later sending him on his way back home. (A return vehicle would be sent later.)

Fritz: You mean the plan that the Air Force colonel proposed involving a one-way trip to stay on the Moon until they developed the technology to bring him back?

Buzz: Yes. Somebody probably straightened that one out and made it "put the stuff there first and then the guy lands."

Fritz: What about early Soviet plans?

Buzz: According to Phil Clark (an expert on the Soviet Space Program), the Soviets had several landing plans. The first one to get there in a hurry was a manned launch just like Apollo. It just took one guy to the surface. Then they revised the plan to try to do some more advanced things. There were ideas about stay times up to 14 or more days, possibly on the Moon's far side, carrying from three to six people.

Fritz: Do you recall any of the early designs for Apollo?

Buzz: One of the initial concepts for Apollo was to not have a lunar module, but to land the command module itself. (See Figure 6-4.)

The Soviet Competition

Fritz: Do you have any comparisons between the astronauts and the Soviet cosmonauts?

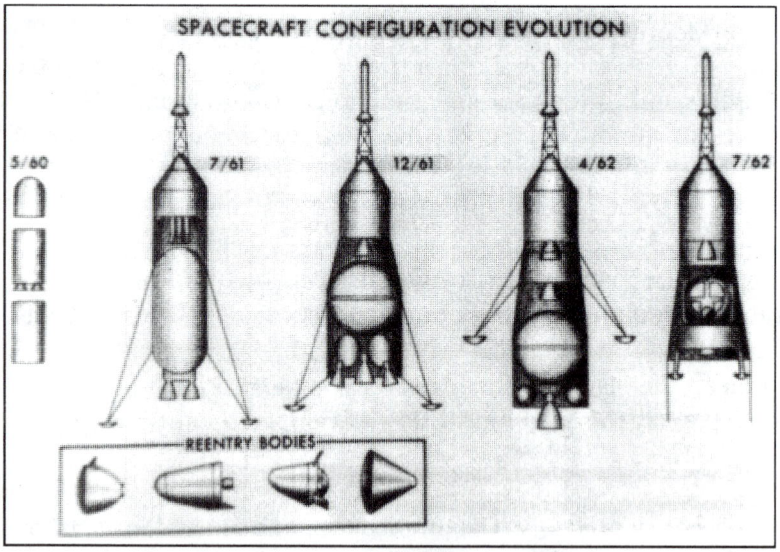

Figure 6-4. Various models of Apollo Direct-Ascent Lander, and Apollo Lunar Module configuration on far right

Buzz: I'm sure the Soviets had a lot more physically and academically demanding testing and more supervision, that is, external supervision rather than internal self-governing supervision.

Fritz: Did either side have an advantage?

Buzz: There are strengths and weaknesses in both systems. We happen to operate on more of an individual freedom basis. But it has its pros and cons. The Soviet spacecraft were capable of always being controlled by the ground, and only at their permission would they allow the guy onboard to begin to do things. We took the opposite approach and tried to make sure that ground control supported the judgment of the crew. Yet the Soviets developed automatic docking very early, at the time of *Soyuz 4* and 5, and we still have yet to develop that type of system. It is still all a manual guidance for our docking.

Fritz: When you met some of the cosmonauts at air and space shows, did they seem like typical military pilots with your same basic background?

Buzz: I think so, but they're brash and overbearing. I think that is most exemplified by Alexei Leonov, the "Bear." I mean, he is obviously an extremely talented person and he was slated to be a key player in their lunar program. I don't know whatever happened to Valery Bykovsky. He was pretty high up in their lunar crew selection listings, but he just faded out of sight. Much has been shared in conversations at space explorers meetings. Georgi Grechko is a pretty open guy.

Fritz: Was it like a fighter pilot reunion?

Buzz: Somewhat, but there is the language barrier and some (cosmonauts) are more extroverted than others.

Fritz: By their training and expertise, do you think they were as competent as our crop of astronauts?

Buzz: No, I don't think so, but it's kind of hard to say why. There wasn't any evidence of their lunar training of course. They wouldn't share much information and they didn't have any real simulators.

Fritz: By the Soviets not having many simulators like we had for crew training, would it have been difficult for one pilot to handle the one-man Soviet lunar module?

Buzz: I don't think it would have been characteristic of the Soviet design to be highly dependent upon manual dexterity and manual input for it to fly.

Fritz: Then you think it would have been highly automated?

Buzz: Yes. And probably it would have been pretty much a brute-force landing! (See Figure 6-5.) I would not have been surprised if in a circular orbit the descent engine would burn horizontally until their horizontal velocity was zero. Of course, while the burn is taking place, they would be losing altitude. Then they would shut off that engine and free-fall while separating from the initial descent stage. At 3,000 feet, the final cushioning engine would ignite for touchdown.

Fritz: The way you are describing the descent, it is pretty dramatic because you have to have nerves of steel!

Figure 6-5. A representation of a Soviet Lunar Module on the Moon. Fruition of this dream never occurred for the Soviets.

Buzz: I don't think it would have been a smooth power descent for 11 minutes the way ours was. We started with full thrust, then a power-down, and finally a pitch-over at these discrete points in the descent profile. Upon reaching 500 feet, we had the option to take over manual control. I do not think they would have trusted their guys to take over manual control. I think they would have made a blind landing and hoped for the best.

Fritz: While you were in the Apollo program, were you closely following the Soviet efforts?

Buzz: I'm sure that Frank Borman (*Apollo 8* astronaut) understood an awful lot more than either Neil or I did about the Zond potential in late 1968. I was not informed at all about the N-1 capability in June and July of 1969.

Tranquility Base Here...the Eagle Has Landed!

Fritz: How difficult was it to land on the Moon?

Buzz: I think people really need to understand that *Apollo 11* had a computer guidance program for landing that was, in my retrospective opinion, significantly inferior to that of *Apollo 12* and beyond. There was a big difference because on *Apollo 11*, when we took over control at 500 feet, there was really no giving it back to the computer.

There were two ways of taking over control. The easy way was where you just take over attitude control while the computer still drives the main engine thrust. If you lose that and have to take over manual control with just an up-and-down thrust lever, it would have been very difficult. I don't think anyone would have had a great deal of success in being able to land with the latter option.

In all missions, once you took over manual control, the rate of descent was controlled by a toggle switch which allowed you to change your rate of descent by one foot per second up or down just by hitting it several times. You hit it three times down and you would increase your rate of descent, or hit it three times up and you decrease your rate of descent. Of course, you had a darn good idea that your rate of descent was in the vicinity of 15 to 20 feet per second upon reaching 500 feet altitude.

It was a relatively simple task to control the rate of descent, but you had to coordinate that with the question of are you pointed in the right direction. On *Apollo 11*, once we took over manually, we still had attitude hold, which was: you let go of the stick and whichever way that engine was pointing, that was the way it was going to continue to point. So, if it was not pointing directly down with zero translation, it would increasingly start moving off to the side because you are unstable standing on the engine thrust. It starts moving to the side in a gradually increasing sideways motion, so you have to correct back. Really, you had your hands full!

Every landing mission after ours, when they (the crew) let go of the attitude controller, not only did they have the toggle switch for the rate of descent, but the computer said to itself "zero translation rates to the best that you know of." So, if you let go of the stick, it stopped you from going forward, aft, left, and right. It was much easier to have a non-drifting translation for all the touchdowns after ours.

It's kind of amazing that on *Apollo 11*, we were able to have just a very small drift on touchdown. Of course, the task that really got to us was the question of how to fly over a boulder field and a crater that you are heading for. Do you dive short, do you veer to the left or right, or fly over it? The prudent way is to fly over it, but that's also the most fuel-consuming way because you essentially stop your rate of descent and stay above the ground considerably longer with the engine burning at a higher rate.

Moon Dust and Moon Shadows

Fritz: When you and Neil got back in the lunar module, and you took your helmet off, did the Moon dust have any kind of smell?

Buzz: Yes, there was a faint metallic gunpowder odor to it. We had just re-pressurized the cabin with fresh oxygen, but it didn't smell like that was all that was there. It wasn't at all overpowering, but it was noticeable. It did not tempt us to stick our nose right up against it and inhale! We were aware of the potential toxicity. This was dust not dirt, and dust might be a lot more metallic-like than dirt.

Fritz: Was it colored, like burned-out charcoal?

Buzz: It was a lot darker than you would perceive it to be. Edmond Scientific Company came out with a little jar of imitation Moon dust after our flight and it had a certain sparkle to it. But the real color was quite dark gray. You would not have guessed that by looking out at the sunlit lunar surface because that was quite light in its reflection. But in the shadow of course, it was pretty dark.

Fritz: With no atmosphere on the Moon, were the shadows extraordinarily dark?

Buzz: You could still see features in the shadow of the lunar module. You could see the features because of the reflected light coming off of other rocks.

A Walk on the Moon

Fritz: You have probably been asked ten thousand times, but when you were walking on the Moon's surface, what was it like?

Buzz: Visually, you had the sensation of being on something that had a much greater curvature than the Earth. I'm looking out the window at the ocean right now, and it's pretty flat looking. But, intellectually, I know that it curves away. On the Moon, there would be absolutely no doubt that you were on anything that curved away, for two reasons. One reason is that there is no atmosphere to fuzzy up the view, so distant vision is very detailed. The other reason is it curves away quite close to you, about a mile away.

Standing on the surface (see Figure 6-6), because of the weight of the space suit backpack, there was a need to counteract that weight by leaning forward. There were very slight foot sensation inputs as you shifted your weight from one side or the other, or fore and aft, kind of hunting around to tell when you are standing upright. You couldn't tell if you were straight by looking at the surface because there was always a slight angle to the horizon. It was hard to look at the horizon and say, "I'm going to align myself perpendicular to that," and know by looking that I was standing up straight. You end up subconsciously using those little cues fed through your feet to tell when you are balanced and equally sharing weight between toe and heel or left and right foot. But those are very obscured on the Moon, so you do not really know when you lean too far forward. But if you do, all you do is put a little pressure on one foot or the other and push yourself up against the weak lunar gravity. Because you don't fall down fast on the Moon, you can just put your foot out and very easily reestablish your stability.

Fritz: Was it fun walking on the Moon?

Buzz: Oh Yeah! There is a freedom. The lesser the gravity, the greater the freedom of methods in moving yourself around. The higher the gravity, the less freedom there is. Everybody walks the same in the zero-G airplane at two Gs. You just shuffle using very, very short steps, because it's hard. You have to put

Figure 6-6. Buzz Aldrin on the lunar surface. Neil Armstrong can be seen in the reflection of Buzz's visor.

all your weight on one foot and then all on the other. But at one-sixth G, there is a lot of freedom. On the Moon, it seems people almost naturally adapted a step like a horse has at a gallop.

Revisiting the Moon

Fritz: If you were planning to go to the Moon today, how would you approach it?

Buzz: My strongly recommended way of going back to the Moon now involves upgrading a PLS (Personnel Launch System). It's sort of an expanded ACRV (Assured Crew Return Vehicle), but capable of being launched on the National Launch System or a follow-on to the Shuttle. Once in orbit, it would be assembled to a propulsion system, landing gear, and drop tanks. Then it would go directly to the Moon, land, and return without rendezvous in lunar orbit.

It does this with one propulsion system that does TLI (Trans-Lunar Injection), LOI (Lunar-Orbital Insertion), descent, ascent, and TEI (Trans-Earth Injection). The system would drop four tanks after TLI and keep two more for LOI and descent. Then it would leave the last two tanks and the landing gear framework on the surface, lifting off with the PLS and a propulsion system for ascent and TEI.

Fritz: You have recently worked on rendezvous techniques for a manned Mars mission, is that correct?

Buzz: Yes. What I had done between 1984 and 1986 was to perfect the concept of cycling orbits between Earth and Mars.

Fritz: Given the chance, would you go back to the Moon?

Buzz: You bet!

CHAPTER 7

A Chronology of Space History

The practical thinkers of today wouldn't exist without the dreamers and visionaries of yesterday. The following is a listing of milestones in the development of space technology and the exploration of the Moon.

150-100 B.C. Heron of Alexandria designs the aeolipile, a reaction device.

145 A.D. Ptolemy declares the Sun and planets move around the Earth.

700 A.D. Invention of gunpowder in China.

1543 Copernicus publishes his theory of the solar system with the Sun at the center and the planets orbiting around it.

1609 Johannes Kepler publishes the Keplerian laws of physics.

1678 Issac Newton's laws of motion is the foundation of rocketry.

1783 First manned balloon flight.

1805 Congreve constructs military rockets.

1865 Jules Verne writes *From the Earth to the Moon*, a spaceflight to the Moon. Figure 7-1 shows an illustration from that book.

1901 Konstantin Tsiolkovsky designs detailed rocket and spacecraft designs. The precursor of today's rockets.

1921 N. I. Tikhomirov establishes the first Soviet laboratory for rocket research.

March 16, 1926 American rocket pioneer Robert H. Goddard test-fires the first liquid-fueled rocket engine.

July 5, 1927 A group of German rocket enthusiasts form the Society For Space Travel. The group, led by rocket pioneer Hermann Oberth, will become the nucleus of German rocket development efforts.

September 29, 1931 Dr. Robert Goddard launches the first liquid-fueled rocket in Massachusetts. It flies 180 feet (55 meters) and stays aloft for 9.6 seconds. High altitude tests with larger rockets will later be conducted in New Mexico.

Figure 7-1. Etching of Jules Verne's lunar adventure

November 1932 German rocket pioneer Wernher von Braun begins conducting rocket propulsion experiments for the German Army. He heads the team developing German liquid-fueled rockets.

August 17, 1933 The Soviet researchers test-fire a rocket that uses a hybrid combination of liquid and solid fuel.

Autumn 1937 The Germans attempt three flights of the A-3 experimental rocket. All go out of control and are not totally successful.

October 3, 1942 First successful test of the German V-2 rocket from the secret rocket research base at Peenemunde. This test is the first supersonic flight of a rocket.

September 7, 1943 First V-2 rocket hits England. The V-2, carrying a one-ton warhead, will wreak havoc on England during the latter part of World War II. There will be 4,320 V-2s fired before the end of the war. A V-2 rocket is shown in Figure 7-2.

May 5, 1945 The Soviet Army captures the Peenemunde rocket research facility from the Germans, only to find that most of the research team already fled to the Americans. The Soviets nevertheless ship 12,000 tons of captured rocket production equipment and thousands of V-2 rocket assembly technicians to the Soviet Union.

Figure 7-2. The German V-2 rocket (U.S. Army)

September 20, 1945 Wernher Von Braun and over a hundred other German rocket scientists are smuggled into the United States under "Operation Paperclip."

April 16, 1946 The United States begins test-firing captured V-2 rockets at the White Sands Proving Grounds, New Mexico. These tests continue for six years and greatly facilitate development of the American space program.

July-August, 1946 Soviet rocket pioneer Sergei Korolev begins testing a Soviet modification of the V-2 rocket developed without the assistance of captured German technicians.

October 28, 1946 Captured German rocket technicians finally begin experiments on behalf of the Soviets, but are quickly phased out and replaced by Soviet scientists.

May 11, 1950 The United States launches a Viking rocket at sea from the U.S.S. *Norton Sound*. Reaching an altitude of 106 miles (170 kilometers), the rocket studies magnetic fields and cosmic rays.

March 21, 1951 The Soviets begin repatriating the German rocket technicians they captured at the end of World War II.

May 31, 1955 The Soviets begin construction of initial ICBM test facilities at what will become the Baikonour Cosmodrome, the place from which all Soviet manned spaceflights will be launched.

August 1957 First full range test of the Soviet ICBM developed by Sergei Korolev.

October 4, 1957 The Soviets launch *Sputnik 1*, the first satellite to orbit the Earth. The 184-pound (83.6-kilogram), two-foot (58-centimeter) diameter spherical satellite is launched into a 142-by-588-mile (228-by-947-kilometer) orbit and is large enough to be seen by observers all over the globe.

November 3, 1957 The Soviets launch the 1,118-pound (508-kilogram) *Sputnik 2* satellite carrying the dog Laika. Biological data is returned for one week before Laika's oxygen supply runs out.

January 31, 1958 The United States launches its first satellite, *Explorer 1* on a Juno rocket, from Cape Canaveral.

May 15, 1958 The Soviets launch *Sputnik 3*. Weighing a record 2,922 pounds (1,327 kilograms), it is the first geophysical laboratory satellite.

October 1, 1958 The National Aeronautics and Space Administration (NASA) is formed.

December 18, 1958 The United States orbits an Atlas ICBM carrying the Project Score payload, America's first communications satellite.

January 2, 1959 The Soviets launch the 795-pound (361-kilogram) *Luna 1*, the first probe to make a lunar fly-by. While intended to impact the Moon's surface, the craft misses the Moon by 3,700 miles (5,955 kilometers) and ends up in a 450-day orbit around the Sun.

March 3, 1959 The United States launches the Moon probe *Pioneer 4*. After the 13-pound (six-kilogram) craft passes within 37,000 miles (60,000 kilometers) of the Moon, it goes into solar orbit.

April 2, 1959 NASA announces the selection of the seven Project Mercury astronauts, who are shown in Figure 7-3.

Figure 7-3. NASA's original seven astronauts

September 12, 1959 The Soviets launch *Luna 2*, which becomes the first man-made object to impact on the Moon.

October 4, 1959 The Soviets launch *Luna 3*, which photographs the back side of the Moon.

April 1, 1960 The Americans launch *Tiros 1*, the first meteorological satellite. It returns 22,952 photos during its three-month life.

May 15, 1960 The Soviets launch *Sputnik 4*, the first orbital test of the Vostok manned spacecraft.

August 10, 1960 The United States performs the first successful recovery of a satellite payload when Navy frogmen rescue the *Discoverer 13* re-entry capsule from the Pacific Ocean.

August 12, 1960 The United States launches *Echo 1*, the first true communications satellite. The 100-foot (30-meter) diameter aluminized Mylar balloon was the first satellite brightly visible to the public.

August 18, 1960 The United States launches *Discoverer 14*, the first successful reconnaissance satellite. After photographing the secret Soviet ICBM test center at Plesetsk, the film return capsule is retrieved by a recovery plane as it parachutes towards Earth.

August 19, 1960 The Soviet *Sputnik 5* was the first to carry living creatures (dogs named Strelka and Belka) into orbit and return them safely to Earth.

December 19, 1960 First test-firing of the American Mercury-Redstone rocket which will later be used for manned missions.

April 12, 1961 The Soviets launch Yuri Gagarin, the first man in space, on a one-orbit flight aboard *Vostok 1*.

May 5, 1961 Alan Shepard becomes the first American in space during the 15-minute suborbital flight of *Freedom 7* (shown in Figure 7-4), also known as *Mercury-Redstone 3*.

June 29, 1961 The U.S. Navy Transit 4-A navigation satellite is the first nuclear powered spacecraft.

Figure 7-4. Suborbital flight of Alan Shepard's *Freedom 7*

August 6, 1961 The Soviet *Vostok 2* carrying cosmonaut Gherman Titov, spends one full day in space before returning to Earth. Titov is the first to experience "space-sickness" or vestibular disorientation.

February 20, 1962 John Glenn becomes the first American to orbit the Earth after a three-orbit flight in *Friendship 7*.

April 26, 1962 The American *Ranger 4* spacecraft reaches the lunar surface, but fails to return data.

April 26, 1962 The United Kingdom is the third country in space with the launch of *Ariel 1* on an American Delta rocket.

April 26, 1962 The Soviets launch *Cosmos 4*, their first military reconnaissance satellite.

July 17, 1962 U.S. Air Force pilot Robert White qualifies for astronaut wings by flying the X-15 rocket plane to an altitude of 59 miles (95 kilometers).

July 19, 1962 U.S. Air Force pilot Joseph Walker flies the X-15 rocket plane to an altitude of 65 miles (104.5 kilometers) and qualifies for astronaut wings.

August 11 and 12, 1962 *Vostok 3* and *4*, launched by the Soviets, perform the first joint manned space mission. Andrian Nikolayev pilots *Vostok 3* and lands after four days. Pavel Popovich pilots *Vostok 4* and lands after three days.

September 28, 1962 Canada becomes the fourth national space power when an American Thor-Agena launches *Alouette 1* to study the ionosphere.

June 16, 1963 Soviet Valentina Tereshkova becomes the first woman in space, spending the days aboard *Vostok 6* while flying a joint mission with *Vostok 5*.

July 31, 1964 *Ranger 7*, the first successful American lunar probe, returns 4,316 high resolution photos of the Moon's surface. (See Figure 7-5)

August 19, 1964 The United States launches *Syncom 3*, the first true geosynchronous satellite. The spacecraft relays television coverage of the Olympic Games from Japan to the United States.

Figure 7-5. Ranger photo of lunar surface

October 12, 1964 The Soviets launch *Voskhod 1*, the first multi-man spacecraft, carrying Vladimir Komarov, Konstantin Feoktistov, and Boris Yegorov on a one-day flight.

March 18, 1965 The Soviets launch Pavel Belyayev and Alexei Leonov aboard *Voskhod 2*. Leonov steps out of the cabin to become the first man to walk in space.

March 23, 1965 The United States launches Gus Grissom and John Young aboard *Gemini 3*, the first manned spacecraft to change orbits.

April 23, 1965 The Soviets launch *Molniya 1*, their first communication satellite.

June 3, 1965 The United States launches James McDivitt and Edward White aboard *Gemini 4*. White becomes the first American to walk in space with a 21-minute EVA.

December 1965 The Americans launch *Gemini 6* and *7*. Frank Borman and James Lovell, in *Gemini 7*, spend 14 days in space, setting an endurance record that will stand for five years. They also rendezvous with Wally Schirra and Tom Stafford in *Gemini 6*.

January 31, 1966 The Soviet *Luna 9* makes the first soft landing on the Moon and broadcasts television pictures from the surface.

March 16, 1966 *Gemini 8*, carrying Neil Armstrong and David Scott, performs the first docking of two spacecraft, joining an unmanned Agena rocket. A faulty thruster then sends *Gemini 8* into an uncontrolled spin, forcing a mission abort after 6.5 orbits.

March 31, 1966 The Soviets launch *Luna 10*, which becomes the first satellite of the Moon.

June 2, 1966 *Surveyor 1* becomes the first unmanned American spacecraft to soft-land on the Moon.

August 14, 1966 *Lunar Orbiter 1* becomes the first American satellite to orbit the Moon.

November 11, 1966 *Gemini 12*, carrying James Lovell and Ed Aldrin, performs the first totally manual rendezvous when the spacecraft computer fails. Aldrin also performs the first totally successful space walk.

January 27, 1967 Gus Grissom, Edward White, and Roger Chaffee burn to death in the *Apollo 1* launchpad fire.

April 23, 1967 The Soviets launch *Soyuz 1* carrying Vladimir Komarov. Upon recovery the following day, the craft's parachutes foul and the cabin crashes. Komarov becomes the first person to die during a spaceflight.

October 28 and 30, 1967 Soviet *Cosmos 186* and *188* satellites are the first automated spacecraft to successfully rendezvous and dock.

November 10, 1967 The American *Surveyor 6* Moon probe, shown in Figure 7-6, performs the first lunar liftoff when it fires its landing engines briefly, propelling it ten feet (three meters) off the Moon.

September 18, 1968 The Soviet *Zond 5* is the first spacecraft to circumnavigate the Moon and return to Earth.

October 11, 1968 The United States launches *Apollo 7*, the first manned test of its manned lunar spacecraft. Walter Schirra, Don Eislee, and Walter Cunningham make a successful 11-day flight.

Figure 7-6. Lunar probe *Surveyor*

December 21, 1968 The United States launches *Apollo 8*, the first manned flight of the Saturn 5 Moon rocket. Three days later, Frank Borman, James Lovell, and Bill Anders aboard *Apollo 8* become the first men to orbit the Moon.

January 14 and 15, 1969 The Soviets launch *Soyuz 4* carrying Vladimir Shatalov, followed the next day by *Soyuz 5* carrying Boris Volynov, Alexei Yeliseyev, and Yevgeny Khrunov. After Shatalov docks with *Soyuz 5*, Yeliseyev and Khrunov transfer to *Soyuz 4* via EVA.

February 1969 The first unmanned launch attempt of the Soviet N-1 Moon rocket results in a catastrophic explosion.

July 4, 1969 The Soviets again attempt an N-1 launch, but the Moon rocket explodes above the launchpad.

July 16, 1969 The Americans launch *Apollo 11* with Neil Armstrong, Michael Collins, and Ed Aldrin onboard. Four days later, Armstrong and Aldrin pilot the lunar module *Eagle* to the first manned landing on the Moon.

October 11, 12, and 13, 1969 The Soviets simultaneously fly a total of seven men on three spacecraft, *Soyuz 6, 7*, and *8*.

April 11, 1970 *Apollo 13* is launched toward the Moon by the Americans with James Lovell, James Swigert, and Fred Haise aboard. An oxygen tank explosion scrubs the lunar landing, forcing an emergency return to Earth.

June 1, 1970 The Soviets launch *Soyuz 9* (see Figure 7-7) with Andrian Nikolayev and Vitali Sevastyanov on an 18-day medical investigation mission.

September 16, 1970 The unmanned Soviet Moon probe, *Luna 16*, is launched toward the Moon. Eight days later, the craft's return capsule returns lunar geological samples to Earth.

November 10, 1970 The Soviets launch *Luna 17* carrying the *Lunikhod 1* unmanned lunar rover which carries out an 11-month exploration on the Moon's surface.

June 6, 1971 The Soviets launch *Soyuz 11* with Georgi Dobrovolsky, Vladislav Volkov, and Viktor Patsayev. They dock with and spend 24 days aboard the *Salyut 1* space station. The crew was killed during reentry by accidental cabin depressurization.

Figure 7-7. Soyuz rocket

July 26, 1971 *Apollo 15* is launched with David Scott, Al Worden, and James Irwin on board. After landing on the Moon, Scott and Irwin use a lunar rover to becoming the first men to drive on the Moon.

December 2, 1972 The launch of *Apollo 17*, the last manned mission to the Moon. Lunar module pilot Harrison Schmitt, a geologist, is the first scientist to be sent to the Moon.

May 14, 1973 The United States launches the *Skylab* space station.

May 25, 1973 The *Skylab 2* mission is launched. The crew, Charles Conrad, Joe Kerwin, and Paul Weitz repair the damaged *Skylab* space station and set a new orbital endurance record of 28 days.

November 16, 1973 NASA launches the third crew to the *Skylab* space station. Gerald Carr, Ed Gibson, and William Pogue spend 84 days aboard, which to this day remains the American space endurance record.

July 15, 1975 The United States and the Soviet Union launch the Apollo/Soyuz Test Project. Two days later, *Apollo 18* carrying Tom Stafford, Don Slayton, and Vance Brand rendezvous and docks with *Soyuz 19*, carrying Alexei Leonov and Valery Kubasov.

July 20, 1976 The United States spacecraft *Viking 1* makes the first successful soft landing on the planet Mars.

PART II

Installation and Game Strategies

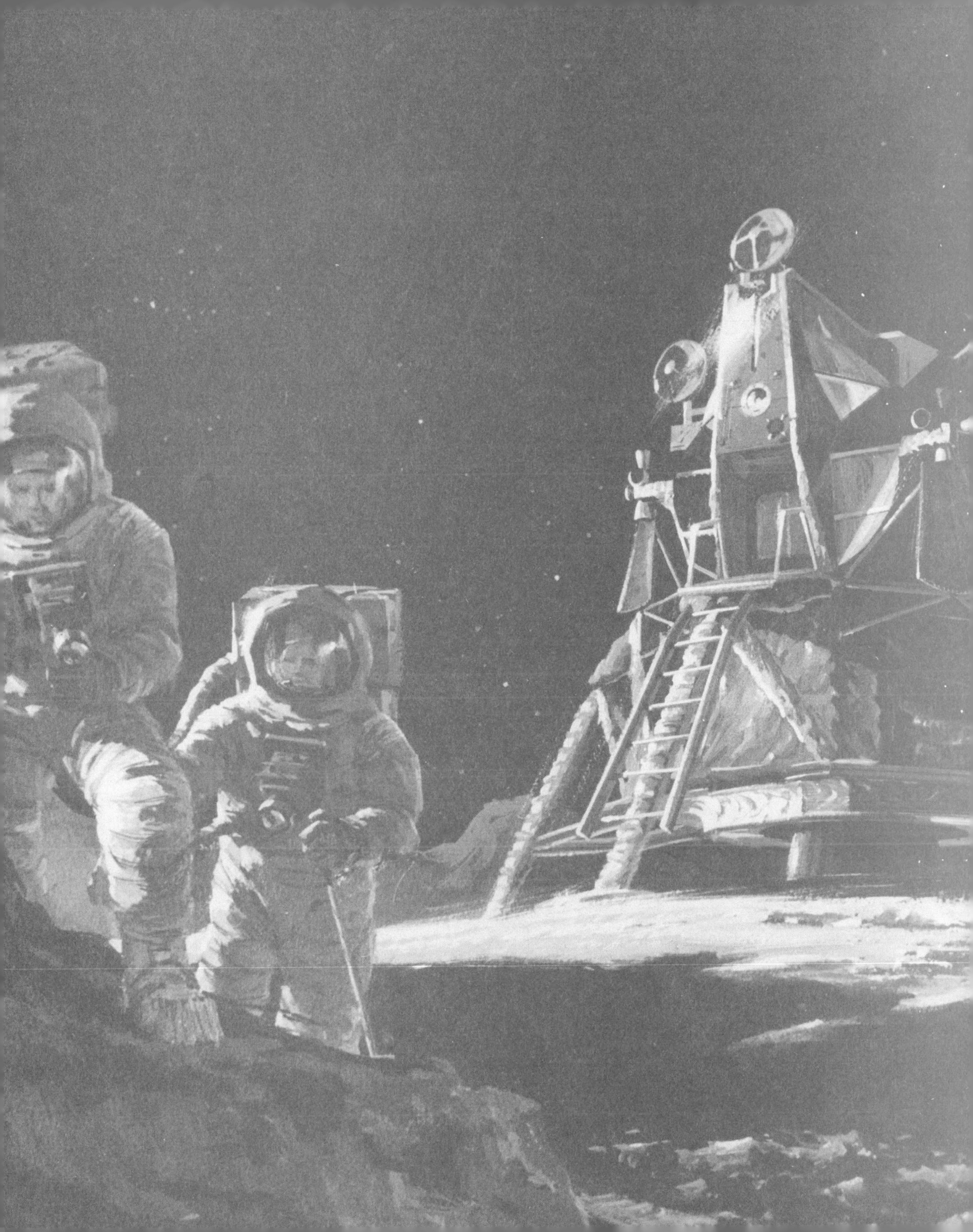

CHAPTER 8

Installation, System Requirements, and Sound Cards

Since "Buzz Aldrin's Race Into Space" is an advanced, multimedia-format simulation, playing the game will make extensive use of your computer's capabilities. Here are some instructions to help you get started, and a few tips to help you get the most out of your system and your game play.

Installation

To play the game, you must first install it on your hard drive. Place the first disk in the drive, and at the DOS prompt, type **INSTALL**. Insert the other disks when prompted, and try to answer the space trivia questions in the meantime. Extra effort was made to make the installation process as quick and painless as possible; however, due to the large amount of data that must be copied, the installation process will take approximately 15 minutes.

NOTE: A DOS subdirectory entitled BUZZ will automatically be created during installation.

Once the installation is complete, you can see what version of the game you have by going to the title screen and typing **V**.

Playing Buzz in Windows

"Buzz Aldrin's Race Into Space" can run in Windows as a DOS application if you want, provided that you disable the sound option in the game. You can disable the sound option at the Preferences screen in the beginning of the game, or, while you're playing the game, by selecting Preferences from the Administration Building screen in your spaceport.

Saving Game Information

Any game information you wish to save will be saved from the beginning of the following turn or season. For example, work done in Spring 1962 will not be saved until the beginning of the next season, Fall 1962.

All saved games are stored in the SAVEDAT directory. There is a limit of 100 saved games. If you exceed this number, not all your saved games will be displayed.

Saving the first game requires 30,000 bytes of hard disk space; additional saved games will each require at least another 25,000 bytes. Some saved games will require more bytes and some less; 25,000 bytes is the average.

Quick Exit Feature

It is possible to perform a quick exit from the game, but this exit option should be used with caution. From any place in the game, press SHIFT-TAB to exit the game. However, the game will not verify that you wish to quit, nor will it ask you if you wish to save game information.

System Requirements

An IBM or IBM-compatible PC is required to play "Buzz Aldrin's Race Into Space." The hard disk drive should have a minimum of about 17 megabytes of free space, and a VGA graphics card is also required. You will need 570K, or 583,680 bytes, of free RAM in the lower 640K block of memory to install the game. Any additional RAM will be beneficial for speed and smoothness of animation.

"Buzz Aldrin's Race Into Space" is designed to work properly with an 80286 processor of 12 MHz speed or greater. However, users of such machines are cautioned that even the minimum animation capabilities of the game will tax the ability of the processor. Animations may play at slower than the designed three-frames-per-second rate, and movement of the mouse pointer will tend to be sluggish and erratic.

Software Information

DOS 3.1 or higher, or Digital Research's DR-DOS 5.0 or 6.0 operating systems is required to play "Buzz Aldrin's Race Into Space."

The following sections discuss the use of extended memory managers and compression programs in conjunction with the game.

Extended Memory Managers

If your computer doesn't have sufficient memory to run the game, you may need to resort to an extended memory manager such as EMM.SYS or Quarterdeck's QEMM. Use of such a memory manager will not conflict with the game.

Stacker and Superstor Compression Programs

While "Buzz Aldrin's Race Into Space" will execute from a drive that contains either the Stacker or Superstor compression software utilities, using them is not recommended for two reasons. First, because the game already uses a sophisticated compression routine on all its files, the use of an additional disk-compression technique will be of little value and will even deceive the compression program into believing that there is more free disk space than actually exists. Second, digitized sounds on most sound boards will not work and may even cause the program to crash.

If you are running Stacker or Superstor with the Sound Blaster, Thunderboard, 8-bit Pro Audio Spectrum, or compatible cards, it will be necessary to turn off all digitized sounds for the game to run reliably. To turn off digitized sounds, select NONE in the Digitized Sound Device Selection menu, and the music will continue to work fine. To avoid the problem entirely, install the game on a drive that does not use these compression programs, if possible.

> **NOTE:** Unlike the other cards mentioned here, the Pro Audio Spectrum 16, with its built-in cache, can tolerate the additional disk access time imposed by Stacker and Superstor, and will play digitized sounds normally when used on a drive running one of these programs.

Using a Mouse or Joystick

A Microsoft or compatible mouse, or a trackball is recommended for play but is not required; you can play using only the keyboard.

Sound Cards

While "Buzz Aldrin's Race Into Space" will play music using its PC speaker driver, its multimedia sound design practically cries out for a sound card of some type, and the Music Device Selection menu is appropriately packed with choices. You may want to consider purchasing a sound card in order to fully enjoy using your computer with this product and many other multimedia products.

To reconfigure the sound driver, type **SETUP** from the DOS prompt (make sure that CAPSLOCK is off during this operation). This will bring you to a configuration menu, the Music Device Selection menu, which lets you select what driver will be used to play music and tones. Select the driver from the choices available by pressing the appropriate number and then pressing ENTER:

0: None
1: PC Speaker
2: Adlib or Adlib Gold
3: Sound Blaster or Thunderboard
4: Sound Blaster Pro Series
5: Pro Audio Spectrum Original
6: Pro Audio Spectrum Plus/16
7: Gravis Ultrasound
8: Roland MT-32 or LAPC-1
9: Roland SCC-1

After you make your selection, you will proceed to the Digitized Sound Device Selection menu. Select the driver to be used for digitized sounds and speech:

0: None
1: Sound Blaster or Thunderboard
2: Sound Blaster Pro Series

3: Pro Audio Spectrum Series

4: Gravis Ultrasound

Again, press the appropriate number, then press ENTER. This returns you to the DOS promt. To start the game, type **BUZZ**.

Multiple sound cards can be used with the program—just select the appropriate driver for each function. The best combination is to configure your system with a Roland for music and a Sound Blaster or Pro Audio Spectrum or compatible for digitized sounds; this configuration maximizes the use of the program's sound drivers.

The program's configuration drivers will support any Sound Blaster-, Pro Audio Spectrum-, or Adlib-compatible sound card, in addition to the choices listed in the configuration menu. However, digitized sounds for the AdLib Gold are not supported. The Covox, although advertised as "Sound Blaster compatible," is not supported by the program. The Microsoft sound board, older Tandy built-in sounds, and any sound hardware that attaches to the printer's parallel port are not supported either.

Reconfiguring Music and Sound Devices

In the event you have a sound card that is either not listed, or that the program's drivers do not access properly, or if you simply wish to reconfigure your music and sound devices for optimum performance, the program can be reconfigured to properly utilize your sound card. Bring up the various configuration menus by typing **SETUP** from the DOS prompt.

As always, remember to check the READ.ME file in the BUZZ directory for the latest updates on the usage of the game.

Sound Blaster/Thunderboard Card

The game's software looks for the BLASTER environment variable to configure any Sound Blaster or compatible card. Set this variable according to your sound card instructions.

Advanced Gravis Ultrasound (GUS) Card

If you select Advanced Gravis Ultrasound (GUS) for either music or sound, "Buzz Aldrin's Race Into Space" will use it for both music and sound in the game. The game relies on the GUS to be set up according to its installation program. Currently this device supports only the game's digitized sound.

Environment Variables

If you want different settings than the game's default settings for your sound devices, you may want to set the game's two environment variables, which are BUZM for the music device, and BUZS for the digitized sound device. These environment variables will inform the game's software which port addresses, interrupts, and DMA channels to use. For example, typing the following command at the DOS prompt

```
SET BUZM=220,5,1
```

will set the music device to Port Address 220h, Interrupt to 5, and DMA Channel to 1. Typing this command at the DOS prompt

```
SET BUZS=240,7,3
```

will change the digitized sound device to Port Address 240h, Interrupt to 7, and DMA Channel to 3.

If a device, for example the SCC-1, does not have a DMA channel, use a 1 as a placeholder.

CHAPTER 9

Defining Your Missions and Goals in the Game

When Neil Armstrong and Buzz Aldrin set foot on the Moon, it was the culmination of the will of a nation. It took many years of research and development costing billions of dollars and was driven by the fear of an enemy, the communist Soviet Union. "Buzz Aldrin's Race Into Space" will give you an idea of what it was like to manage such a lofty ambition. No one would ever have successfully completed such a project without a well defined and carefully planned agenda. Careful planning, clear objectives, and a little bit of luck are all ingredients to a successful Moon ascent.

In the next few chapters we'll provide many hints and tips that will help you win in "Buzz Aldrin's Race Into Space" (as shown in Figure 9-1). For the sake of brevity and readability, we will use American hardware for our examples except where there is a specific reference to Soviet hardware. Note that the techniques described in the following pages will work equally well with equivalent American and Soviet hardware. For instance, the American one-man capsule Mercury is equivalent to the Soviet one-man capsule Vostok.

Figure 9-1. Title screen for "Buzz Aldrin's Race Into Space"

The Mission Objective

Your ultimate goal in this game is to land a man on the Moon and bring him back safely to Earth before your opponent does. Never forget that this is your primary task. All the other missions are a means to an end as far as the goal of the game is concerned. While you may not win the prestige point race, you will win the game if you beat your opponent to the Moon and back. Of course, this would be impossible if we did not start with smaller accomplishments. Balancing patience, a budget, and the desire to be first is a tough act.

You must remember to be patient and persistent at the same time. Nothing comes easily in the space race. If you fail, try and try again. Sometimes failing a mission a couple of times is preferable to skipping it. A major catastrophe during the later stages of the race is more detrimental to your program than one occurring during the early ones. During the early stages of the game, you still have time to change the direction you want to take without affecting the method you've decided to take to get to the Moon. During the later stages, a change in method will cost too much and take too long—you'll probably lose the race. If you stick with the planned method despite failed missions in the later stages, the time required to recover could also cost you the race.

First Foray into "Buzz Aldrin's Race Into Space"

The first few times you play the game, play it at Level 1, use the Basic Model/Historical Roster, and set the opponent to another human. Decide which side you're going to play and just skip right on by the opponent's turn. Playing the game this way may not seem as exciting, but it will take the heat off trying to beat a computer opponent. Getting to the Moon without having to race an opponent is a hard enough goal by itself. Playing at Level 1 without an opponent will give you a stress-free opportunity to famliarize yourself with the different screens, resources, and the space program in general. Since you will not get fired at this level, you can experiment freely. Try playing the game this way with different approaches to get to the Moon. You will then begin to see the various failures that can occur. The strengths and weaknesses of the individual methods will surface.

When playing against a computer opponent, set Player 2 to a higher level than yourself. The higher level setting will actually make it a harder game for the computer. For example, set yourself to Level 1 and set the computer opponent to Level 3 (see Figure 9-2). Setting your opponent's astronaut level higher will provide you with an advantage. These are settings you should use when you first start out with "Buzz Aldrin's Race Into Space."

There is a significant difference between the Basic and Historical models. In the Basic model, both sides' hardware and budget are equalized. In the Historical model, American hardware can be researched to a higher level of safety, but cost more to research than do their Soviet counterparts. The weights of your payload will also change to resemble the actual realistic historical weights. The payload capabilities of the rockets are also affected in the same way. A Titan rocket with strap-on boosters may work for a particular payload in the Basic model, but may not be adequate in the Historical model—you may need a Saturn V with strap-on boosters for the same job.

Read and understand the rules of the game. Chapter 13 provides several viewpoints and good plans of attack for getting a man on the Moon. All of these approaches can work. I recommend that you give them all a try. Many would choose to just jump in with both feet, not realizing that a space program is no piece of cake to manage. There is much at stake, including invaluable human lives. History is a good precedent and starting point—learn from the mistakes and try to duplicate and improve upon the successes.

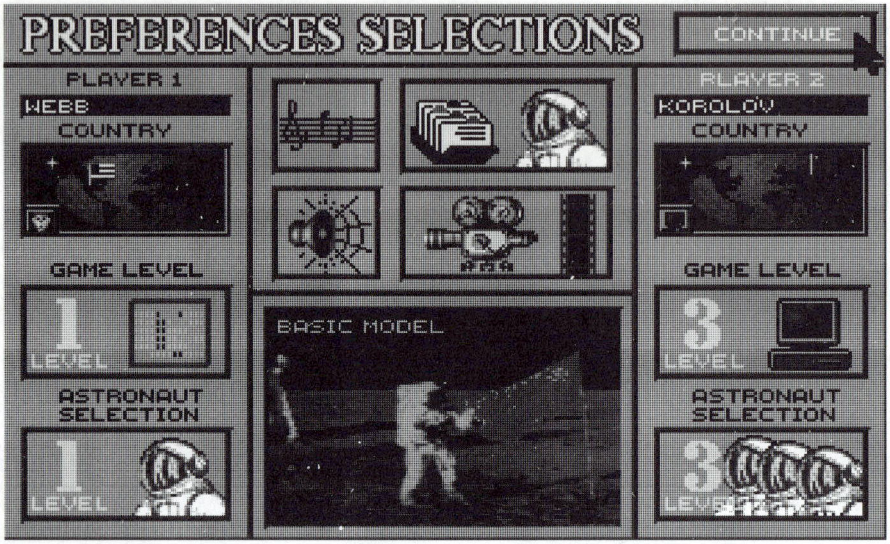

Figure 9-2. Setting difficulty levels and models

Getting to the Moon

It is a good idea to decide early on just exactly how you want to get to the Moon. Are you going to attempt a direct ascent? Or are you aiming to go the historical landing route? Exactly how you're going to get to the Moon will have a direct effect on what missions you should plan. Keep the big picture in mind when you're deciding what program to purchase next and how much to R&D (research and develop) on different hardware.

Aside from choosing the method for getting men on the Moon, the other important factor is which capsule or spacecraft you're going to use. Together, the capsule and the mission type you will use for your Moon landing will set the tone for your space program. From these two factors, you can determine the number of astronauts you'll need to ready, what rocket program to invest in, and so on.

There are four (five if you're playing the Soviets) different missions that get your men on the Moon in "Buzz Aldrin's Race Into Space." However, except for the Direct Ascent Lunar Landing and the Soviet-specific Soyuz Lunar Landing missions, there are dozens of combinations of hardware that can be used instead. Each

	Two-Man Capsule (Gemini/ Voskhod)	**Three-Man Capsule (Apollo/ Soyuz)**	**Four-Man Capsule (Jupiter/ Kvartet)**	**Mini Shuttle (XMS-2/Lapot)**
Historical Manned Lunar Landing	1. Saturn V/N-1 or Nova/Vulkan 2. Cricket/L-3 LM 3. Kicker B 4. Docking Module	1. Saturn V/N-1 or Nova/Vulkan 2. Cricket/L-3 LM or Eagle /Duet LM	*Not Possible*	1. Saturn V/N-1 with boosters or Nova/Vulkan 2. Cricket/L-3 LM or Eagle/Duet LM 3. Kicker B 4. Docking Module
Direct Ascent Lunar Landing	*Not Possible*	*Not Possible*	1. Nova/Vulkan	*Not Possible*
Joint Lunar Landing EOR *(Earth Orbital Rendezvous)*	1. Minimum of two Titans/Protons with one booster 2. Cricket/L-3 LM 3. Kicker B 4. Docking Module	1. Minimum of two Titans/Protons with one booster 2. Cricket/L-3 LM or Eagle/Duet LM	*Not Possible*	1. Minimum of two Titans/Protons with one booster 2. Cricket/L-3 LM or Eagle/Duet LM 3. Kicker B 4. Docking Module
Joint Lunar Landing LOR *(Lunar Orbital Rendezvous)*	1. Minimum of two Titans/Protons 2. Cricket/L-3 LM 3. Kicker A (two) 4. Docking Module	1. Minimum of two Titans/Protons with one booster 2. Cricket/L-3 LM or Eagle/Duet LM 3. Kicker A 4. Docking Module	*Not Possible*	1. Minimum of two Titans/ Protons with one booster 2. Cricket/L-3 LM or Eagle/Duet LM 3. Kicker A (two) 4. Docking Module
Soyuz Lunar Landing LOR *(Lunar Orbital Rendezvous)*	*Not Possible*	1. Soyuz Only 2. Two Protons with boosters 3. Kicker C	*Not Possible*	*Not Possible*
Note: Astronauts and EVA suits are required for every mission				

Figure 9-3. Hardware requirements guide for basic model

method will require a variety of hardware and a different strategy when planning missions. Use Figure 9-3 as a guide to some of the minimum requirements for the Moon ascent.

Figure 9-3 shows only the minimum requirements for the different missions. For instance, in the Joint Lunar Landing LOR approach with a two-man capsule, the minimum rocket requirements are for two Protons. You may opt to use two Novas instead, or one of each type. Keep this in mind when deciding which hardware to purchase and build. Some hardware are more reliable than others, so they might be worth investing in even if it is overkill for the type of mission you've selected.

There is a lot of variety in the combinations of hardware that can be put together to accomplish the ultimate goal of getting someone on the Moon. Further study of the table will show, for instance, that Kicker A may not really be a good investment; it can only be used for one type of mission—Joint Lunar Landing LOR. If you encounter a catastrophic failure during the later stages of the game, it would be impossible to switch to another lunar ascent method. Having to either purchase a Kicker B and then R&D the program, or rebuild after the failure could take two years and cost you the game. The Kicker B offers a lot more flexibility.

Explore the Different Approaches

Play both sides of the race. Interesting differences between the two space programs exist. You'll discover that the Soviet program has rockets that will carry heavier loads. However, this is offset by the fact that Soviet cargo (satellites, landing modules, etc.) is generally heavier than the American counterparts.

The computer opponent is actually quite "intelligent." He will react dynamically to your successes and failures. If you take a commanding lead, it will usually start taking more risks and adjust its strategy to keep up with your pace. It is highly unlikely that any two games will play alike, even if all your game settings are identical.

"Buzz Aldrin's Race Into Space" contains many painstakingly put together full-motion video sequences for the different mission types. These clips are actual

Figure 9-4. A Gemini-Titan at launch

footage from historic films. There's a wealth of video footage for you to discover, so go ahead and try different approaches. Figure 9-4 shows one of the many possible mission animations in the game. Every mission type presents a different challenge. While you may want to avoid failures, rocketry is not a perfect science. Therefore, a gamut of failure mode animations is included for each hardware failure.

The next few chapters will provide you with the necessary tools for administrating a space program. Through many experiences, you too will find your favorite way to be the first to conquer the Moon.

So get ready and set your sights high—we're going to the Moon!

CHAPTER 10

Scheduling and Running Missions

You've memorized the prime directive: Get a man on the Moon and bring him back safely before any other. You're raring to go; you're ready to make tracks. You've taken care of first things first—you've decided on a Lunar Landing mission and the space capsule to use. The funding has been appropriated, the facilities have been built. Looks like it's a green light.

This chapter will be your guide to understanding the different missions you will decide to perform or forgo. The success of each mission will have a profound effect on your ultimate mission—the one that gets you to the Moon. The comprehension of all mission types is critical to your planning and success in "Buzz Aldrin's Race Into Space."

Mission Milestones

A baby usually learns to crawl before it stands, stand before it walks, and walk before it runs. He may fall many times along the way, but he'll get up and keep going. If holding on to a chair on rollers won't work, maybe holding on to a big table leg will. The process of getting a man on the Moon is similar in many respects.

When first starting on your quest, many technological obstacles stand in the way. At this point, no data exists on either machines' or man's abilities to work in space. Through experimentation by running various missions, you can begin to learn about this great body of unknowns. In time, you will have learned enough to attempt a safe Manned Lunar Landing.

A milestone mission is one that represents the discovery and successful conquest of an obstacle in the space exploration race. For example, a Manned Earth Orbital is a milestone step because it is our first lesson about spacecraft reliability and man's adaptability to space.

Milestone steps can be executed out of order, or skipped completely. A penalty on the safety factor of subsequent missions is usually incurred whenever you decide to skip a milestone. The penalties are cumulative. This is the reason why you can't just recruit some astronauts, build a program to purchase and research a Saturn, an Apollo capsule, an Eagle landing module, EVA suits and head for the Moon. Without completing any milestones before attempting a Manned Lunar Landing, you would incur at least a –50% penalty (usually more) on your safety factor!

There are milestones that affect space travel in general and those that affect only the lunar landing. The Lunar Landing milestone is independent of the general milestones and do not affect them adversely. On the other hand, general milestones are imperative to the Lunar Landing milestone. You can't expect a safe Manned Duration D mission without at least sending a man up into space first on a prior mission. Yet a failed Manned Duration mission will not affect the success of a Lunar Satellite Fly-by.

Some of the milestones that affect the lunar landings may be skipped depending on your selected Lunar Landing mission. When a docking module is not required for the final mission, Docking milestones can be skipped. In the Direct Ascent approach, for example, you can leave Docking tests out. Landing module tests are also not required because a landing module isn't required for the mission.

Primary Milestones are steps that must be completed in sequence. Durations B and C must be completed in alphabetical order, after Manned Orbital and before Manned Lunar Pass, Duration D must be completed before Manned Lunar Orbital. LM tests must be completed anytime after Manned Orbital and before Manned Lunar Landings. The following are some examples of penalties you could incur:

Example 1. If the first mission that is attempted is Manned Orbital, the penalty is –9% on all aspects of the mission.

Example 2. A Manned Orbital B has been successfully completed. If a Manned Lunar Pass were attempted, it would be –14% on all aspects of the mission.

Example 3. A Manned Orbital Duration D has been successfully completed. If a Manned Lunar Landing were attempted, it would be –12% on all aspects of the mission.

The table below outlines the necessary steps for a successful completion of a Manned Lunar Landing and the penalties you will incur for skipping milestone missions. Note that EVAs and Docking tests can be performed jointly or separately; there is no penalty for skipping them, but you will have low docking reliability if no docking tests are attempted.

Primary Milestone Steps	Penalty	Secondary Milestone Steps	Penalty
Orbital Satellites	–3%		
Manned Suborbital	–3%		
Manned Orbital	–3%	Duration B	–5%
Lunar Satellite Fly-by	–3%	Duration C	–5%
Lunar Probe Landing	–3%	Duration D	–5%
Manned Lunar Pass	–3%	LM tests (min. 2)	–3%
Manned Lunar Orbital	–3%	(each LM test skipped)	
Manned Lunar Landing	–3%		

Tackling the Missions

The variety of missions in "Buzz Aldrin's Race Into Space" can generally be broken into three categories: robotic, manned, and unmanned missions. You can accomplish these missions in very different manners. They are listed within their categories in the following section. If the mission is a milestone, the penalty incurred when skipping it is also shown.

Please note that not all missions need to be executed chronologically or in the listed order. Repeating successful missions several times is a good way to increase the reliability of the hardware used in the mission. Typically, you can send two lunar probes to the Moon on separate missions and improve the photo reconnaissance by 30%!

You can also combine tests, such as the Lunar Orbital LM test and EVA. Safety points and prestige points are only awarded for the parts of the mission that are successful. In a Duration D, EVA, and Docking mission, you will be rewarded for completing each step successfully. So if out of the three only the Docking test is unsuccessful, you will still get points for the two other successful steps.

The Robotic Missions

The robotic missions refer mainly to satellite and probe programs. Missions that are in this category are the Orbital Satellite, Lunar Fly-by, Lunar Probe Landing, and the Interplanetary Fly-by.

Orbital Satellite (Prestige: 12 points; Milestone Penalty: –3%)

This milestone step will help your program understand more about the hazards of space. You will discover what types of protection and shielding is required for manned missions. A man-made object will have successfully escaped Earth's gravity and actually orbited the Earth from space. Imagine the possibilities we can better predict the weather, spy on the enemy, map the Earth from orbit, transmit radio signals, etc. We've realized many of these dreams as evidenced by the many satellites orbiting the Earth today, performing some of what was impossible back in the 1940s and early 1950s.

This is almost a must-do step. It is a cheap and easy mission to complete and should be your first successful mission. Figure 10-1 shows a screen for this milestone step.

Lunar Fly-by (Prestige: 5 points; Milestone Penalty: –3%)

Here is where you learn about the range of the Van Allan Belt, heading towards the Moon, and also more about deep space conditions. This mission helps you discover the strength of solar radiation and how it would affect humans that far out into space.

With on-board cameras, images of potential future landing sites are sent back to Earth. Other images show the yet uncharted regions on the Moon. With each successful Lunar Fly-by, photo reconnaissance of the Moon increases by 5%. Figure 10-2 shows a screen for this milestone step.

Lunar Probe Landing (Prestige: 16 points; Milestone Penalty: –3%)

A single successful probe landing will increase your photo reconnaissance by 15%. By far, this is the most efficient method for mapping the Moon. Even a failed landing will result in a 5% increase in photo reconnaissance if the probe manages a lunar de-orbit burn.

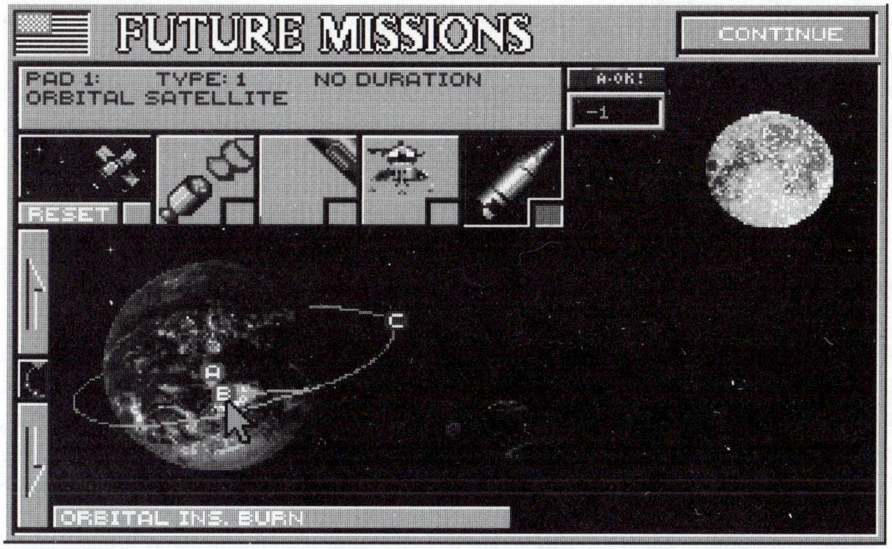

Figure 10-1. An Orbital Satellite mission

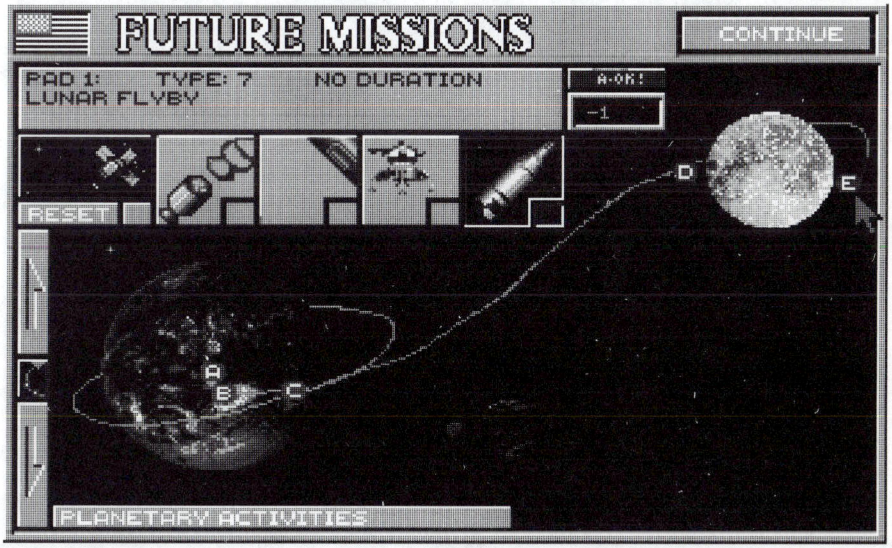

Figure 10-2. A Lunar Fly-by mission

A well-mapped Moon will ensure a safer landing site for the lunar party. The penalty for not having a highly reliable photo reconnaissance can have a direct effect on Lunar Probe Landings and Manned Lunar Landings. What might appear as a safe landing site several miles up in orbit can have devastating consequences if the terrain is rough and cratered. Some of the failure mode possibilities are crashed probes and dead spacemen littered on the lunar surface. Figure 10-3 shows a screen for a Lunar Probe Landing.

Interplanetary Fly-by (Prestige: 5 points)

These missions are used primarily for prestige points. The best use of an Interplanetary Fly-by mission is when you have an unused launchpad for a season. Launching a satellite to Mars, for instance, is a good mission when you have nothing going on that season. By gaining some prestige points, you may squeeze more funding for the following seasons, as well as keep yourself on good terms with the president.

Remember that this is a long-term investment. Some of these fly-bys will take two to three years to complete. The farther out you send the satellite, the longer it will take. Mars, Mercury, and Venus are fairly close, but Saturn and Jupiter are much farther out. You gain the prestige points at the completion of the fly-by, not when you launch the satellite. Figure 10-4 shows a screen for a Venus Fly-by.

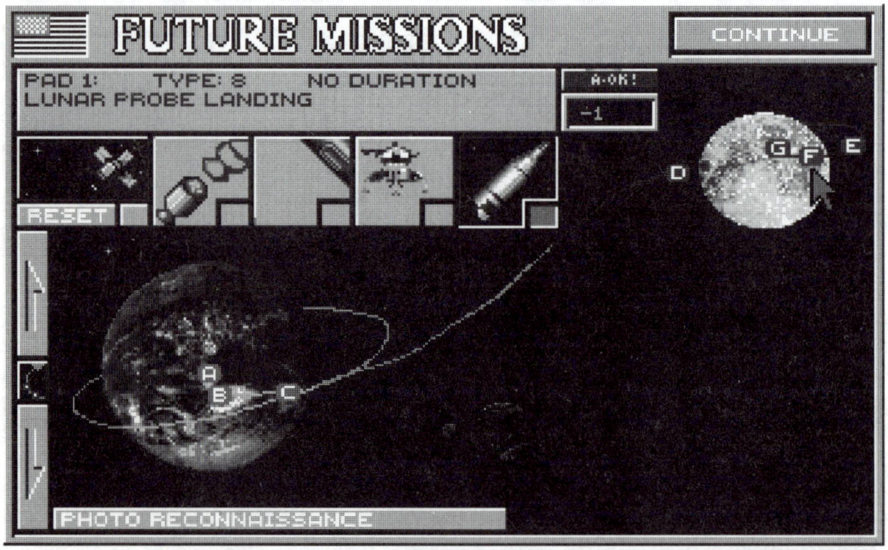

Figure 10-3. A Lunar Probe Landing mission

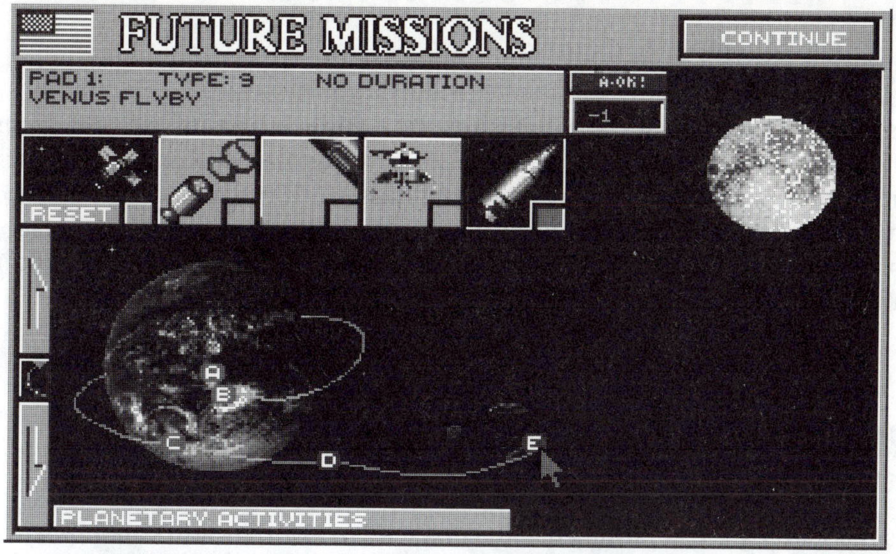

Figure 10-4. An Interplanetary Fly-by mission to Venus

The Unmanned Missions

Unmanned missions are equivalent to certain manned missions minus the astronauts. For example, you can perform Duration tests and Docking tests without astronauts. Successful completion of such missions will increase the safety level of the hardware involved, but do little for you in the way of prestige points. However, they are appropriate whenever you want to avoid a catastrophic failure.

You could use a manned mission to get the prestige points and then use unmanned missions to improve the reliability of the hardware on subsequent tests. No prestige points are given for unmanned missions; they also do not qualify as a milestone mission—it will not clear a milestone hurdle. The astronauts will miss out on their skill improvements on these follow-up tests as well, but this is more desirable than having dead astronauts on your hands if a mission fails.

Since unmanned missions are actually similar to manned missions, they are explained in the following Manned Missions section of this chapter.

The Manned Missions

This category encompasses both the manned and unmanned missions but does not include the satellite programs. These missions are performed in preparation for the all important final Manned Lunar Landing mission. If the mission is a milestone step, you are required to accomplish the mission with astronauts to complete the step. You can run a similarly configured mission unmanned, but it would not qualify as a milestone completion.

Manned Space Mission/Suborbital (Prestige: 15 points; Milestone Penalty: –3%)

This is a startling accomplishment—the first man to exit the Earth's atmosphere and step into the vastness of space. The spacecraft's ability to sustain life in the harsh environment of space is tested. This step may be skipped in favor of the Manned Orbital mission. Figure 10-5 shows a screen for a Manned Suborbital mission.

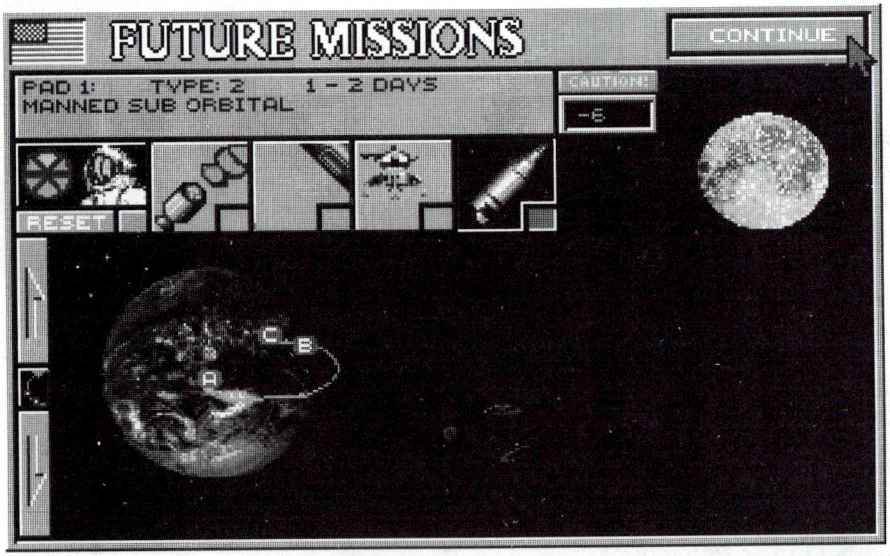

Figure 10-5. A Suborbital mission

Manned Orbital (Prestige: 15 points; Milestone Penalty: –3%)

This mission sends a man around Mother Earth in space. Since the astronaut remains within the confines of the spacecraft, the special EVA suit is not required. (During actual flights, the astronauts wore pressurized suits and were totally reliant on the spacecraft for lifesupport; i.e., air, temperature regulation, pressurization, etc.) More data is gathered in a Manned Orbital mission regarding spacecraft reliability and man's adaptability to space, paving the way for longer missions.

Both this and the Suborbital missions are fairly easy ones to execute. Use the astronauts with the highest capsule rating for these missions. All the other skills are not applicable to these early missions. The astronauts' pilot skill rating is almost critical for the one-man capsules since the capsules themselves are inherently unreliable at only a maximum R&D safety of 76%. Having an astronaut with a capsule rating of 4 would bring the safety factor for a one-man capsule up to a tolerable 80%. Mission experience and dummy tests (unmanned missions) can increase the capsule saftey factor by 1% on each successful mission.

An alternative is to skip the one-man capsule completely and invest in the safer two-man capsule for these missions. This will stretch your early budget, especially since you'll need either boosters for the Atlas or the bigger and costlier Titans. You could make up for it by planning for a more ambitious mission including sending a female up in the same capsule and including an EVA to garner additional prestige points. Figure 10-6 shows a screen for a Manned Orbital mission.

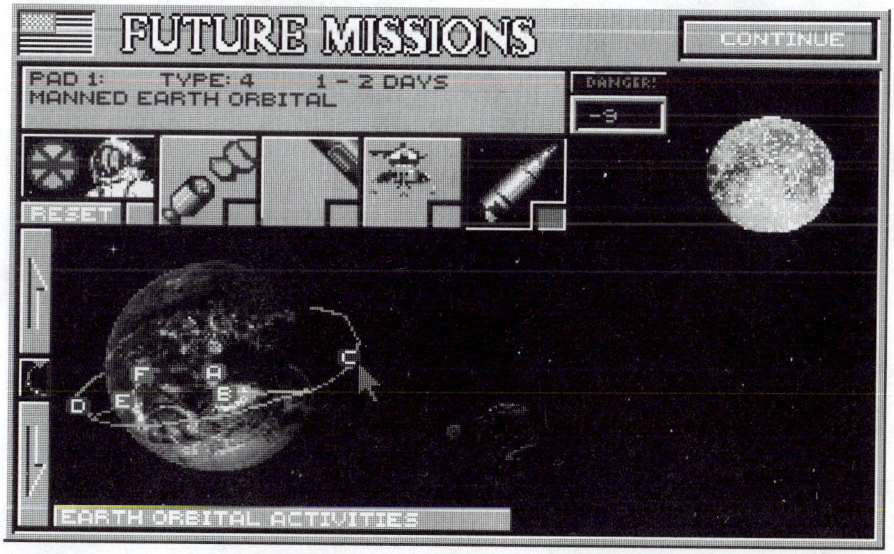

Figure 10-6. A Manned Orbital mission

EVA (Prestige: 8 points)

This is an important accomplishment that demonstrates man's ability to work freely outside the confines of the spacecraft in space. The EVA is usually combined with other missions. Figure 10-7 shows a screen for a Manned Orbital EVA.

Duration Steps (Prestige: 1 point; Milestone Penalty: –5% to –8% per duration skipped)

The object of Duration tests is to ensure that both the spacecraft and astronauts are capable of sustaining and working for extended periods of time in space. A trip to the Moon and back will take approximately 8 to 12 days.

Every successful step translates to more scientific data taken to analyze the effects of spending time in space, which makes the next step more attainable. The penalties for skipping steps is calculated by averaging the flight crew's endurance skills and the number of steps skipped. They range from –5% to –8% for every step skipped.

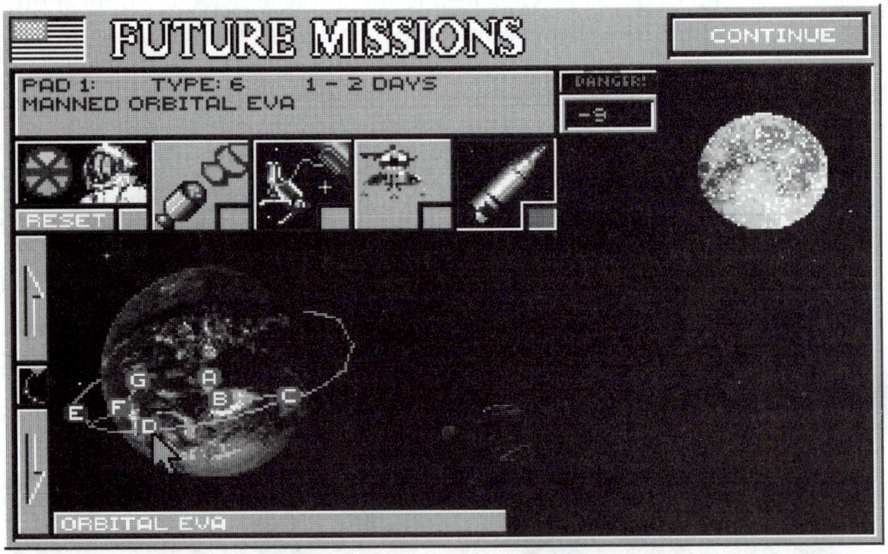

Figure 10-7. An EVA mission

The steps are named alphabetically as following:

- Duration A 1 - 2 days This is achieved with an Earth Orbital mission
- Duration B 3 - 5 days
- Duration C 6 - 7 days
- Duration D 8 - 12 days
- Duration E 13 - 16 days
- Duration F 17 - 20 days

The penalties for each step are cumulative, so if a Duration E mission was planned after a successful Duration B mission, the Duration E mission will be penalized between –10% to –16% for all aspects of the mission.

Ironically, few prestige points are passed out for these critical steps. The penalties for skipping any Duration step is very high. Therefore, we suggest patience in making sure each one of the steps is completed in sequence before attempting a Lunar Landing. Figure 10-8 shows a screen for a Duration step.

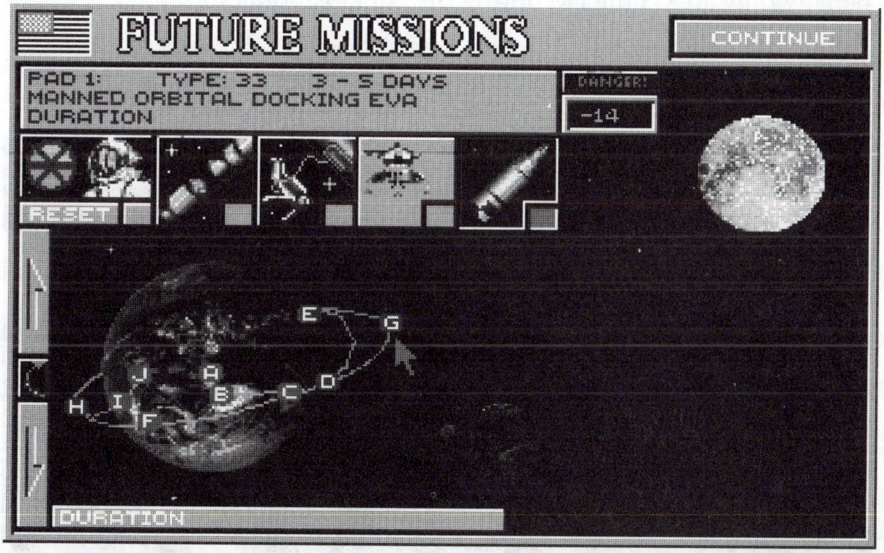

Figure 10-8. A Duration test mission

Docking (Prestige: 8 points)

Docking is another important skill for astronauts to master. A rendezvous in space is an intricate process. Several of the lunar approaches require docking in space. While this is not a milestone step, improved docking skills will decrease the likelihood of failure in this stage of any mission.

A successful Docking test increases the docking skill by 10%. Even a failed mission will improve this same skill by 5%. (The Docking step must be reached during the mission for it to count as an actual attempt.) Docking tests are best combined with other tests such as an EVA. The Direct Ascent method does not require this skill. Figure 10-9 shows a screen for a Manned Orbital Docking.

Lunar Module (LM) Tests (Prestige: 15 points; Milestone Penalty: –3% per test skipped)

LM tests can be conducted with manned missions only. There are two ways to test the LMs: Earth Orbital or Lunar Orbital. A total of at least three Earth Orbital LM tests are required to clear this milestone. A –3% penalty is incurred for every test skipped; skipping all LM tests will result in a –9% penalty. Unlike most penalties, this penalty only affects the LM during the Lunar Landing stage.

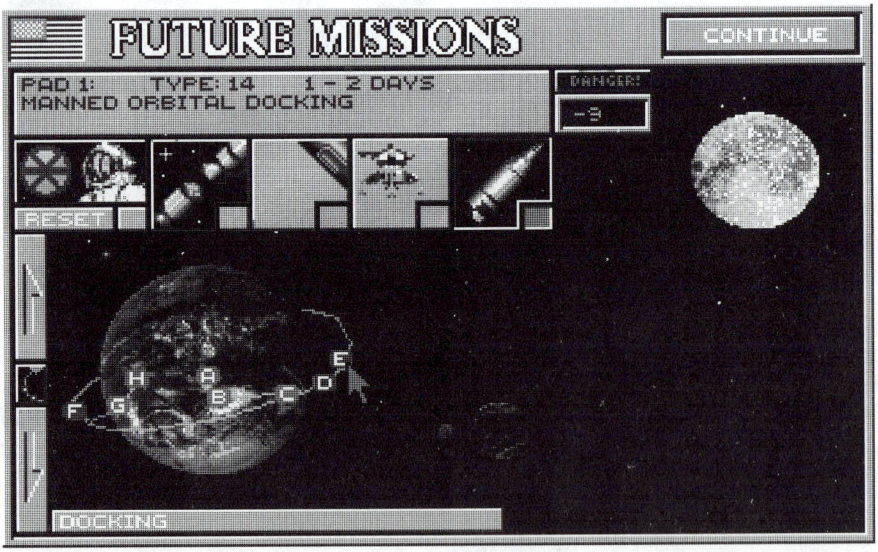

Figure 10-9. A Docking test mission

One Lunar Orbital LM test is equivalent to two Earth Orbital LM tests. Therefore, a combination of one Earth Orbital and one Lunar Orbital LM test is adequate for clearing the three steps and this milestone. This milestone is not a requirement for Direct Ascent missions.

You may use cheaper hardware to complete these tests. Even if you have decided to use an Eagle LM for the lunar landing, you can run all the tests with the lighter Cricket which can be sent into Earth orbit with a lesser and cheaper rocket.

Important as these tests are to the lunar program, the ignorant general public won't give you much accolade for completing them. There are no prestige points given for these missions. Figure 10-10 shows a screen for an LM test.

Manned Lunar Pass (Prestige: 12 points; Milestone Penalty: –3%)

The Manned Lunar Pass presents for the first time a view of Earth as a rising moon to an alien world. Many observations are also made of lunar gravity and potential landing sites. Photo reconnaissance is improved 5%. The spacecraft systems are tested in lunar gravity and in the extremes of space far from Earth's atmosphere. Figure 10-11 shows a screen for a Manned Lunar Pass.

This mission helped man conquer a great psychological barrier. Man can explore and traverse space and reach new worlds.

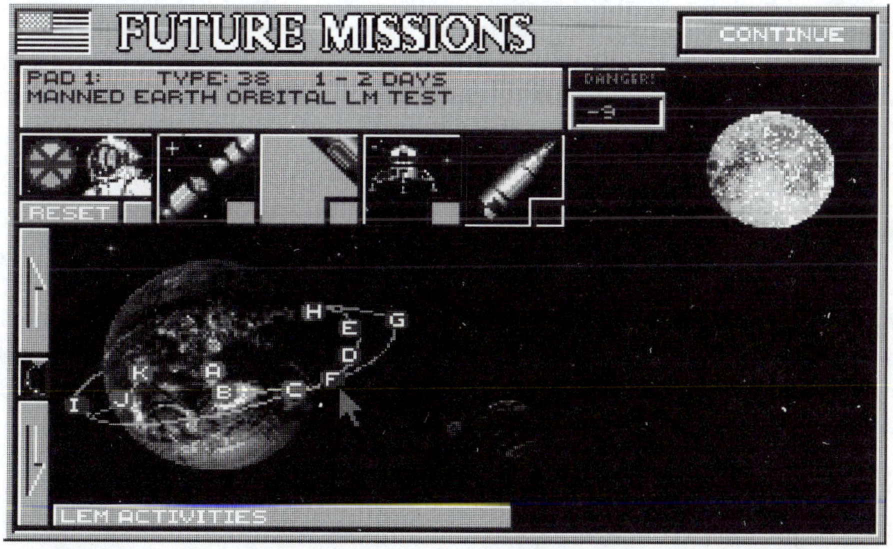

Figure 10-10. A Lunar Module test mission

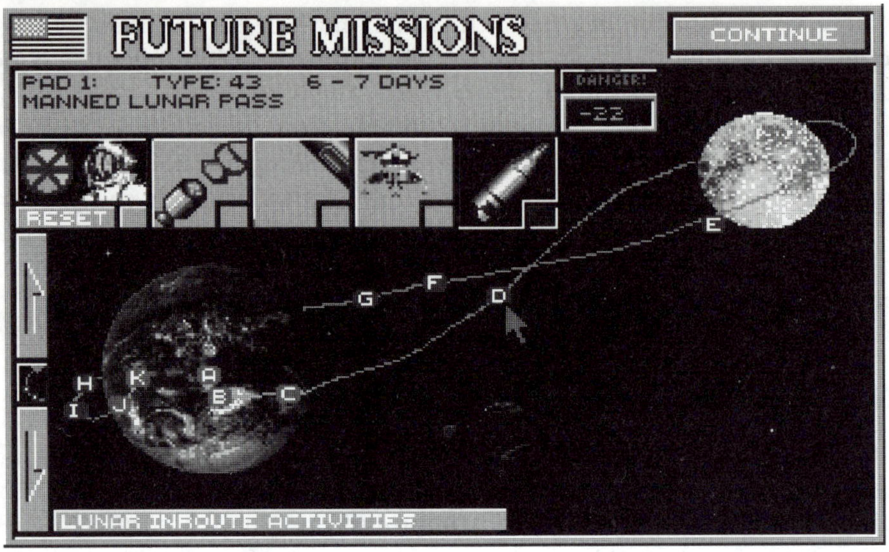

Figure 10-11. A Lunar Pass mission

Manned Lunar Orbital (Prestige: 20 points; Milestone Penalty: –3%)

The Manned Lunar Orbital is a prelude to the final ascent of the Moon. More tests are conducted and photo reconnaissance increases by 5%. Typically, you can include a LM test with this mission. This would take care of one of the last remaining milestones at this stage. Figure 10-12 shows a screen for a Manned Lunar Orbital.

It won't be long now before man sets foot on an alien world.

Manned Lunar Landing (Prestige: 40 points)

"Space, the final frontier...To explore strange new worlds...To boldly go where no man has gone before." These immortal words rang from Captain Kirk from the then not so popular TV series of the '60s, "Star Trek." You're on the edge of this accomplishment. Who would have imagined that in less than a decade and a half since the first orbital satellite, man would set foot on the Moon.

Try all the different routes when you get a chance. Each method requires a slightly different approach requiring different strategies during the middle phase. Historically, there was much debate over which mission would be used to get to the Moon. The Americans ended up using a three-man capsule, the *Apollo 11*, the Saturn

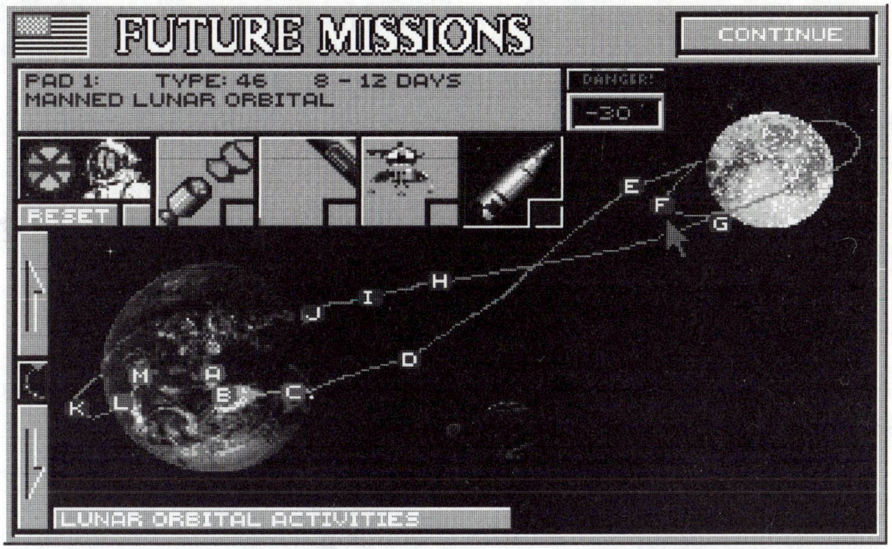

Figure 10-12. A Lunar Orbital mission

V rocket, and the Eagle LM. It proved to be an effective way to get on the Moon. Many will forever remember the spoken words, "The Eagle has landed!"

The Direct Ascent is probably the most straightforward approach, but also the costliest. It is also not very flexible. The only requirements are a Nova rocket with a Jupiter four-man capsule. The hardware can be R&D'ed to a very safe level.

The EOR or LOR methods serve as good backup plans for the Historical method. Since these later missions require lesser rockets, a catastrophic failure of your larger Saturn rocket would not leave you paralyzed if you have the Titans with boosters as backups. The EOR and LOR methods require two launches each. One rocket carries the LM, the other carries the capsule. Then depending on which mission you select, the capsule docks with the LM either in Earth orbit or in lunar orbit. These two missions require more steps which may allow Murphy's Law more opportunity to act up.

It was discovered that the Soviets had a unique alternate plan for getting to the Moon with a Kicker C. This mission has been recreated in "Buzz Aldrin's Race Into Space" for those playing the Soviets as another way to reach the Moon.

This is it! Succeed in this mission and you'll win the game. Nobody will remember if you come in second in this race. All the megabucks and planning boil down to this one mission amidst dozens of others. Don't rush this mission; a failure here could bring about a disaster. If you make it, the congratulations and the praise of a nation are due. Figure 10-13 shows a screen for a Manned Lunar Landing.

Three Phases of the Game

As you play "Buzz Aldrin's Race Into Space," you'll discover that there are three distinct phases to the game. The phases are not exactly based on a timeline, but instead on the state of your space program; i.e., which missions have been completed and which have not.

The early phase should be from 1957 to around 1961, the middle phase should be between 1960 and 1966, and the start of the final phase will vary from 1965 to 1970. The aforementioned schedule is merely a guide and will vary from game to game.

The Early Years

In the early years, the goals are to successfully launch an orbital satellite and get a man into space. These are fairly easy goals to accomplish and is the easiest of all the three phases to complete.

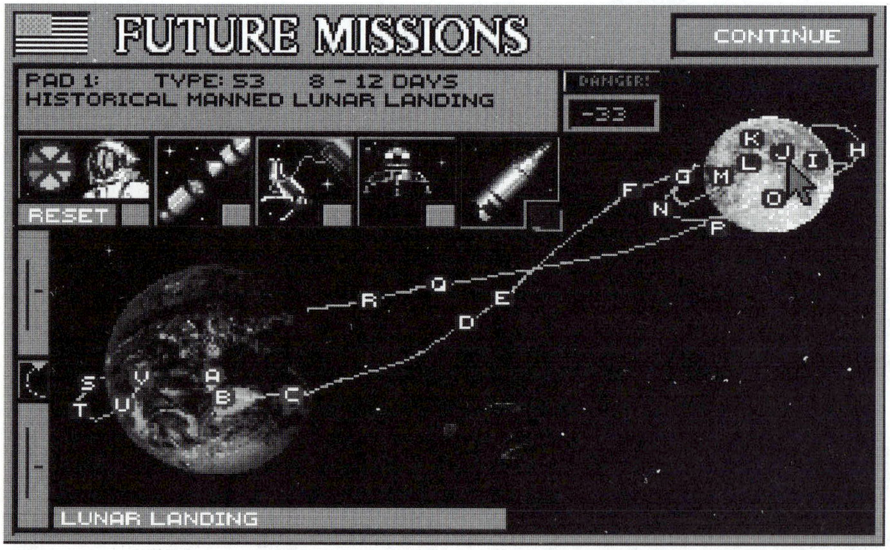

Figure 10-13. A Lunar Landing mission

The early years' goals are straightforward ones, but they should not be taken lightly. A total of up to 47 prestige points are at stake. These points are important because you really want to enter the next phase with lots of megabucks in your budget. And since these missions are the cheapest, they can be easily fixed at little expense if a failure occurred. Beware of killing your astronauts in the manned missions; those are considered to be major catastrophes that can set your space program back by as much as three or four seasons.

Chances are that you won't be able to successfully send an orbital satellite up into space until the third or fourth season of play. The earliest possible season you could complete this mission is the fall of 1957—but you'll need a large dose of luck. Since this is only the second season in the game, your rocket will be very unreliable. Taking a risk here won't cost you too dearly. At worst, your rocket will explode on the launchpad, destroying it and setting you back a couple of seasons. Even then, you should accomplish this mission by 1959 or 1960.

As for getting a man into space, a Suborbital mission will suffice. There is no reason to acquire astronauts as soon as they're available in the spring of 1958 because the capsule program is probably not close to the highest level of R&D safety. There is no use for astronauts if you don't have a capsule program. In fact, you may just be acquiring the capsule program at this point. The 20 megabucks could be used for important R&D on the rocket, satellite, and capsule. It takes longer to develop a capsule program than to train astronauts, so start with the capsule first.

Do recruit the astronauts two seasons prior to the planned first manned mission. Take enough astronauts out of Basic Training immediately for a primary and backup crew and assign them to a capsule program. You will then be able to plan a future mission in the following season. By the third season, all your ducks should be in a row (maximum R&D safety on all hardware used for the mission) and ready to go.

It is very unlikely that anyone from the first group of astronauts will end up on the Moon. In real life, Alan Shepard was the only Mercury astronaut to reach the Moon. By the end of the middle phase, many of them would have retired or left the program for other reasons. Don't worry too much about what's left of this group of astronauts later on in the game. If any of them manage to hang on, so much the better.

You could go for the gusto and try a Manned Orbital EVA mission as your first attempt at launching a man into space. On the other hand, you could be conservative and run unmanned Suborbital and Orbital missions to improve the capsule and rockets prior to sending an astronaut into space.

Completing a Duration B mission and an EVA mission is a good idea during this phase. You want to clear these foundation milestones and steps before reaching the critical middle phase.

Everything in this phase becomes obsolete very quickly and won't have any long-term consequences. What you acquire will be replaced by superior hardware, and is expendable. What happens to the Atlas rocket program, the one-man capsule, and the orbital satellite isn't going to affect your lunar landing hardware at all. What is important is that they will help you complete those early missions to build up the prestige points, which ultimately puts more megabucks in your hands. With more funds available, you have greater flexibility during the middle phase to choose the type of programs you want.

If you're unsuccessful here in achieving firsts, don't despair. It is the start of a very long race. Your opponent is in no better position than you are, because they too have to complete the remaining steps, and they face the same problems and potential for disaster. Remember the story of the tortoise and the hare—it's not who leads the race, but who crosses the finish line first.

The Mid-Game Crisis

This may be the toughest part of "Buzz Aldrin's Race Into Space" for most players. There is good reason for concern at this stage. Here's where we're going to shift gears and try to maintain the momentum from the first phase. At the same time, we're starting to handle more money, new programs and equipment as we build towards the final configuration of hardware required for the final phase.

Part of what makes this phase tough is that you're making the biggest commitment in "Buzz Aldrin's Race Into Space" at this point—what hardware you're going to use to get on the Moon. During this phase, you'll have to contend with missions that have the widest variety of possible combinations of hardware and steps. All of a sudden you're faced with having to allocate the right resources to the right programs and have them all going at the same time. You may have multiple launch facilities and multiple missions every season. Docking tests, Duration tests, and Fly-bys are all types of missions you have to complete in this phase.

Always have a secondary plan in case your primary plan encounters an insurmountable roadblock. Plan for this during the middle phase. Not having a backup plan in the final phase when a major catastrophe occurs will spell disaster for your program. You'll probably end up playing second fiddle to your opponent.

It is extremely important that you keep focused on your final hardware configuration. What mission have you decided to conquer the Moon with? What rocket, spacecraft, and other equipment will you need? Keep this final configuration in mind when deciding on what hardware to purchase and R&D. You have to decide on whether to achieve all your early milestones with cheaper hardware before investing in the more expensive programs required for the final phase. Even though it could be expensive, you could alternatively purchase and develop the final components during this stage, and use them for testing as well.

Funding and its management are clearly more difficult at this stage. You need to invest in the right areas but at the same time depend on the tried and true so that you'll continue to gain prestige points. One mistake that is frequently made is to allocate funds to R&D during a season, but to forget to leave enough money to purchase the hardware for the mission planned for the same season. Having to scrub the mission could be costly, especially if during that same season, your opponent beats you to the prestige points for being first at something or other.

You'll recruit the second and third batch of astronauts during the mid-game stage. A few lucky astronauts from these batches will likely end up on the Moon. Matching astronauts' personalities and their skills becomes more difficult. You've got many astronauts on your hands to keep happy. Fortunately, they're all professionals. No matter how incompatible their personalities are, they will perform to the best of their abilities during a mission. However, they may quit or retire on you before you get a chance to send them on a mission. You can find out more about astronaut management in Chapter 11.

Towards the end of this phase, you'll finally purchase a Kicker and a landing module. Again, premature purchases of these programs will mean less money for other programs. Since you won't really need these two items until the final phase, it would make sense to purchase them only towards the end of the middle phase.

The Final Thrust

Your program may have gone through its ups and downs, but it all boils down to the final mission in the final phase—the Manned Lunar Landing. In final preparation for that ultimate mission in "Buzz Aldrin's Race Into Space," the Manned Lunar Pass and Manned Lunar Orbital are performed to further improve Moon reconnaissance and lunar landing equipment, and to test astronauts.

Continue to launch other missions to improve your hardware to new levels of safety. Every percent in safety is important because the later missions are very involved and require many intricate steps.

There is a possibility that you may encounter a disaster in your primary program. In that case, your secondary program should become your primary program, and your failed primary, the new secondary program. Don't forget about that failed program though; you may need it again if your second primary program fails! Continue to R&D and test your secondary programs throughout this stage.

This stage is the make or break stage. If you're behind, you may consider skipping some milestones, or combine a milestone step with some tests. For instance, you can schedule a Lunar Orbital and LM tests on the same mission. Kill two birds with one stone if you can. And since this is your last chance, be aggressive with your megabucks. If you get to the Moon, the game ends. All the megabucks that are hoarded will be useless.

Is the Cup Half Full or Half Empty?

Managing a space program requires many compromises in the safety of all your resources, among other things. Most players will view the safety factors from an optimist point of view. Is the cup really half full or half empty? It is, of course, both. No matter how high the safety factor is for any component, there is a chance that it will fail if that safety factor isn't at 100%. The 100% safety factor is for the most part unachievable in "Buzz Aldrin's Race Into Space," as in real life. Rocketry is not a perfect science, and the space program is a high risk endeavor. Failure in the harsh environment of space can be quick and deadly.

Some of you may look at a safety factor of 65% and conclude that there's a very good chance the hardware will work. What is more important however, is that there is a 35% chance for failure and that's pretty high. That's not counting the percentage lost if you're skipping a milestone step. How much would you bet on it raining if the weatherman told you that there is a 65% chance of precipitation today? Very quickly, we can see that the launch actually isn't all that safe, and that it may not rain at all today.

Generally, it is best to R&D all hardware to maximum safety factors before using them. At least then you'll be satisfied that you did the best you could to prevent any potential failures. Sounds like shirking your responsibilities? On the contrary, you're taking responsibility and ensuring the highest integrity and safety in your program in areas where you have control. This may not be possible all the time, but watch out for the hidden factors that could reduce your safety levels further than what is showing.

In this game at least, it is preferable to look at the cup as half empty, and start from that premise.

A Case for Multiple Launchpads

Having multiple launchpads opens up a whole new world of possibilities and gives you much more flexibility in mission planning. You can launch missions more often and earlier, without rushing penalties. The extra maneuverability is almost like having a queen as opposed to a bishop in a chess game.

If you're planning an EOR or LOR Lunar Landing, having at least two launchpads is a requirement. All joint launches also require multiple launchpads. Only missions on adjacent launchpads can be used for joint launches. Either missions on Launchpads A and B, or B and C can be linked. Joint mission cannot be launched from Launchpads A and C.

Usually, a scheduled mission is launched toward the end of any season. You can launch the mission up to two months earlier, but a rushing penalty usually applies, as well as extra costs. If you had extra megabucks to spare, it may not be a bad idea, especially if you know your opponent is pretty close to launching a similar mission and you only have one launchpad. The difference in prestige points between being first or second can certainly be worth what money you spend on an early launch.

But with multiple launchpads, your missions are now staggered. You can now assign a more important mission to launch earlier without incurring any penalties or additional costs. Of course, if one of your pads got destroyed in a prior mission by an explosion or a storm, you still could use the other launchpad.

In the middle phase where the density of missions is at its highest, having multiple launchpads provides you with the ability to speed up the program by allowing multiple missions per season.

All of the above adds up to a lot of value for only 20 megabucks!

Downgrading and Scrubbing Missions

It has been oft said, "Better live to fight another day." Such is the case sometimes with your missions. Each cancelled mission will decrease the prestige points for your space program by 3%. Again, –3% is a small price to pay compared with launching a risky mission that may end up costing you even more prestige points and megabucks. Nothing else is lost when scrubbing missions, which is a very important factor. The penalty in prestige points reflect the public's letdown whenever a mission is announced and then later cancelled. Following are a few reasons why you may want to cancel a mission.

Sometimes you may assign a future mission based on the assumption that the current year R&D will bring the safety factor of a component up to an acceptable rating. However, you may be dealt a card which lowers the current year's R&D. Since the results of the R&D are lower and does not meet your expectations, the launch may be in jeopardy. Consider seriously the risk for a catastrophic failure, especially during the final phase. A 5% increase in safety factor may not seem significant, but could prove to be the margin between success and failure on any particular mission. Losing 3% in prestige points, conserving the hardware and megabucks, and postponing the launch may be the better route.

During a season, you may have to unexpectedly spend megabucks on unforeseen expenditures. For example, let's say you get hit with a 20 megabuck cost to repair a damaged launchpad. You may end up not having enough megabucks for both R&D'ing a program and purchasing the hardware for a mission. If the R&D that season is to affect the chances of a successful mission, then the mission should definitely be canned and the monies applied instead to just the R&D program.

In other cases when you have more than a launch planned for any season, the second or third launch may depend on the success of the previous launches. There is no sense in launching the second part of a joint mission if the first one failed. It is a waste of time, megabucks, and other resources.

Sometimes a catastrophic failure may occur in a particular mission which will affect subsequent missions. For instance, you may be using the Titan rocket for two separate missions in a season. Unfortunately, during the first launch, the Titan blows up on the pad and the safety factor for the Titan is dropped to 5%. With this in mind, it is almost guaranteed that the following launch will fail—it makes sense to cancel the second launch.

Note that the Orbital Satellite and Fly-by missions are not worth scrubbing. Since these are unmanned and relatively inexpensive, you may as well take the chance. After all, a failed mission would still only cost you 3% in prestige points. Also, failures that occur inflight, after all missions are launched for the season, will not affect other flights in space at the time. So if the first capsule loses its heat shields on re-entry, the safety factor for that capsule will only affect all subsequent launches. It will not affect the second identical module that was launched at the same time as the first.

Downgrading a mission is also a viable alternative when you want to lower the risks involved with any mission. It will still cost 3% in prestige points whenever a mission is downgraded. Some missions cannot be downgraded; for example, an Orbital Satellite mission cannot be downgraded.

CHAPTER 11

Managing the Space Program and Its Buildings

Part of doing a good job as the director of the space program is to understand your duties and responsibilities. Your spaceport grows substantially as time passes and as new programs are birthed. The group of few essential buildings that you start with mushrooms into a metropolis of buildings and facilities during the middle and final phases of the game, as shown in Figure 11-1.

The rule book for "Buzz Aldrin's Race Into Space" explains in detail what tasks are performed in each building. It is recommended that you review these functions so that you can efficiently run your space program.

Figure 11-1. The bustling American spaceport in the final phase

Recommendations for Visiting Buildings

You will develop your own turn sequence within each season as you become more familiar with the spaceport. The following is a good turn sequence to adopt, although you could customize these steps in any order you wish:

1. Visit the Vehicle Assembly (American) or the Vehicle Integration (Soviet) building to purchase and assemble all hardware required for the season's launches. This should be the first thing to do to make sure that the funds are primarily spent on the current mission. If there was an oversight in your calculations the previous season, leaving you with less than enough megabucks to purchase the required hardware, now would be a good time to scrub a mission.

2. Review the budget for the following season in the Administration building so that you can plan ahead. Knowing what to expect for a budget is important for planning and disbursing funds into the various programs. You can also compare your budget to your opponent's and discover what he can or cannot accomplish with the funds he has available.

3. Check out the intelligence reports in the Pentagon/KGB building. They may provide hints on the current status of your opponent's programs. Review the Prestige Firsts screen in the Administration building while you're there. By observing your opponent's prestige points, you can tell whether he's had a major failure or a major success. This information will help you decide to take a risk if you're behind, or to relax if you're ahead.

4. Purchase new hardware and space programs in the Administration building. Once an initial purchase has been made in any program, you are not required to have the hardware in hand to perform R&D. If you've used up your inventory of Atlas rockets, for instance, and there are no launches scheduled for this season, you can still R&D the Atlas without having to purchase one for inventory. The money is better spent elsewhere on new programs. In "Buzz Aldrin's Race Into Space," it doesn't make sense to have anything in inventory. Spend your budget wisely by purchasing only what you need for the current missions.

5. Research and develop space hardware in the R&D Complex. First, always R&D programs that will be used in the current season. Priority should be given to manned spacecraft to ensure the highest reliability and to avoid catastrophic failures.

6. Plan future missions by going into the Administration building. If you have multiple launchpads, you can schedule the more important mission first by assigning it to Launch Facility A (upcoming launch announcements are shown in Figure 11-2). Even if you're short on megabucks, you may assign a dummy launch to the other launchpads. By doing so, you will move the mission on Launch Facility A forward several months without incurring any rushing penalty. You can then scrub the other missions just before they're launched. This is a good tactic when you're short on funds, but do keep in mind that you lose 3% in prestige points every time you cancel a mission. On the other hand, cancelling a mission may be the only choice if you're short of funds for paying the rushing penalty.

7. Visit the Astronaut complex and assign ready astronauts to capsule programs or send them off for further training. Check the moods of the crews in the different capsule programs to make sure everything is stable. Reassign crews if necessary. Astronauts who are not assigned to a flight crew in any capsule program are returned to the Astronaut complex. Also, if part of a two- or three-man crew retires, gets killed, etc., the crew is broken up and the remaining astronauts end up in the Astronaut complex until reassigned.

   It is a good idea to have at least four flight crews in a capsule program you intend to use in the next turn. This ensures that there are enough

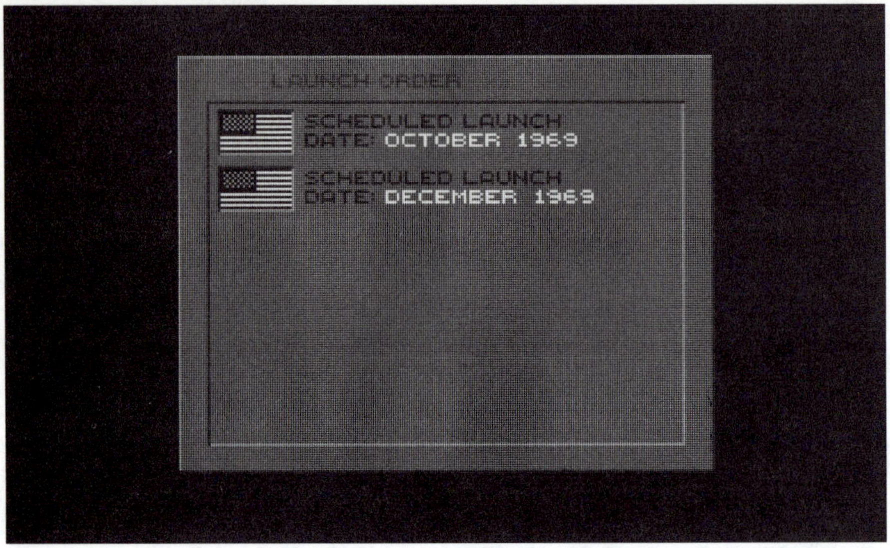

Figure 11-2. Two scheduled launches in Mission Control

crews around (primary and backup) if one or two of them gets broken up during the turn in which you want to assign the mission.

8. Review presidential approval in the Capitol/Politburo building. When playing the game at a higher difficulty level (Level 2 or 3), a low approval rating could threaten your job security. Reviewing this feature might alter your short term plans by assigning missions to improve your job performance.
9. Review the launches for the current season at the Launch Facility, shown in Figure 11-3, and determine whether you need to rush or scrub them.
10. Click on your flag and end the turn.

Additional Recommendations for Successful Play

Following are a few other steps that you may want to make a habit of completing during the later stages of the game:

- Click the Moon in the Spaceport screen to see your current photo reconnaissance reliability. Photo reconnaissance starts at a base of 55% reliability. A lack of photo reconnaissance will lower your chances of success for a safe Moon

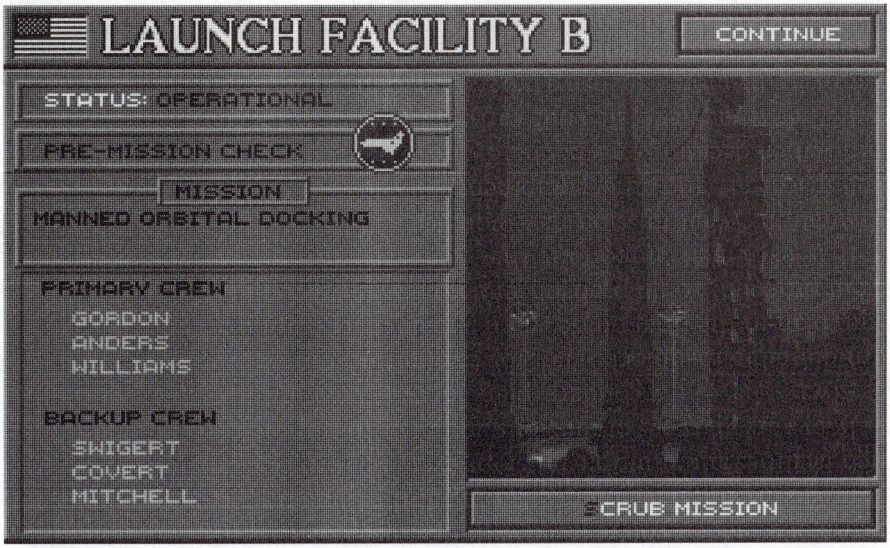

Figure 11-3. Launch facility showing a planned mission

landing. If you fail photo reconnaissance, it will affect the LM landing portion of your Moon ascent mission.

- The Viewing Stand is a repository of all the previous news that has had a direct effect on your space program. It is the place to visit to refresh your memory as to what penalties or bonuses are in effect for the current turn. You can review all the news and event cards to remind yourself of what happened to different astronauts, events that affected your R&D, additional funding that you received, and historical events. A most useful list of the last three missions and their results is also displayed in the Viewing Stand.
- The Space History screen in the Museum building lists all the missions you've launched. You can review details of all previous successful and unsuccessful missions here. This is a good place to visit to review the many animation sequences in the game. Be sure to click the space hardware icons and mission patches for specific information. The information helps you review your launch history in order to determine your short-term goals.
- Further in the game, you may want to check the LM building. This is where you can find out how many LM points you have. You want to acquire at least three points total to avoid any safety deductions from your lunar landing mission. Each missing point below the desired total carries a penalty of

3% in safety factor. For example, if you only have 1 LM point and you attempt a Moon landing, the LM safety factor will be reduced by 6%.

➤ The Satellite building gives you an immediate update on all your unmanned programs. These programs include orbital and interplanetary satellites, docking modules, and the lunar probe. This screen is helpful because it displays not only your statistical success in each category, but also which countries, if any, have succeeded in each of these categories or prestige milestones. It is useful for analyzing your opponent's successes (prestige first) in the unmanned programs. The Satellite building screen is shown in Figure 11-4.

➤ During game play, you may want to know where all your astronauts are located. The Astronaut History screen, shown in Figure 11-5, provides background information about the astronauts, including tenure (duration of active duty), total prestige points (cumulative total from all missions), current skills, space duration (total days in space), and location.

The Astronaut History screen in the Museum building also displays your astronauts' accomplishments. All astronauts who complete at least one space mission will get an astronaut wing. Mission patches and spacecraft icons on the left side of the screen display the number of missions the astronaut has flown and the corresponding text provides data about what type of missions they were. You will notice that some of your astronauts may

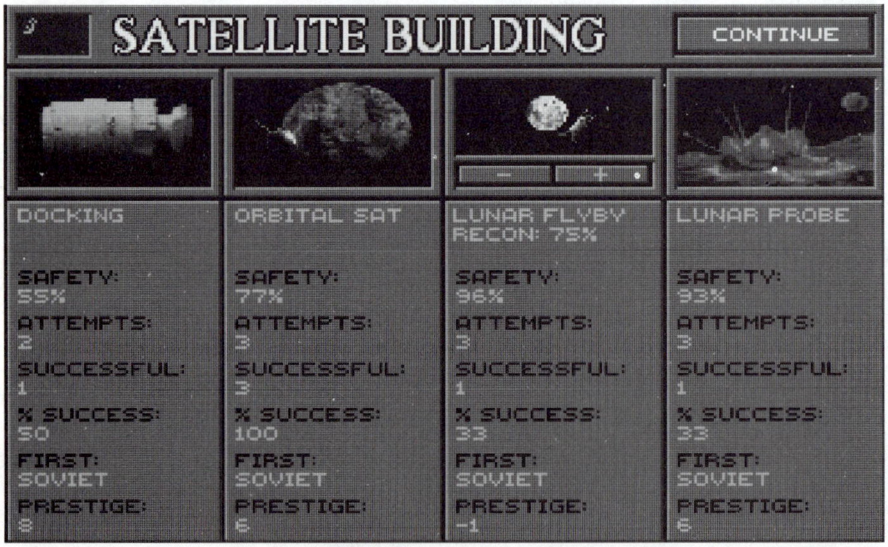

Figure 11-4. Data shown in Satellite building

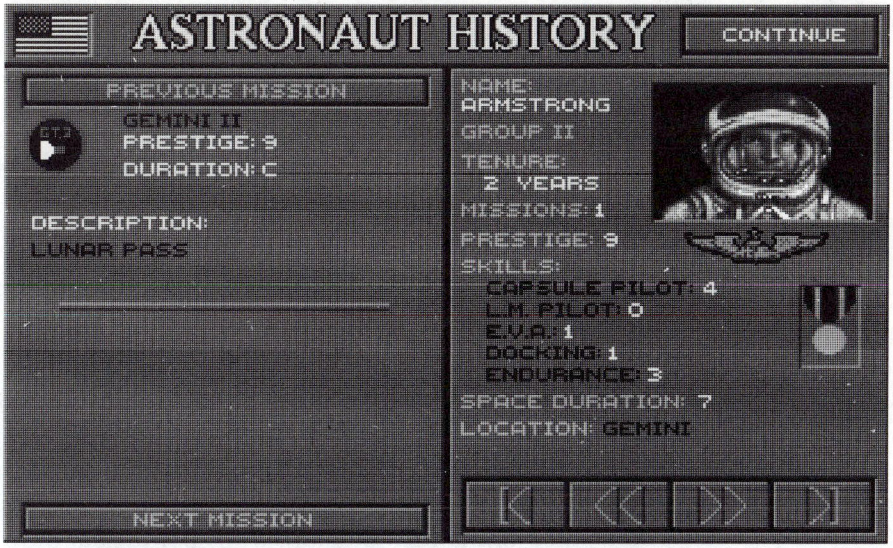

Figure 11-5. The Astronaut History screen showing individual accomplishments

have medals while others do not. A medal is handed out when the astronaut is the first to complete a milestone mission or when the astronaut completes four successful space flights.

Gathering Intelligence Data

Knowledge about what your opponent is up to is part of your responsibility as the space program manager. This information can help you make many critical decisions, such as whether to schedule risky missions or play it safe. A close assessment of your opponent's capabilities is tremendously helpful in planning your strategy and judging your own capabilities.

The most obvious source for intelligence information is, of course, the Pentagon and KGB buildings. There, the CIA (or KGB) reports on the possible missions or space hardware development your opponent may be planning. The CIA (or KGB) also informs you of what it suspects are the programs being researched and tested by the opponent. From the Intelligence Statistics and Intelligence Briefing screens within the Pentagon and KGB buildings, you can get a comparison of the opponent's safety factors (represented with bar graphs) and mission plans for the different hardware to that of your own. Safety factors for the various rockets are shown in Figure 11-6.

Assess the information in the Intelligence screens and the Prestige Summary screen found in the Museum, and you'll be able to figure out the method your opponent is planning to use to beat you to the Moon. If you see your opponent complete a lunar orbital mission with an LM test, chances are he's ready to land on the Moon in the next couple of seasons. Or if you see your opponent complete a first EVA with docking tests, that probably means he's going to the Moon using the methods that include docking.

Check your opponent's budget in the Administration building. This will help you determine what is feasible for him to acquire and what type of missions he can afford. It will also give you an indication of how he's faring in his government and before public eyes.

Study the "space firsts" in the Prestige Summary screen in the Space Museum. Keeping track of your opponent's prestige points will give you a clear indication of his success or failure the previous turn. If his prestige point total drops three points or more, he probably had a failed unmanned mission. If his point total drops 10 points, he probably had a serious failure on a manned mission. The Prestige Summary screen is shown in Figure 11-7.

The Director Ranking screen in the Space Museum can also tell a story. For example, if your opponent's performance or ranking has been flat for several

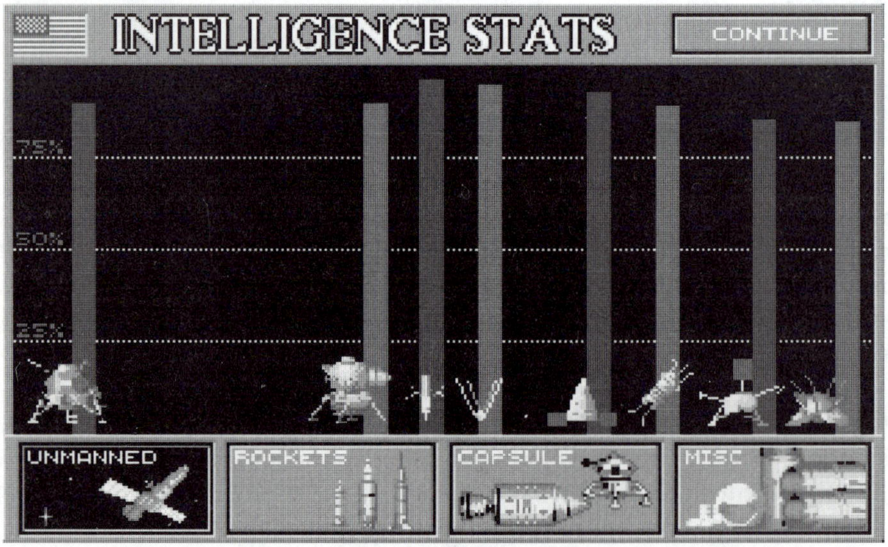

Figure 11-6. The Intelligence Statistics screen comparing safety levels on space programs

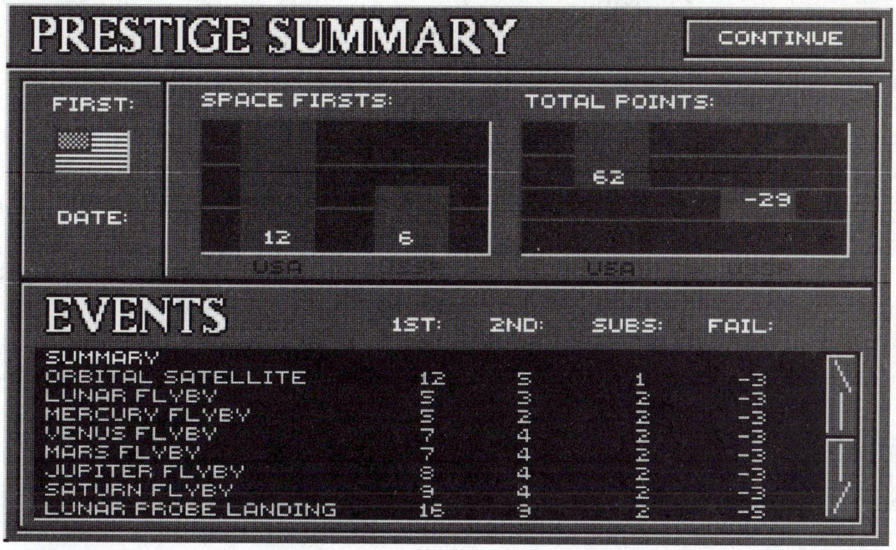

Figure 11-7. A useful tool: comparing and tracking prestige points in the Prestige Summary screen

seasons, it is very likely that he's going for the Direct Ascent. Since his ranking has not decreased (failed missions), or increased (successful missions), you can be fairly certain that the president knows that he's spending the money on purchasing and R&D'ing this costly method.

All of the data you gather will be important to understanding your opponent's capabilities and strategies. Use it to your advantage by planning your missions appropriately to win prestige points. You should be able to determine what some of your opponent's successes or failures are from these reports. Do you want to compete on getting a first? Or maybe you'd prefer to go for other firsts if your adversary has a definite advantage. Should you rush your mission? Careful examination of all available information regarding your opponent will help you answer these questions wisely.

Hopefully in the end, all the intelligence information will prompt you to adjust your strategy so that you can ultimately beat your opponent to the Moon.

Avoiding Potholes

Throughout "Buzz Aldrin's Race Into Space," there are a few precarious potholes you don't want to fall into. Of course, any pothole you can fall into is not good and should be avoided. Two of these involving astronauts are described here.

Astronauts in "Buzz Aldrin's Race Into Space" have some very human traits. If you have a Mercury and Gemini program operating simultaneously, the astronauts in the Mercury program will eventually grow dissatisfied and lose morale more quickly because they feel like their program is obsolete. This drop in morale begins two years after the more advanced program has been purchased. Naturally, they want to be assigned to the more advanced program. If you want to keep some of the Mercury pilots, make sure to break them up and reassign them to the new program before too long.

Pulling an astronaut out of any advanced training program early will cost three megabucks, and the astronaut will not have improved on the skill which you sent him to train for. Normally, upon completion of a two-year advanced training program, the astronaut stands to gain one or two skill points in the area of training.

Game Replay

"Buzz Aldrin's Race Into Space" has several successful approaches for achieving victory in the game. Along with these long-term game winning approaches, there are numerous short-term methods that lead to the final victory lap. There are over 12,000 different types of space missions in the game that can be scheduled with over 50,000,000 possible failure mode consequences. The event cards add another layer of variation so that no two games will ever be the same. You have a better chance of winning the lottery twice!

Another challenge for the more advanced player is breaking the space records in the Space Museum. There are 53 different categories of space records to compete in. Your initial installed game is supplied with actual historical records in some of the categories. You have the option of deleting these records, or attempting to break them. Records do not take effect until a player (human or computer) lands on the Moon. If you are fired from the game or do not finish the game, any records from the current game will not be included.

Some of the space records are easier to break than others. Can you beat history? See if you can achieve a better score in the Director Ranking screen. Who will have the record for the earliest successful lunar landing? An individual in the play-testing department at Interplay completed a manned lunar landing in May 1964! Can you beat this? Which human will be the first into space? What woman will perform the first EVA? All of these milestones are glorious achievements to be proud of.

CHAPTER 12

Walkthrough of a Mission

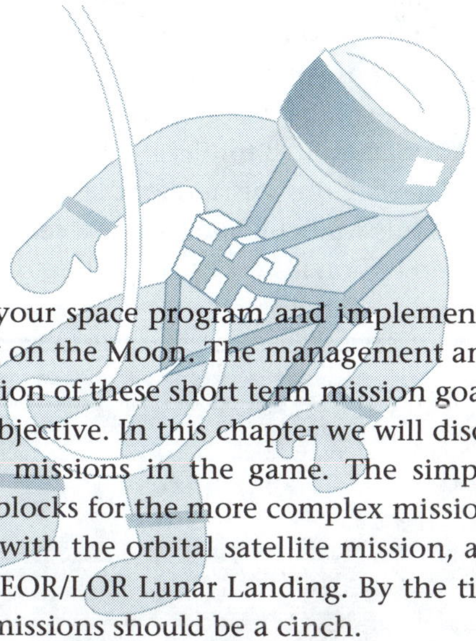

The operation of your space program and implementation of space hardware is crucial for landing on the Moon. The management and planning of future missions and the completion of these short term mission goals are equally critical to achieving the overall objective. In this chapter we will discuss how to plan and execute most of the key missions in the game. The simpler missions are the foundation and building blocks for the more complex missions. Starting from the ground floor, we'll begin with the orbital satellite mission, and build up to more complicated ones like an EOR/LOR Lunar Landing. By the time we're done, even the most complicated of missions should be a cinch.

Anatomy of a Mission

This is a breakdown of the sequence in setting up a mission. By experimenting with these sample missions, you'll easily be able to move on to scheduling and running more complex joint-missions.

The order for scheduling future missions, purchasing programs, and visiting other buildings can be altered to suit your tastes and needs. This is just a recommendation of what areas of the space programs must be visited and implemented.

Orbital Satellite

Launching the Orbital Satellite mission is easy. Starting on the first turn, which takes place in Spring 1957, go to the Administration building screen and then to the Hardware Purchase screen. Purchase either the Explorer or Sputnik satellite program and a rocket (Atlas or A-series) to launch the satellite into space. Leave the Administration building and visit the R&D Complex. Research each program with the maximum of five teams, as shown in Figure 12-1. You can also get to the R&D Complex by clicking on the tunnel button on the left side of the Hardware Purchase screen. This tunnel is also available in the R&D Complex, allowing you to get back to Purchasing. After you have R&D'd the satellite and rocket programs, visit the Administration building again and select Future Missions.

In Future Missions, you'll select a pad to launch from (you start with only one pad in your possession, Pad A), and then be presented with a declarations screen, where you declare your missions. Click the up arrow button to increase the mission type to #1, or just press UP ARROW on the keyboard. Mission #1 is the Orbital Satellite, which is shown in Figure 12-2. If you wish, click the Draw key (or press SPACEBAR) to see the flow of this mission. Move the mouse pointer over each mission bubble, which will display each mission: A for Launch, B for Orbital Insertion Burn, and C for Hardware Power-On. To finish and declare this mission for the next turn, just click the Continue button. A-OK! You have a mission scheduled for next season! It's not necessary to view the mission flow; that's an optional step, as you will see later.

Figure 12-1. R&D of a satellite

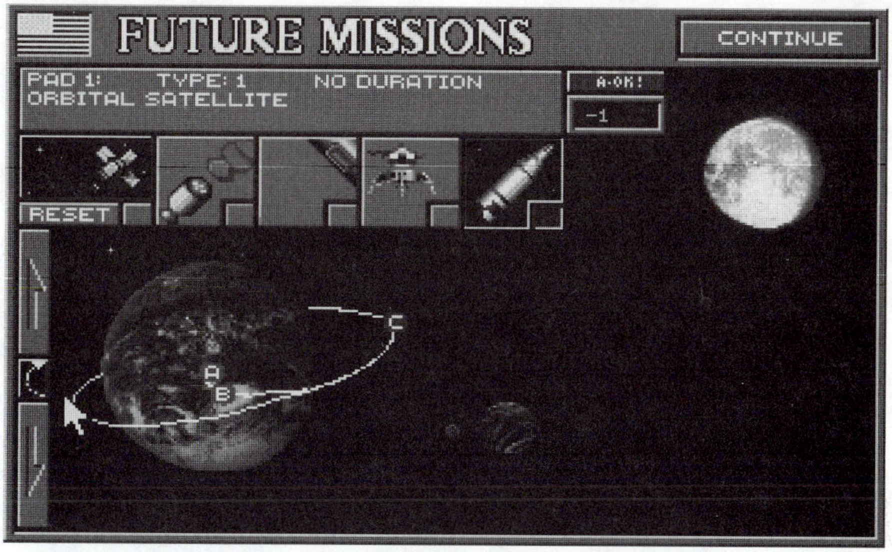

Figure 12-2. In Future Missions, scheduling an Orbital Satellite

Proceed to the next turn, which takes place in Fall 1957. After reviewing Event News in the Spaceport, go to the R&D complex and R&D the rocket and satellite programs.

Next, visit the VAB (Vehicle Assembly building), shown in Figure 12-3, or the VIB (Vehicle Integration building). Select Pad A, and the screen will show the hardware to be assembled, its safety factor, and the number of vehicles bought. For this mission, everything is already set properly; just click the Continue button. Your vehicle (payload and rocket) is assembled! Take note of the weight of your payload and the payload lift capability of your rocket in the upper left portion of the screen. On more complex missions, this ratio of weight to lift capability can be critical.

At this point in the mission, the exact launch date for this season will not be set yet. Since it is possible to rush a mission, you can choose to launch as early as September, or you can wait until October or November. Proceed to Mission Control, where a prompt will ask you for a date. For now, leave the "Nov" button pressed so that the launch will take place in November, and just leave the Downgrade button alone. (In this case it would be ineffective; it's impossible to downgrade from this mission. We'll get into the use of the Downgrade button later in this chapter.)

Leave the Mission Control building. The mission is now ready to launch! The mission will be launched after you finish your turn, so you can declare additional missions, buy other systems, view the museum, and so on, before the mission is launched. Once the flag is lowered, however, your turn will be over.

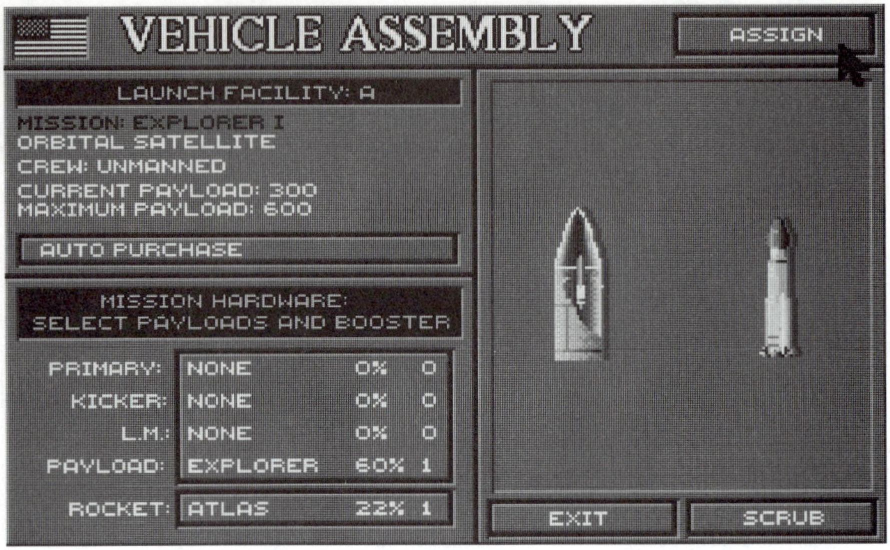

Figure 12-3. VAB, assembly of components

> **NOTE:** Just before every launch, you are presented with the opportunity to scrub a launch. While this is advantageous if you lose your nerve, the intent of this option is to allow scrubbing after a catastrophic failure that kills astronauts, or after the safety factor of a defective space hardware program is lowered. This is of no concern in this satellite mission, so click Continue and monitor the launch!

Lunar Fly-by

This mission is almost identical to the Orbital Satellite mission. The following discussion assumes that you have played the game long enough for you to have purchased a Ranger or Luna program and a Titan or Proton to launch the satellite with. Additionally, it is assumed that you have R&D'd each of these programs to an acceptable level, but we won't worry about that here. (By the way, you should do these things sometime between 1962 and 1964 if you're playing a competitive game.)

In the Future Missions screen, there are two ways to get to the Lunar Fly-by mission, which is shown in Figure 12-4. Since the Lunar Fly-by is mission #7, one way is to click the up arrow button (or press UP ARROW on the keyboard) seven times.

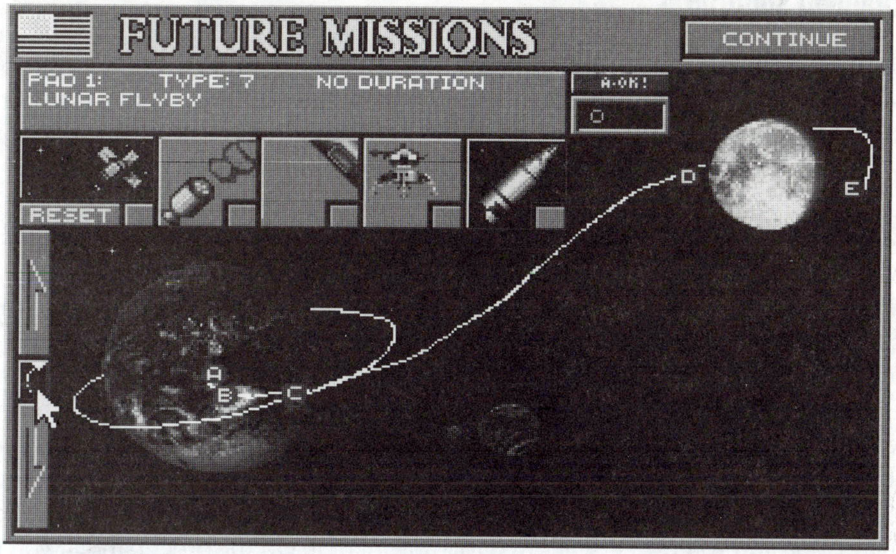

Figure 12-4. Lunar Fly-by

Another is to lock the Manned/Unmanned selector key on Unmanned. To do this, find the key that corresponds to the Manned/Unmanned selector. (It looks like a satellite when on Unmanned.) Clicking the selector once will show an Astronaut's helmet (indicating Manned) and one pie slice out of six possible (indicating the shortest duration level, Duration A). Click it again, and it will show Duration B. Continue clicking it until the satellite reappears, indicating Unmanned missions. Click the small red box in the lower-right corner of this button; this locks the Unmanned selector into place. Henceforth, the up and down arrow buttons will scroll through *only* those missions that are unmanned. As it happens, clicking the up arrow button four times will do it, past Orbital Satellite, Unmanned Suborbital, Unmanned Orbital, to Lunar Fly-by. With that done, examine the mission path, or just click Continue to get out.

The next turn, the VAB will be used as before. Again, the hardware should already be appropriate for the mission. Click the Auto-Purchase button if there isn't sufficient hardware to fly the mission; it will buy what is needed. Then leave the VAB.

Visit Mission Control as before to assign the launch date and end your turn by figuratively lowering the flag.

Unmanned Suborbital

This mission is only a little more complicated than Orbital Satellite. For the sake of discussion, we'll assume you have a Gemini capsule with an Atlas and strap-on boosters to launch it with. This will also illustrate the use of strap-ons.

In Future Missions, the Unmanned Suborbital is mission #3. Select it any way you wish; it's an unmanned mission. When you click Continue, however, you'll be asked what type of capsule is to be launched on this mission. Capsule programs you haven't yet bought won't be accessible; their buttons will be unavailable. Click the Gemini button; this is the capsule you'll be launching. You'll then be ejected from Future Missions (you pressed Continue, remember?), so if you made any mistakes, or want to see the mission flow, you'll need to reassign the mission and declare it again.

In the VAB, you'll need to do something additional because of the strap-on program. The technicians will have already prepared the Gemini capsule for you, but they'll need instructions to decide when to use strap-ons. The VAB technicians will suggest using the Titan program. If you haven't yet bought this program, it will be obvious from the red display color, and you won't be allowed to implement it. On the other hand, if you have already purchased the Titan program, but it's not yet safe to fly, it's very easy to allocate it for use. Nothing will warn you otherwise. Be careful of this trap!

To change the rocket so that the Atlas with strap-ons is used instead of the Titan, click the Rockets button (the one that should now say Titan at the bottom left of the screen). This will toggle through the rocket combinations. Leave it on "Atlas w/B," which stands for Atlas with boosters (shown in Figure 12-5). The safety factor represents the average of the two systems' (rocket and booster strap-ons) safety factors, and the number available represents the number of complete Atlas and booster sets available. If you have an Atlas but no strap-ons, the number available will be zero. Click the Auto-Purchase button if necessary, and exit.

Manned Suborbital

The procedure for this mission is exactly the same as for the Unmanned Suborbital, except that it'll be necessary to get our astronauts or cosmonauts ready before they can fly this mission. Let's briefly go over how that's done.

Astronauts are bought in the Administration building when available, which can be as early as Spring 1958. To recruit them, click the Astronauts icon in the Administration building screen, and choose which astronauts to recruit. Highlight a name on the right, and click the Recruit button (shown in Figure 12-6). If you recruit any accidentally, you can Dismiss them by selecting their names on the left, and pressing the appropriate button. The top right of the screen will inform you how many astronauts you must recruit. Group I is a total of seven Astronauts. When all astronauts have been recruited, click Continue.

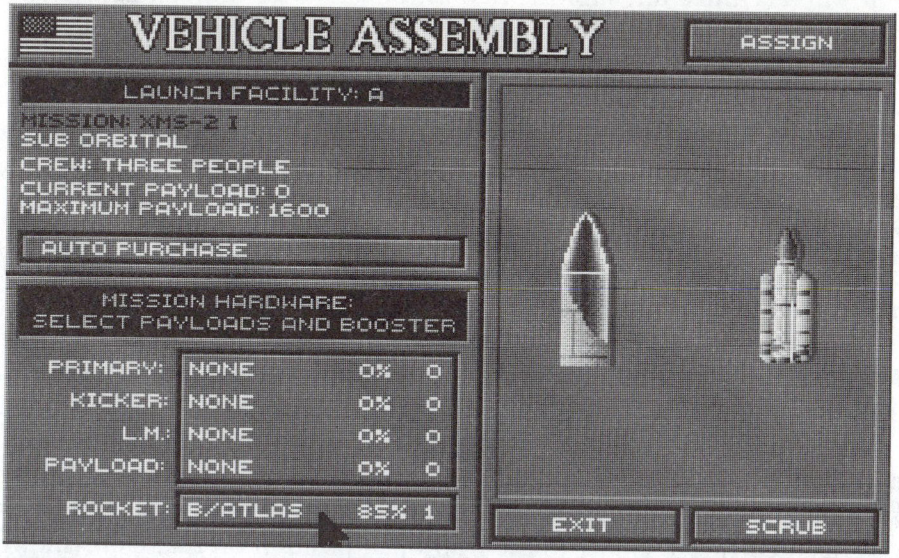

Figure 12-5. Atlas with boosters in VAB

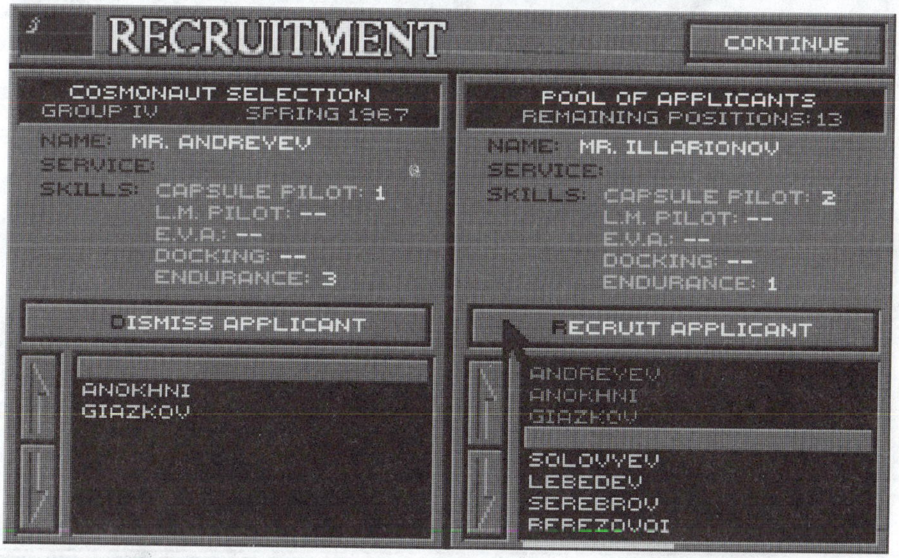

Figure 12-6. Cosmonaut recruiting

It will take three turns for the new astronauts to go through Basic Training, so you'll probably want to remove some of them early. To do that, go to the Basic Training building, highlight the appropriate name, and click the Withdraw From Training button. This decision is irrevocable, so you'll be asked to confirm it. For the first group, pulling four astronauts out early is a good idea. Once that's done, they'll once again be available in the Astronaut complex.

Go to the Astronaut Complex. A list of all the non-allocated astronauts will be shown. Since astronauts get demoralized when just sitting around, it's best to assign them to programs, or at least send them to advanced training in the Astronaut Complex screen, shown in Figure 12-7.

Make sure that the Assign To button is pressed (not the Visit Facility button, which would send *you* and not the highlighted astronaut to a facility). Then, click the Gemini program four times. This sends all four astronauts to that program.

Visit the Gemini program by clicking the Visit Facility button, and then the Gemini button. Once inside the Gemini Program building, you can assign crews by pressing the flight crew number, then pressing the Assign Crew button to assign the highlighted crew member to the next open position on that crew (shown in Figure 12-8). Click the highlighted name to examine an astronaut's skills to determine if the astronaut is suitable for the crew position. Assign all four astronauts to positions,

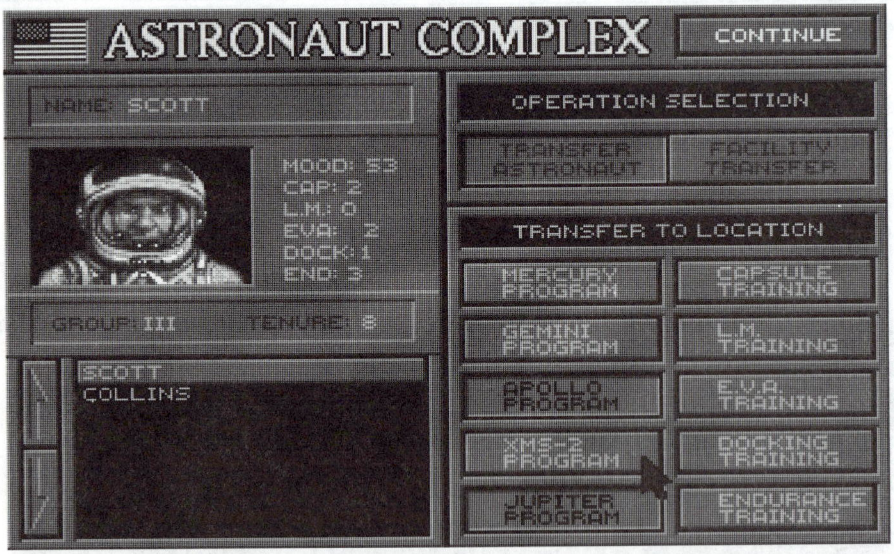

Figure 12-7. Astronaut Complex; waiting for assignment

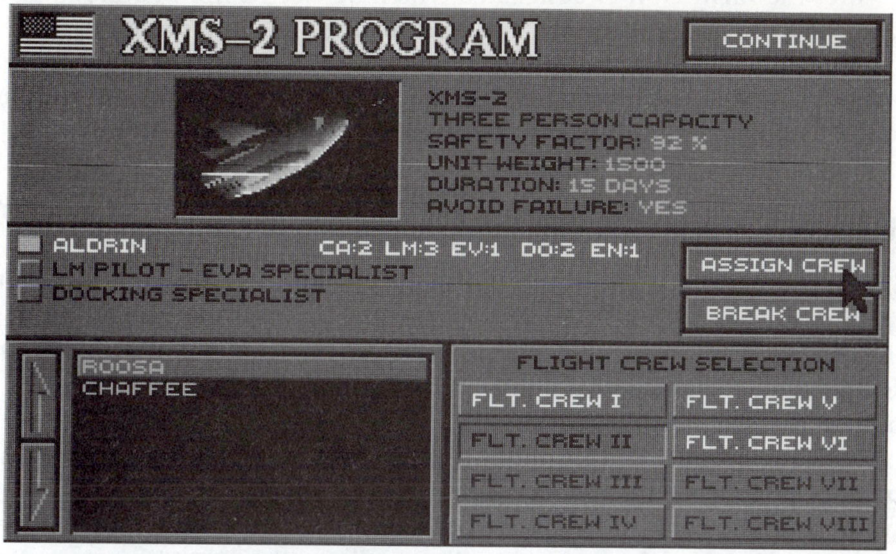

Figure 12-8. Assigning astronauts to flight crews

then click Continue to leave. The crew is now assigned to the program! Another way to visit the Gemini Program building or any other manned spacecraft program is by clicking on the specific program building while in the Spaceport screen.

Even though your astronauts are assigned to flight crews, they are not yet ready to fly. It'll take another turn (one season) for them to familiarize themselves with the systems aboard their new capsule, and receive orientation with their crewmates (if its a more complex multimanned spacecraft). At the end of that time, they can be used as crews—in Future Missions. That means it takes a full year from the time astronauts are sent into their capsule program and are assigned as flight crews, for them to qualify to fly a mission. So plan ahead!

In Future Missions, the mission declaration will proceed almost exactly the same as for Unmanned Suborbital above. However, this is a *manned* mission (Duration A), so the lockout keys will behave accordingly.

After the screen which asks you which capsule will be used, you'll be asked which flight crews will train as the primary and backup. Designate one crew as the primary; who will fly the mission unless they get sick. Designate a second crew as the backup; they will be unavailable for other missions as they're training for this specific mission. This backup crew will not fly the mission unless the primary crew is incapacitated in some way.

After the flight crew is designated, you'll be ejected from Future Missions. Proceed with the mission as usual. Your first manned mission is ready to fly!

The more advanced missions are the orbital, missions that have increased duration, and EVAs. The procedure is about the same for these missions, although hardware requirements might vary somewhat.

Joint Manned Earth Orbital, Docking, EVA, and Duration

Looks like a mouthful, doesn't it? If you can do this mission, you can accomplish anything. Let's get it out of the way now, since you'll need to accomplish it fairly soon. This mission was historically performed by the *Gemini 8, 9, 10, 11,* and *12* missions. It was instrumental in building up our docking experience, so it's important that we learn how to perform the same task!

For starters, you'll need a multimanned capsule, since a Mercury or Vostok cannot dock. You'll need astronauts or cosmonauts trained in that program as flight crews (at least two flight crews), and you'll need a rocket to launch it. You'll also need the Docking Module program, and something to launch it into space. You'll need a second launchpad, too.

Visit Future Missions, and declare this mission by locking down the Manned key with an appropriate duration. Lock the Docking, EVA, and Joint keys on. Lock the LM key off. Scroll up (there aren't many missions that meet these requirements) and go to number 33, the Joint Manned Earth Orbital Docking EVA Duration (shown in Figure 12-9).

After you've looked at the mission flow (if you desire), click Continue. You'll be asked which capsule and which flight crews will be flown. Designate them the usual way and end the turn.

The next turn, you'll need to go to the VAB and assemble hardware on *each* pad, since it's a joint launch. In joint launches, the unmanned portion is always launched before the manned portion. On Pad A, the Primary Mission Part will be a docking module launched on an Atlas or A-series rocket, which is very similar to a satellite launch assemble. Click Assign and then Continue, and assemble the Secondary Mission Part on Pad B. This should be the assembly of the rocket and spacecraft which includes the crew. Click Assign.

Proceeding on to Mission Control, we find that even though we're using two pads, there's only *one* launch date to set. That's because this is a joint launch. The launches occur with in a few hours, instead of being separate missions. Set the launch date.

Let's do one other thing before we go on; we'll play with the Downgrade button. Click this once, and the mission drops its EVA. This might be useful, for example, if the EVA suits program isn't ready. Click again and the duration is dropped, but the EVA returns. Click again and there's no duration and no EVA. And

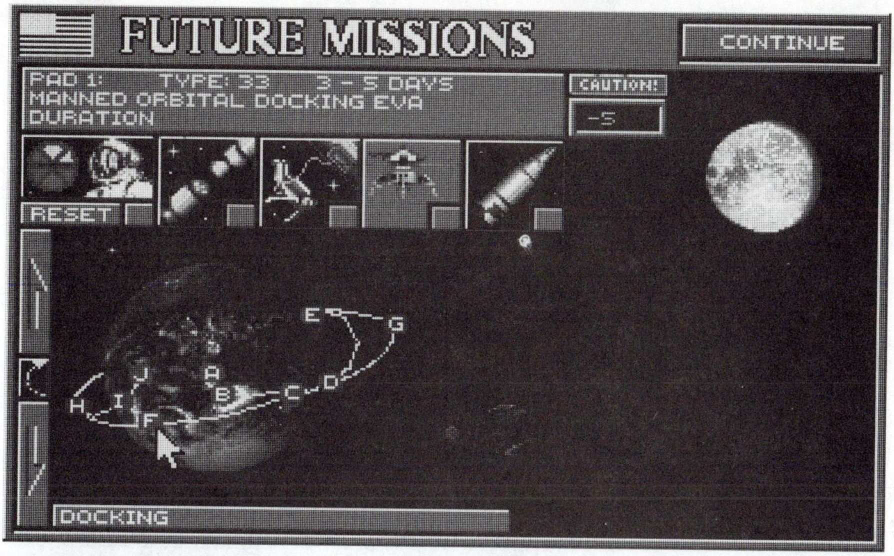

Figure 12-9. Mission 33, Joint Manned Earth Orbital, Docking, EVA, and Duration

so on. Eventually, you'll be returned back to where you started—with a Joint Manned Earth Orbital Docking EVA Duration. Let's stay with that one.

When you launch the mission, the primary part will launch and orbit, then the secondary part will launch and join it, all in the same mission! A failure of either part, even the unmanned part, will abort the mission. Unmanned parts won't be launched, and manned parts will come home by the most expeditious way. Historically, this happened when *Gemini 9* found that their Agena docking target vehicle had failed its power-on step.

Manned Lunar Pass

After the last mission, this may seem easy. For purposes of this discussion, we'll assume that a Gemini with Kicker B is being used for this mission. If an Apollo or Soyuz is used instead, no kicker would be required.

In Future Missions, the fast and easy way to do this is to lock in a C duration, then click the down arrow button (or press DOWN ARROW on the keyboard). This scrolls backwards through the mission list, starting with the highest mission number for which a duration C is acceptable. This is number 43, the Manned Lunar Pass (shown in Figure 12-10).

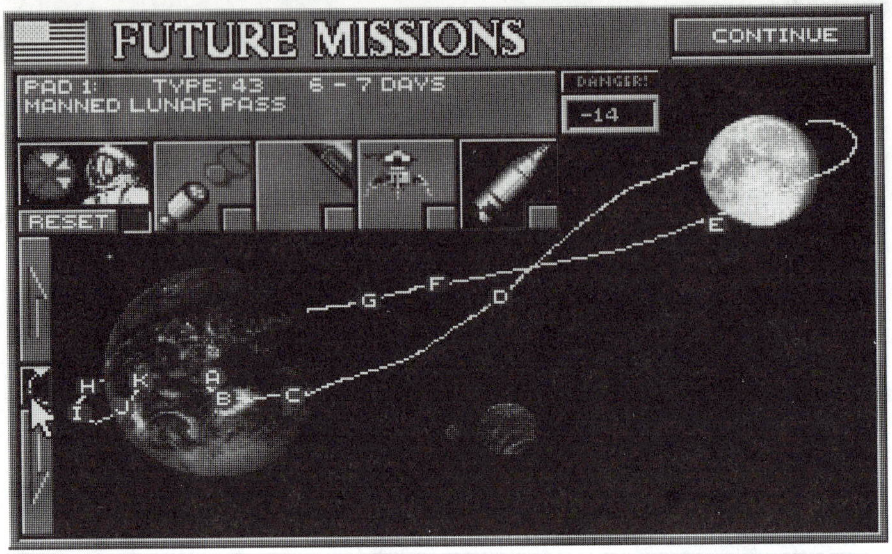

Figure 12-10. Manned Lunar Pass

On the next turn, visit the VAB. As usual, the technicians need to be instructed as to what rocket program is most suitable, so you may want to change it from "Saturn V" to "Titan w/B," if your program warrants it. You already know how to do this from the Manned Suborbital discussion.

The technicians will also need to be told what kicker program is best. It will select the Kicker A by default, but if the Kicker B is the one you need to use, you'll have to select it. That's easy: just click the large button with the mission hardware on it, shown in Figure 12-11. It's the one just above the rocket button you just used, and it works in a similar manner: it toggles through the allowable combinations. In this case, there are only two combinations: the Gemini with Kicker A, and the Gemini with Kicker B. Choose whichever is more appropriate. Finally, choose the launch date as always and monitor the launch!

Historical Manned Lunar Landing

The big one! The Earth Orbital and Lunar Orbital LM tests are very, very similar, by the way.

The Lunar Landing mission is pretty easy to schedule. In Future Missions, just scroll backwards until you come to this mission, or use the lockout keys. Click Continue, and select the capsule and crew. Note that a "Historical" Manned Lunar

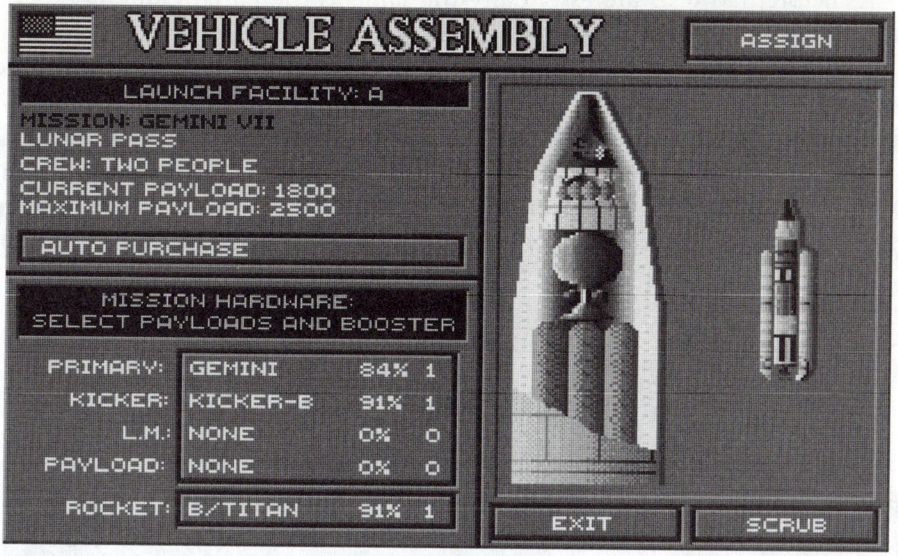

Figure 12-11. Gemini with Kicker B

Landing can be performed with a Gemini, Voskhod, Apollo, Soyuz, or even a Minishuttle.

VAB assembly proceeds the same as always. You may have to click the rocket buttons to get the correct rocket, or the hardware button to get the right combination of kicker and LM. Remember, your VAB crews may be superb at assembling rockets, but they don't have the big picture. They don't know whether the mission needs a one- or two-person LM, and they're liable to make the wrong decision. You'll have to check to make sure.

Set a launch date in Mission Control as you've always done, then sit back and enjoy the flight! If this is successful, you've won the game!

Joint EOR/LOR Manned Lunar Landing

These are sort of a cross between the joint docking mission discussed previously, and the historical landing described immediately above. This mission is used when a rocket as powerful as the Saturn V or N-1 is not available. A Titan or Proton with boosters can launch the same hardware (capsule, LM, and kicker or docking module if required), but it has to do it in *two* launches, which then rendezvous in Earth orbit and proceed to the Moon in the LOR mode we're used to.

As before, LM tests are essentially the same as the Lunar Landing described here, just with a different mission number.

Lock down the Joint Missions button, then scroll through the possible choices. Pick one, click Continue, and choose the capsule and crew. In the VAB, choose the proper rocket and hardware as always, for *each* mission part on *each* launchpad. Set the single launch date in Mission Control. Then, sit forward and watch nervously. Joint missions are notoriously prone to failure at some point. But if you're successful, you've won the game!

CHAPTER 13

Winning Strategies

In this simulation, there are numerous successful approaches for a manned landing on the Moon. All of these methods do indeed work, and depending on the circumstances, some are more reliable than others. Whether the historical approach is better than EOR is up to you to determine. You must decide when to initiate a specific approach. Each of the strategies presented here is an alternative philosophy on selecting space hardware and methods of how to win the game. Here are some my favorite methods; they are the recommendations from space veterans of both the board game, "LIFTOFF!," and the computer game, "Buzz Aldrin's Race Into Space."

Simply Put, Direct Ascent, by the Rocketman

Maybe those old sci-fi flicks had the right idea. Imagine a powerful, sleek rocket soaring to the Moon and landing upright on the lunar surface. The "spaceman" climbs down the extended ladder and collects some lunar rocks and after a brief stay, gets back into his rocket-ship and fires up the engine for the journey home.

In the movies, Direct Ascent worked. In real life, it would have required some technological hurdles. Likewise, if you attempt it in the game you'll need a healthy budget, good R&D, and some luck. But with some early successful missions, this approach can work.

Direct Ascent is the single launch of the megarocket (either the American Nova or the Soviet Vulkan) and the Lunar Landing of either the American Jupiter or the Soviet Kvartet spacecraft. After a walk on the Moon, the crew returns directly to the

Earth. This giant rocket and spacecraft are both very expensive programs and can take several years of costly R&D to implement.

The advantage of Direct Ascent is that it is not complicated. It's a good plan for the inexperienced player because it is straightforward and doesn't require concurrent development. You don't need to worry about docking technology or Lunar Modules or performing any LM tests. If it works, the baby just goes straight to the Moon.

Early Goals

To ensure the best conditions for Direct Ascent, you should accomplish some preliminary milestones. First, go for the gusto and get some early prestige points by attempting the first Orbital Satellite. Maybe take a risk and rush the launch a month. Because the Orbital Satellite is an unmanned mission, rushing the launch will not cause a catastrophic setback. In 1958, purchase and begin R&Ding the Mercury/Vostok, shown in Figure 13-1. Recruit astronauts and put a primary and backup crew immediately into the capsule program. When that one-man capsule program is greater than 55%, announce a manned mission for the next season. Schedule a Manned Orbital mission. Yes, it's risky, but if it's successful, you're sure to take an early lead.

Figure 13-1. Research and development of the Soviet Vostok

Hopefully you did not suffer any event setbacks and were successful in getting the first man into space. So far, with the space milestone steps that you've completed (Orbital Satellite and Manned Orbital), you still have nearly all of the lunar approaches available for going to the Moon. This is the point that you decide whether or not you want to pursue the Direct Ascent option. The conditions you should take into consideration are whether your budget is healthy (85 megabucks or greater) and whether you got the first satellite and man into space. You should also check the Prestige Firsts screen in the Museum to get an idea of what successes or failures your opponent has had. If your opponent has had a manned failure and you have the lead, then it's time for Direct Ascent.

Schedule one more one-man capsule flight for a Duration B (a 3-5 day mission) and an EVA. Don't worry about earning a Milestone First for the Duration B because the one-man capsule will already be obsolete when you have completed this mission. In level 2 games against the computer, try to get a second launch facility (when the funds are available) so that Orbital Satellites can be launched. (Figure 13-2 shows the screen where you can purchase your second and third launch facilities.) That way you'll keep gaining prestige points to maintain your job security. (It's embarrassing to get fired.) When playing level 2 or 3 against the computer, always try to have an Orbital Satellite launched every season; your prestige and budget will always increase.

By 1961, purchase and develop the Interplanetary Satellite and the Titan/Proton rocket program. Start sending Lunar Fly-bys (shown in Figure 13-3) as soon as

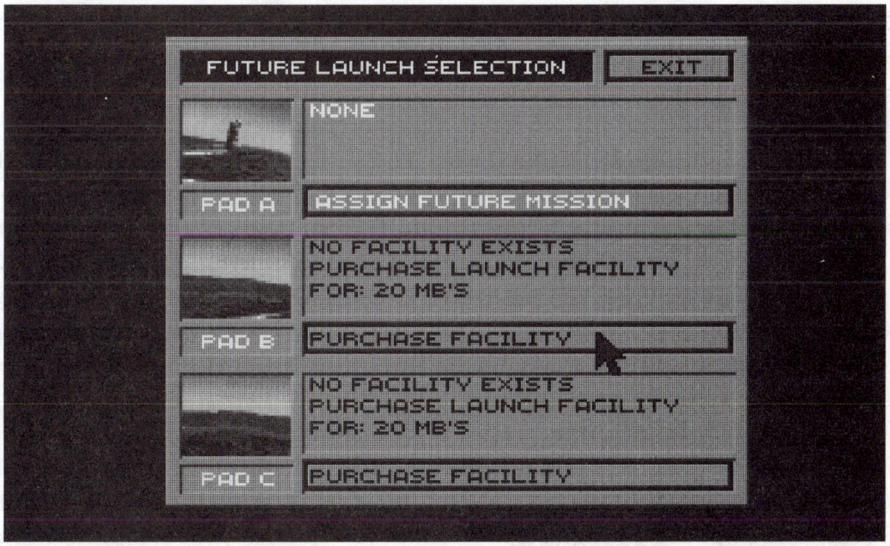

Figure 13-2. Purchasing a second launch facility

Figure 13-3. In real life, the Ranger satellite failed on its first six attempts!

the technology (the safety factor) is near the maximum. It may take several attempts before a Lunar Fly-by is successful, but it's an important milestone step (some Photo-Recon doesn't hurt either) for future manned lunar missions. As soon as the funds are available, purchase Group Two astronauts and when available, Group Three astronauts. You'll need lots of fodder for the cannon (Nova/Vulkan). This plan requires fewer overall manned launches, which will obviously disappoint a lot of astronauts, so it's a good idea to have plenty of volunteers.

By 1962, '63, or '64, begin the Lunar Probe, shown in Figure 13-4, another important milestone step. By this time, you should have scrimped and saved enough funds to buy the Jupiter/Kvartet Module. From now on, every season always give it some R&D funding, even if it's only one team of scientists. It's an expensive program and will probably take four to five years to develop. Keep assigning flight crews, even though a lot of the astronaut corps will probably get frustrated and retire. Better to have flight crews ready and available for when that big bird (Jupiter/Kvartet) is finally ready to fly.

In the mid-1960s, the only missions you should be scheduling are Orbital Satellites, Lunar Fly-bys, and Lunar Probes. Keep launching Lunar Probes until at least one of them is successful. It's another milestone step that you don't want to skip. (You'll be skipping enough steps later!) This will raise your Photo-Reconnaissance to a safer level and gain you additional prestige points.

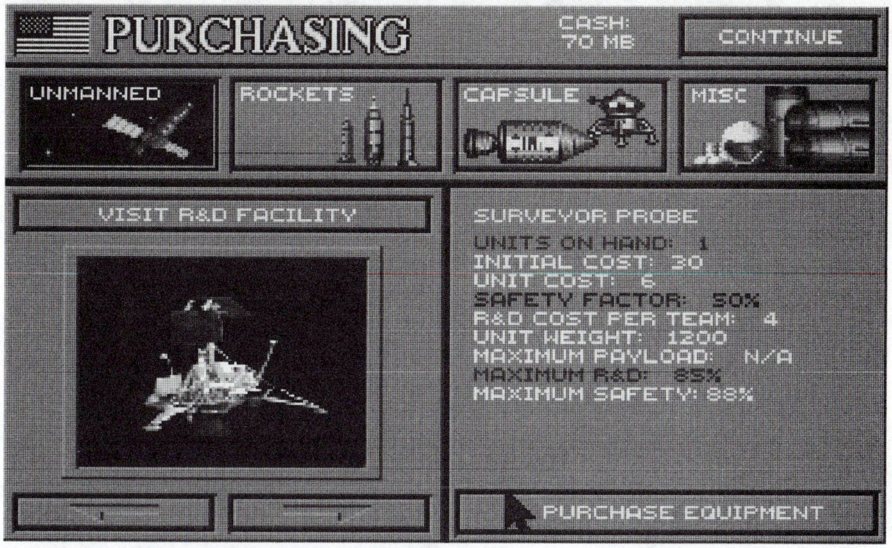

Figure 13-4. The Lunar Probe, Surveyor

The challenge now is to save money so that by 1967 or '68 you can afford the big gun, the Nova/Vulkan. Since your Titan/Proton has a high reliability, you'll get a break on the giant rocket safety factor. (When you purchase your initial rocket or capsule programs, you are given a technology bonus on other, more advanced programs, if the initial safety factor is 75% or greater. So if the Titan/Proton program has a safety factor greater than 75%, the safety factor of the Nova/Vulkan will start at a higher, more reliable percentage.) It should only take two years to "man-rate" the cannon.

Now you should be ready for your first four-man mission. Each Jupiter/Kvartet flight is going to cost at least 60 MBs, so be sure to have R&D'd it to the maximum safety factor since it's too expensive to let it fail. A catastrophic failure would cause an additional two-year delay in your program. Depending on the progress of your opponent, you can pursue a risky plan or a conservative approach with your missions. If you assess that your opponent is a few years away from a Lunar Landing (he has completed no LM Tests or Lunar Pass yet), then you can gradually build up your manned Jupiter/Kvartet duration missions. Hopefully you achieved a Duration B during the one-man capsule program several years earlier. That way you can begin the Jupiter/Kvartet with a Manned Orbital C Duration, shown in Figure 13-5.

If your opponent is threatening to reach the Moon in a year or less, I recommend that you let them take the first risk and maybe they'll botch the

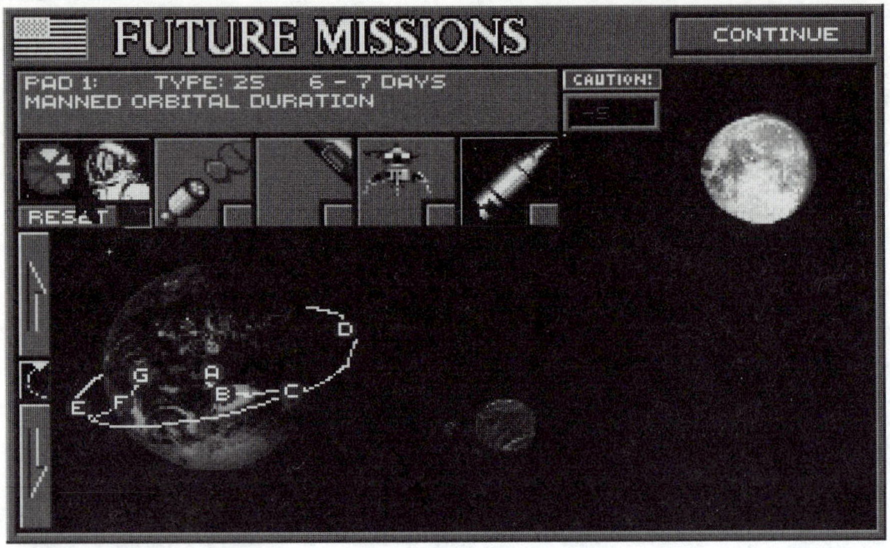

Figure 13-5. Manned Orbital C Duration

mission. (In any case, don't skip two or more milestone steps; to do so is foolhardy.) Follow the methodical approach. After the C Duration, do a Lunar Pass and then the Lunar Orbital. The next mission is the big one and this should happen between 1968 and 1971. Direct Ascent can work and win the game, but it depends on the circumstances of the game and a little luck. Destination Moon here we come! Good luck Commander!

The Glorious One-Cosmonaut-on-the-Moon Approach, by Comrade Kawato

The beauty of "Buzz Aldrin's Race into Space" is that there are so many ways to reach the Moon. An inexperienced player might think that the "Americanski" historical approach of Mercury/Gemini to Apollo is the best way. Well comrade, this couldn't be further from the truth.

My simple method calls for skipping (not purchasing) the one-man capsule and the A-series rocket and going directly to the two-man spacecraft and Proton

rocket. Later on, the one-man LM, Kicker B, and the boosters must be purchased and developed.

Just imagine, comrade, the two-man Voskhod spacecraft launched atop the glorious Moon rocket N-1 lifts off to the Moon. In lunar orbit, one of our great cosmonauts enters the one-man LM and makes a solo flight down to the surface. After the historic walk, our great comrade docks and joins his crewmate in lunar orbit and the pair heads home for a stunning victory!

The advantage of our road to victory over that of the capitalists is simple, comrade: it will cost fewer rubles. Why waste 168 megabucks with three different rocket programs, and another 78 megabucks on three different spacecraft? Our efficient socialist plan only requires 60 megabucks for the Proton rocket, 24 megabucks for the Voskhod spacecraft, and 18 megabucks for the Kicker B. Only a foolish capitalist would fail to the see the cost benefit of using the one-cosmonaut-on-the-Moon approach. The savings of rubles increases even more when you tally up the research and development costs.

Some might argue that skipping the A-series rocket will allow the Americans the first Orbital Satellite. My answer to that objection is, "So what, comrade? Historically the great Soviets got the first satellite, Sputnik, up first, but failed in beating the Americans to the Moon." If we use five-year planning methods instead of quick results, our gains will be far greater and we will be glorious in our overall objective.

We must strive for quality by researching and developing all programs to the maximum levels. Numerous dummy tests must be conducted on the Proton and Voskhod programs, shown in Figure 13-6, until they reach typical Soviet perfection. Only at this zenith of perfection do we send our brothers and sisters to the stars. Caution is recommended by starting first with a Manned Suborbital flight.

Our policy is not to skip any milestone steps and to fly the manned missions with increasing duration until a Duration C has been achieved. While conducting these missions, be sure to include space walks and docking attempts. (These added mission events can be selected in Future Missions. By pressing the appropriate button you can choose Docking, EVA, LM Test, and Joint Launches.) It is also wise to prep the boosters for possible use on these docking missions. (Depending on which game you're playing, Basic or Historical, boosters may be required.) By using boosters on the launch, their reliability will increase. To speed up our docking missions, try to have additional launchpads. Another option is to attempt joint docking missions, because having more launch facilities creates more options for your primary and backup lunar missions.

After our cosmonaut comrades have flown an Earth Orbital mission, the Cosmos Interplanetary Satellite is the next goal. After the Cosmos has at least an 85% safety factor, it is time for Lunar Fly-bys. We should continue these Lunar Fly-bys until the mission has been successful. Try to repeat this several times to increase your Photo-Reconnaissance. (The Photo Reconnaissance Summary screen

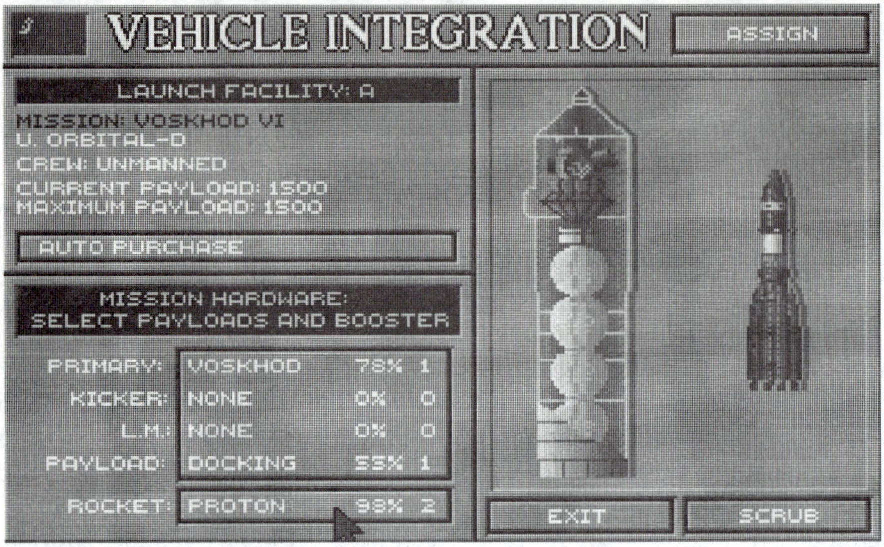

Figure 13-6. A Proton/Voshkod combination in VIB

is shown in Figure 13-7.) Imagine the Americans' reaction when they see the success of our Lunar Fly-bys. It is likely that they will panic and attempt riskier missions. Do not waste precious resources on Cosmos flights (Interplanetary Satellite Missions) to Mars or Venus, as they contribute nothing towards our objective, landing on the Moon.

During these initial Lunar Fly-bys and the continuing Manned Docking Duration missions, the Luna Probe should be finishing R&D. As soon as a Lunar Fly-by has been successful and the Luna Probe is ready, schedule a Lunar Probe Landing. Even with our reliable Soviet technology, this will probably require a couple of attempts.

When the Luna Probe is nearing completion, it will be time to go for the Kicker B and one-man LM. After a successful Lunar Probe landing, the scheduling of a Manned Lunar Pass is the next step. Only a greedy capitalist would be tempted to jump ahead to a Lunar Orbital.

The next phase is LM Tests. First we schedule an Earth Orbital LM Test, followed by the Lunar Orbital LM Test. The Lunar Orbital will increase our Photo-Reconnaissance to acceptable levels for the upcoming Soviet manned landing. A configuration of this space hardware is shown in Figure 13-8.

Some might think our way of reaching the Moon requires nerves of steel, but it also requires patience, especially early in the game. It will take time to develop the glorious Voskhod. Don't be frustrated if these Americans gather a few early

Figure 13-7. Photo-Reconnaissance summary

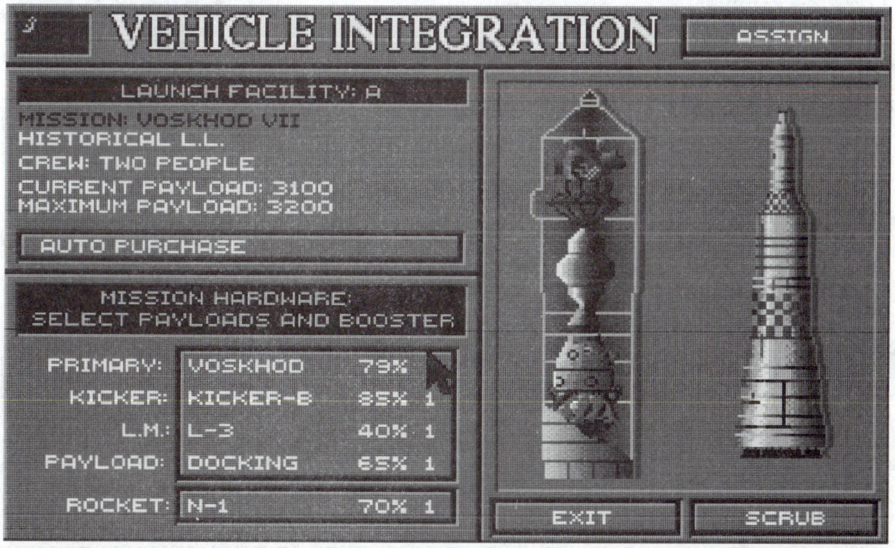

Figure 13-8. The Voskhod "historical" lunar landing hardware in VIB

achievements. For those who cringe at letting the capitalist take the lead, let us all recall the old Russian tale of the hare and the tortoise. The hare took an early lead, but then became complacent which allowed the socialist tortoise to win.

The key to our strategy is traditional conservative Politburo planning. Take no foolish risks. By these methods, we should avoid losing cosmonauts. Let the Americans send their astronauts to Arlington Cemetery.

A toast to our cosmonauts on the Moon!

The Winged Gospel, by Erik "Bucko" Anderson

In the race to the Moon, a breakneck pace will result in great risks and several casualties unless, of course, the XMS-2 or Lapot minishuttle is used.

The minishuttle, shown in Figure 13-9, is the most reliable manned system in the game. It deserves this rating because it is a true, piloted spacecraft, as opposed to vehicles with *Spam-in-a-can,* a term for an astronaut crammed into a small capsule. A space capsule is a converted, man-rated artillery shell. A minishuttle, on the other hand, is a real, human-controlled spacecraft, with a lot of flexibility, reliability, and abort options.

Figure 13-9. The minishuttle on final approach

Of course, the minishuttle comes with a price—a rather hefty one, in fact. As we'll see, however, it pays for itself by the end of the game.

My basic game plan involves launching the Orbital Satellite, then following it up with Mercury/Vostok launches to get a Suborbital, Orbital, EVA, and perhaps a Duration B. That's when the fun begins.

The minishuttle is expensive, so it's best to purchase it early and work on research and developing it a while. In fact if possible, it should be bought while the Mercury is still being R&D'd. When it becomes ready for use, it should be flown continuously, improving the Duration and Docking skill for the middle of the game. Along with development of a heavier rocket, a Kicker B, and an LM, it's the ideal system for a Moon trip.

Sure, you've heard that the minishuttle is expensive, but it does pay for itself. Consider that the budget bonus for flying the first one is 20 MBs, more than anything else except the Jupiter/Kvartet, and it can be flown on an Atlas or A-series with boosters! Additionally, it's cheap to fly. Whereas an Apollo capsule would cost 12 MBs per flight, a minishuttle costs only 2. This adds up to a great savings by the end of the game.

A minishuttle saves money in two other ways. First, its use assures you'll only need a two-person module, so it won't be necessary to build the more expensive (and less reliable) one-person version. Second and above all, the reliability of the minishuttle makes it unnecessary to build a backup system. This is a large expense saved. The reliability of the minishuttle also means that astronauts won't be killed, missions wasted, and budgets busted as a result. This means that the minishuttle's expense will have been more than offset by its benefits when the game is over, and it assures that the game will end with *your* minishuttle orbiting above *your* two-person LM on the Moon! The minishuttle program is located in the aircraft hanger, shown in Figure 13-10.

Other advice? It's probably okay to skip the Lunar Probe mission to increase speed. If that's the only mission you skip during the entire game, it's a good risk to take. Be careful not to take too many other risks, though. Skipping both the Fly-by and the Probe, for example, and then going on to, say, the Manned Lunar Orbital...well, even the legendary reliability of the minishuttle has its limits! Don't push them.

EOR joint missions are probably okay, as the minishuttle's reliability can probably carry off even the most complex of missions. In this way, only a few of the steps are unmanned. There's nothing more frustrating than watching the unmanned portion of a joint mission fail an easy roll, ruining the whole mission and a year of planning.

Of course, if you'd rather have a Saturn/N-1 program to boost the whole mission stack in one launch in historical fashion, don't forget that you'll need strap-ons for the all-up launches!

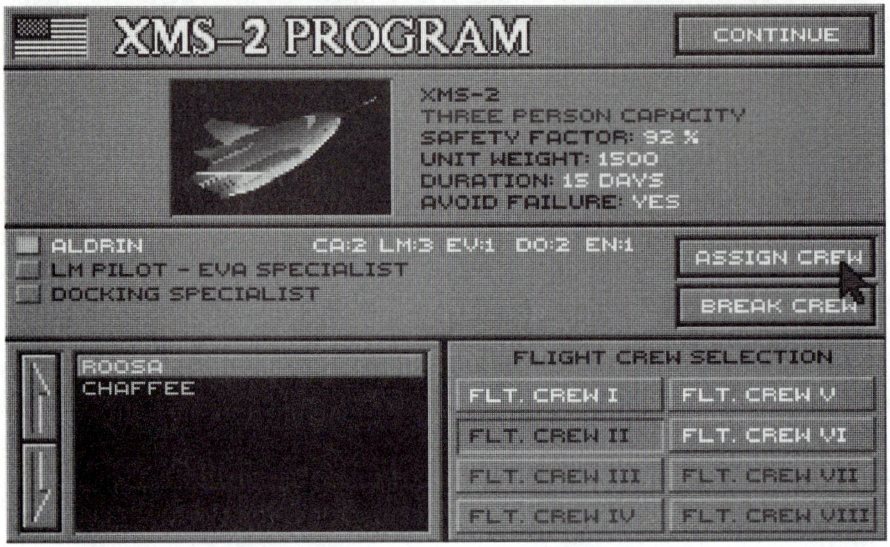

Figure 13-10. Minishuttle astronaut training and flight crews

Skipping the Mercury/Vostok and going straight for the minishuttle is possible, but not recommended. The high up-front cost and the large number of astronauts required mean that, even with an early start, the program will take a long time to get going full-swing.

Doing the Gemini/Voshkod instead of the Mercury/Vostok as a precursor is probably a waste of time and money. It's more expensive, and you'll need that money for the shuttle. Besides, it's got duplicate capabilities (docking, long durations, lunar missions) that you won't want to make use of because the minishuttle is cheaper to operate. So why do it?

Getting the Titan/Proton and later the Interplanetary Satellite worked into the budget somewhere is a trick, but one worth mastering. Ideally, you should finish the minishuttle, followed in a turn or so by the rocket, followed in another turn by the satellite for lunar photography. And all this should come off by '64-65. After that, it's a short trip to the Moon.

Get the Kicker B for the minishuttle program every time. The Kicker A means the mission *has* to be joint, and the unmanned part *has* to fly to lunar orbit before linking up. It's just not likely to work consistently.

In short, this game stands as proof that we did things the wrong way with *Spam-in-a-can*, and that the X-20 (the XMS-1, if you were curious) and other piloted spacecraft were the right way of doing things. Best of luck with them! Keep the shiny side up, and the rubber side down, and keep 'em flying!

Kicker C, The Red October Surprise, by Comrade Wheeler

So, comrade, you wish to know the fastest way to land our cosmonauts on the Moon. Some of my collegues may try to convince you to build huge expensive rockets and fly there directly. Others undoubtedly will argue for a terribly risky Lunar Orbital rendezvous, which will require not one but two dockings a quarter of a million miles from the Rhodinia. I propose a path that has the advantage of simplicity, thriftiness, and timeliness!

Our magnificant Soyuz spacecraft flys to lunar orbit, rendezvous, and docks with the previously launched Kicker C. After docking, the joined craft lands on the lunar surface. After a long climb down the ladder, our cosmonauts become heroes of the universe! The Soyuz then launches from the lunar surface and sends our three brave cosmonauts home.

Begin with the basics, comrade, by purchasing the Sputnik and A-series rocket programs and R&D them to perfection. When the A-series rocket and the Sputnik programs are well advanced, it will be a great benefit to strive for getting a small manned capsule into space as soon as possible to find early those surprises that the cosmos has in store for us. Schedule one-man Suborbitals and Manned Orbitals. Another option is to begin the first manned spacecraft program with the more advanced Voskhod or Soyuz. Be aware of the weight requirements of these payloads. Boosters will need to be purchased and R&D'd.

After successfully completing manned missions with the one-man Vostok or two-man Voskhod spacecraft, it will be time for the three-man Soyuz.

When you begin the three-man Soyuz program use boosters on the A-series rocket. You can launch, test, and refine Soyuz until you are reasonably sure it and our cosmonauts are capable enough to go to the Moon. The cosmonaut pool may not be large enough for many manned flights early on, so in the meantime your technicians can test out systems and try automated docking without risking your pilots.

Ah yes, the Moon comrade! What is needed is the Proton rocket to launch our lunar robot probes, with the Soyuz not far behind that. The Proton, with strap-on boosters, can also put the large Kicker C stage into lunar orbit. Once there, all a Soyuz has to do is to follow, dock with it, and land. (A display of the Soyuz atop the Kicker C is shown in Figure 13-11.) The whole program requires nothing larger than a Proton rocket. Other than a single docking, the mission is almost as simple as Direct Ascent. If the docking fails, then the Soyuz returns without stranding the

Figure 13-11. Soyuz with Kicker C on the lunar surface as a cosmonaut collects soil samples

cosmonauts. If the Kicker C stage fails, then you can either delay the Soyuz launch until another stage is prepared, or orbit the Moon again.

Simple and effective, yes? The cost of this approach is low—you know how to build rockets comrade, like sausages! The key to success is to make the Soyuz capsule as rugged and reliable as possible. The only way to do that is to start the program as soon as you can, so all the inevitable problems are fixed in time. The Kicker only has to work getting to the Moon, and doesn't require this nonsense about testing it a bunch of times because it has all new control systems.

As with all approaches, management skills are vital. A delay in astronaut training could set the program back, as can ill-timed major purchases. Even when you are the first to put a man around the Moon, the Politburo's increase of your funding for the final few steps will be too late except for the purchase of a few extra rockets for future trips. Your bosses want results yesterday yet promise resources tomorrow. Your engineers want resources yesterday and promise results tomorrow. I don't envy you your job.

Glory to Mother Russia!

Your Winning Approaches

There are additional winning approaches, with varied mission milestones and hardware combinations that are not listed in this chapter. I encourage you to try to discover some of these methods. I also would be interested in hearing of your successful exploits and novel approaches. Please send in a written account and description to the address provided in the back of the book for comments on the game. I would need your permission to be able to reprint your winning approach in articles and other publications. This would be a benefit to other players, students, and teachers.

Playing Tips

The road to the Moon is a long and dangerous one. The following paragraphs lead you through a series of tips for avoiding problems you may encounter when playing "Buzz Aldrin's Race Into Space."

Capsule and Rocket Choice

Here are some methods for choosing the right capsules to negotiate the worst potholes:

One favorite tactic is to start with the Gemini/Voshkod capsule, rather than the Mercury/Vostok. This is more expensive. It's just barely affordable with the budget bonus from getting the first Orbital Satellite. While it may be somewhat slower, Gemini/Voshkod features increased reliability (it's not *nearly* as dangerous as the Mercury/Vostok), more capability (first Manned Docking is practically assured), and it even comes with its own budget bonus of 7 to offset the increased likelihood of being the second to put man in space. Best of all, the Gemini/Voshkod is lunar-capable hardware, which means you can either proceed to the Moon on that very same capsule program, or use it as a backup while proceeding with a better capsule.

That, of course, leads into the next point, method. Undoubtedly, in the final phase of the game, the most versatile method for lunar missions employs an Apollo/Soyuz spacecraft, using the Gemini/Voshkod as a backup. This way, in case of catastrophic failure, each program keeps the safety factor of the other from falling too far, and serves as a backup for the other while the cripled program is R&D'd

back up to health. Of course, this requires the use of the more expensive one-man module (Cricket or L3). The one-man LM is worth it, though.

As for rockets, purchase the strap-ons, shown in Figure 13-12, as soon as the one-stage rocket (the Atlas or A-series) is up to a 75% safety factor. You'll need these strap-ons in order to lift the Gemini/Voshkod spacecraft into space. Getting the Titan or Proton rocket is difficult because of the initial expense, but is worth it since it allows Fly-by and Probe missions, as well as Manned Lunar Pass and Manned Lunar Orbital missions.

Having the Titan/Proton in your inventory is an asset. Besides all of the space mission options, a high safety factor will give you a bonus when you purchase the Saturn/N-1. The Saturn/N-1 safety will start as high as 60% and can be R&D'd up in a single turn. That means you can afford to wait to purchase the big rocket until last. Just remember that you'll need the Saturn/N-1 for LM tests, too. The Nova/Vulkan programs are to be used for the Direct Ascent approach and for no other reason. They're too expensive for anything else!

Choosing or Avoiding Other Systems

Avoid joint missions whenever possible. While it may look tempting to avoid the cost of the big rockets, it adds up with the cost of the second launch facilities, and

Figure 13-12. Strap-on boosters

it also ties you down to exactly one mission per turn. Invariably, one of the launches will fail, setting the whole program back six months. With an Earth Orbital LM test, a Lunar Orbital LM test, and a Manned Lunar Landing, what looks like three turns will probably take five or six. With a large rocket, however, it can be done in one or two turns.

There are other systems you'll need, of course. Purchase and R&D EVA suits and Docking Modules as soon as possible. Buy the EVA suits first, as they need time for R&D (the cost of that R&D is miniscule). Buy a Kicker B about two years before the Manned Lunar Pass, unless you're *sure* you'll be doing everything with an Apollo/Soyuz capsule. The Kicker C is a good option for the Soviets, since it eliminates LM tests. It necessitates a joint mission. Without the need for LM tests, however, that's not such a disadvantage. Only one mission needs to be a joint one: the Lunar Landing itself.

If you're avoiding joint missions, you should also avoid the Kicker A. The Kicker B isn't much more expensive, and it's *much* more useful with more options.

Be sure to buy an LM (one-person is best, so that the Gemini/Voshkod capsule can be used) in plenty of time to use it for LM tests. An LM test mission is shown in Figure 13-13. It'll take over two years to R&D it properly, so it's very easy to put it off until it's too late. Don't!

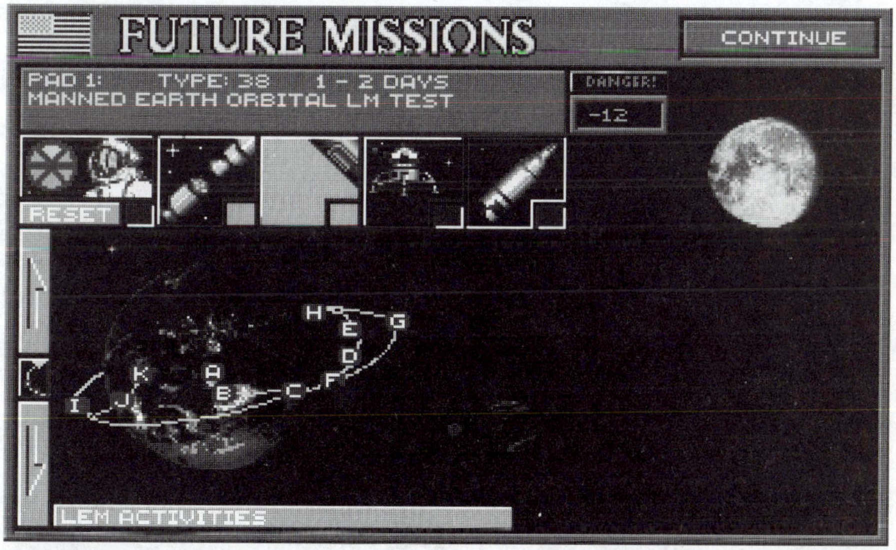

Figure 13-13. LM test mission

Alternatives

There are plenty of differing philosophies on which capsule programs are best. I've already mentioned one favorite, the Gemini/Voshkod and Apollo/Soyuz combination. That's the best way to go when in doubt. If your budget comes out either very low or very high, you might try something different out of desperation or for insurance. Some of these alternatives are listed here.

- **Gemini/Voshkod spacecraft for lunar missions** This is a pure desperation move for the person with a very low budget. You're banking on zero failures here and that's statistically very unlikely. This is truly putting yourself into the hands of fate.
- **Apollo/Soyuz spacecraft for lunar missions** Same as preceding alternative, but the odds of succeeding with this program are much higher. You're still gambling but it should work. Also, if you *start* with the Apollo/Soyuz (it's your first manned program), the budget bonus for getting that will more than offset the budget losses for the slow program pace. It does take a while to collect enough astronauts and rocket power to make the purchase useful. Purchase the two-person module (Eagle or Duet) to save money later in the game.
- **The Jupiter/Kvartet Module** This is the most expensive way to go, but it's no more reliable than the Apollo/Soyuz capsule. Elimination of docking missions and LM tests only mildly offset the high cost and development time. With this system, you must count on never having a catastrophic mission—because if you do, you're out of the race for good. As was concluded in real life, it's tough to get this to work before 1970.
- **XMS-2/Lapot Minishuttle** This is also very expensive, but worth every penny if you can afford it. With the minishuttle, you really can expect to fly the entire program without suffering a single catastrophic failure. This makes up for the increased cost in the long run, and the budget bonus for its first flight is very nice, too. There are really two ways to do this: buy the minishuttle as the first manned program, or develop a small, cheap spacecraft, and use it for docking and/or endurance while the minishuttle is being developed. Either method will work well as long as the budget is above average. With the minishuttle, you can safely buy the two-person LM, which will save a little money.

Generally, go with the mixed two-person/three-person capsule strategy, but if you find that your budget is significantly above average, buy the minishuttle instead of the three-person capsule. Both seem to work pretty well.

Skipping Missions

Another important question to be answered is, what missions should be skipped? The Lunar Probe is undoubtedly the most commonly skipped step, and with good reason. Its expense and unreliability will mean a lot of money will be spent before it is completed. Since the penalty for skipping it is a mere –3% for the Manned Lunar Pass mission, it's generally a good trade. (This is an excellent place to have the backup capsule program ready.) On the other hand, a well-funded program can buy the probe to increase its Manned Lunar Pass reliability, boost its Photo-Reconnaissance, and get a hefty budget bonus (16 prestige!) as well.

Skipping both the Lunar Fly-by and Lunar Probe missions yields a –6% for the entire duration of the Manned Lunar Pass mission. The cost savings may make it look tempting, but a penalty that large will put even a minishuttle at considerable risk. Capsule programs with this penalty will fail with alarming regularity. It is easily possible to lose both capsule programs due to disasters, R&D them back up, and lose them both *again* before completing a Lunar Pass. Therefore, skipping both of these steps is a really bad idea, and should only be considered in the most desperate of situations.

Skipping LM tests can be a good bet. While the –3% may look serious, the LM is used for only a short while (just during the LM mission steps). However, it's not always a good idea to make a habit of it. Therefore, what I would do is plan both LM tests as the Lunar Orbital type. The following season I would schedule the Lunar Landing. If either LM test fails for some reason, then and only then will I take the risk. In this way, I'm not slowed by LM test failures, but neither am I exposing the program to undue risk.

I never skip the Manned Lunar Pass or Manned Lunar Orbital. Near the Moon the missions are too long and the mission steps too numerous to accept that kind of risk.

I find that skipping the Manned Suborbital step is pretty risky, since it's done with a primative, low-reliability capsule. However, another advantage of starting with a Gemini/Voshkod capsule is the increased reliability that comes with it, offsetting this disadvantage. Additionally, the Gemini/Voshkod missions can't be scheduled back-to-back (due to the limited astronauts and flight crews available), so this is a good step to skip in the Gemini/Voshkod approach. I usually never try replacing Suborbitals with Orbitals in using Mercury/Vostok; they're dangerous enough without the added risk.

One not-so-obvious step to skip is the Orbital Satellite. Starting out by buying the hardware for a manned mission will go a long way towards being the first to put a man in space. The rewards for first man in space are the big prestige bonuses. One disadvantage, however, is that you can't buy astronauts until Spring 1958.

You'll have to pull them out of basic training immediately, then wait a turn until they've trained in their capsule. Only then can you assign them, so Fall 1959 is the earliest possible time to launch a manned capsule. However, a good opponent might even be ready at the same time, so the advantage is hardly devastating. You'll probably want to go back and get the Orbital Satellite program sometime, too. That way you can get the budget bonus, the reliability bonus on other satellites, a 95% power-on reliability for docking modules, and the ability to launch an Orbital Satellite when your government orders you to. Don't skip this step or the Manned Suborbital; that's really asking for trouble.

When to Acquire Astronauts

Speaking of buying astronauts, you'll definitely want to do that as soon as possible. It is preferable to purchase astronauts in Spring 1958. Not only will you need them, but each following astronaut group will become available that much sooner. When deciding whether to buy another astronaut group, you must look not only at your needs for the immediate future, but also at when you will need the *subsequent* group of astronauts.

There are only two kinds of astronauts: those with a capsule skill of 4, and everyone else. Always take care to put the specialty astronauts in their proper places, but *never* have a capsule pilot with a skill of less than 4 if you can help it. That skill of 4 helps out in nearly every mission step, whereas the other skills are used only on limited missions or mission steps. A capsule skill of 4 is easily ten times as useful as an EVA skill of 4, for example. Count your pilots carefully, don't waste them, and train more if you have to. There's no good reason to send any astronaut to advanced training—unless it's capsule pilot training.

Dummy Tests

A good, safe way of improving the reliability of rockets and spacecraft is by performing Dummy tests (unmanned missions). When your hardware has been R&D'ed to the maximum levels, schedule Suborbital or Orbital unmanned missions. By testing this hardware unmanned on an actual mission, you won't be risking a catastrophic failure. The worst that could happen is a failure mode and a failed mission, and since it is only a Dummy test, you won't even lose any prestige points. If the mission is successful (through all of the required mission steps), the safety reliability will improve by 1 percent on all of the hardware components on the mission. If you schedule and successfully launch several Dummy tests with the same hardware, you not only improve the reliability, but ensure a safer flight when the mission is eventually manned.

Missions

The mission declaration screen may look intimidating at first, but its selection of 53 possible space missions is the heart of the game. Here, you will decide what to do with all that fancy hardware you've been working on.

Generally, the first thing you should do is to order the launch of an Orbital Satellite for Fall 1957. It's not very likely to succeed, but if it does, you're well ahead. Continue ordering Orbital Satellite missions until one succeeds.

This brings up an important point about launch declarations: You must declare before this season's missions have flown. You must always think six months ahead. Ask yourself what missions will have been accomplished, what missions will need to be flown, and what hardware will be R&D'd and ready to fly? The answers to these questions are crucial.

Let's say, for example, that you're considering a Manned Lunar Pass mission. The capsule is ready, but the Kicker is only at 70% or so. You're launching the first Lunar Fly-by and Lunar Probe landing this season; should you plan the Manned Lunar Pass for next?

In this case, probably not. Consider what might happen if the Lunar Fly-by fails; the Lunar Probe will be at –3% on every mission step, and it wasn't too reliable to start with. If it fails, the Manned Lunar Pass would be at –6%. Additionally, it's unlikely you'll do well enough in R&D rolls to get the Kicker up from 70% to the 90%-plus range that I would consider safe for a circumlunar mission. Of course, you could plan it anyway, and scrub it if indeed any of the above happens. But that comes with a budget hit, and takes up a launchpad, so you only want to do that if it's fairly likely that the launch will go off. In this case, it isn't.

When planning missions, a good trick is to keep launching an Orbital Satellite, every turn. In fact, purchasing an additional launchpad early is worth it for just this purpose. The reason is not for the prestige or the reliability, but simply to keep from being fired.

The way the Presidential Review system works, your rating goes up or down a level if you have gained more prestige points, or less, respectively, during that season (see in Figure 13-14). Thus, it's better to launch a satellite and get 1 prestige point than it is to sit and do nothing. If the opponent also does nothing, then your rating will go up a level. If the opponent knows this trick, you have at least nullified its effectiveness. The computer opponent knows this trick, and uses it.

Another trick to keep in mind when planning missions is that it's always better to get big prestige bonuses in the fall season. This is not just because the budget bonus comes right away, but also because of the way the budget system works. Each season's event card comes with a random adjustment in the budget level, either

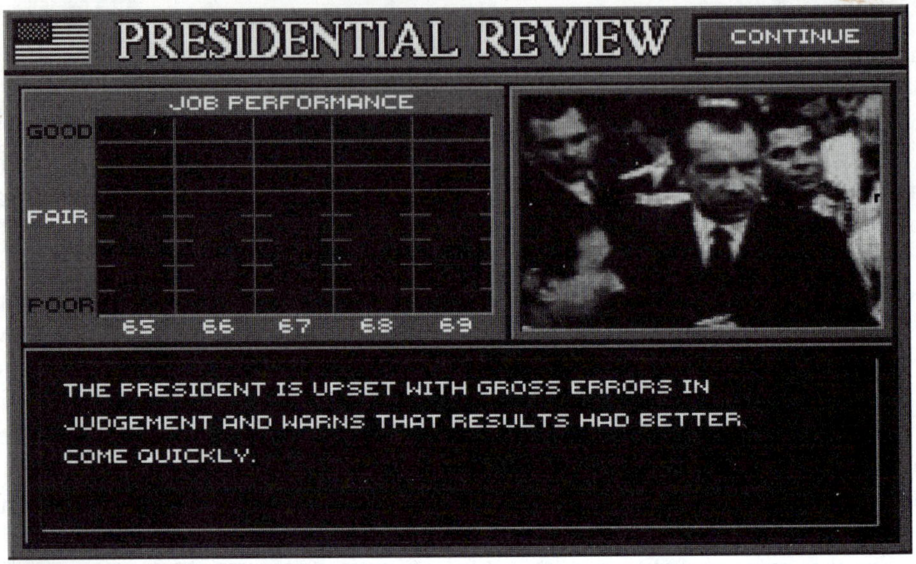

Figure 13-14. Presidential Review

plus or minus. However, higher budgets are biased more towards the negative, for game balance. The system operates using "brackets" of 50 megabucks in size, so throwing yourself into the highest (150+ megabucks) bracket in the spring season can result in a large penalty.

As an example, a player with a budget of 143 has what would be the first Manned Lunar Pass on the launchpad. If it were launched, it would throw the budget into a new bracket, possibly penalizing the budget by as much as (or more than) the 12 MB obtained from the Lunar Pass. Therefore, consideration should be given to scrubbing the mission and rescheduling it for the fall season. Of course, there are plenty of other considerations, such as what the opponent is doing, what missions are planned for the next season, the number of launchpads, and so on, that would make the risk worth taking anyway. Ideally, this player should have been ready six months *earlier,* so that the bonus would have come the previous fall.

How much Photo-Reconnaissance is enough? It's always good to send a few extra interplanetary satellites to the Moon to gather data on safe landing sites. A minimal program will include one successful Lunar Fly-by (boosting Photo-Recon from 55 to 60), a Manned Lunar Pass (worth another 5), and a Manned Lunar Orbital (another 5), for a total of 70%. This is adequate, but risky. A Lunar Probe brings another 15%, for a total of 85%, and that's probably more than enough. Without it, it's probably a good idea to send another interplanetary satellite or two towards the Moon. Best of all, this is a risk-reducing step that can be saved until near the end, and skipped if time becomes critical. The Photo-Recon Step is shown in Figure 13-15.

How much docking is enough? If it's needed during the Lunar Landing mission, then get it as high as you can. When using the Gemini/Voshkod or more advanced spacecraft, every mission can be scheduled with Docking, EVA (a two-man docking EVA mission combination is shown in Figure 13-16), and an additional Duration level. (Mercury/Vostok can't Dock.) A successful docking on the first try and subsequent will yield a Docking level of 60%. Success on the third and fourth mission would increase the skill to 80%. This is probably sufficient to try an LM test with, and of course two LM tests will bring the docking level up further. An LM Test has two seperate docking attempts. Failing the first docking in an LM test is embarassing and expensive, as the whole flight is wasted. Failing the second docking necessitates an emergency EVA for crew transfer, which is actually not much of a problem. If it fails, however, the results will be catastrophic. I therefore will not try an LM test with a safety factor of less than about 70% under any circumstances. If the LM is late in development, additional time and money can always be spent on additional, less-risky docking missions with Docking Modules.

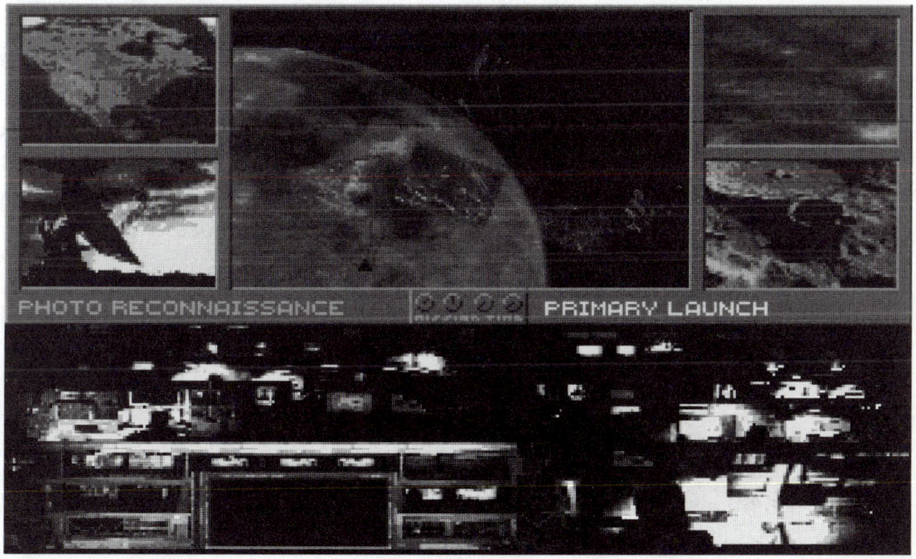

Figure 13-15. Photo-Reconnaissance mission screen

Figure 13-16. Historically, most of the Gemini missions and one Voskhod mission performed EVAs

Expose your astronauts or cosmonauts to risk when it will help, but never make a habit of it. Astronautics is a dangerous business, at least in the twentieth century, and the consequences of killing a crew are far-reaching and uniformly negative (see Figure 13-17). However, sitting still and accomplishing nothing is even worse. Nothing ventured, nothing gained. Knowing what is a "venture," and what is just plain suicide, and what is just wasting time, is the key to victory in the space race.

Good luck!

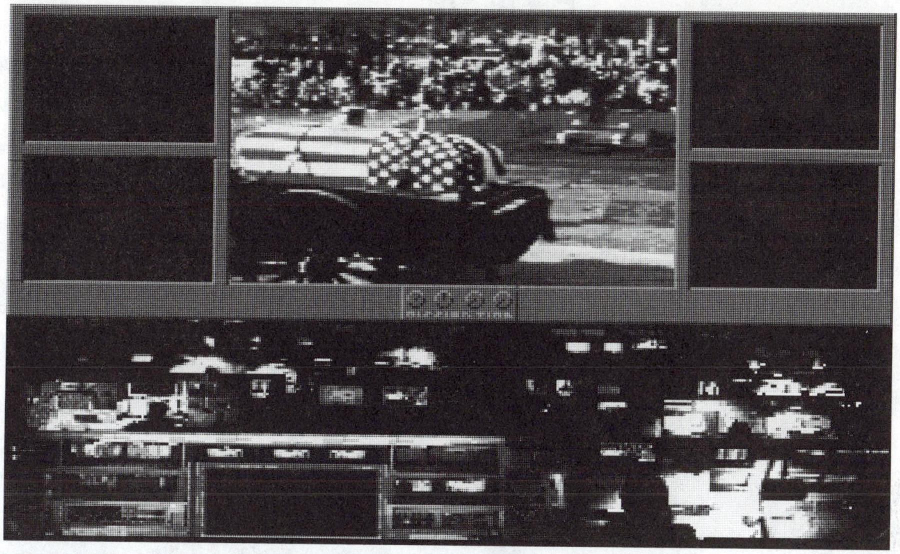

Figure 13-17. Too many funerals and your fired!

PART

Debriefing

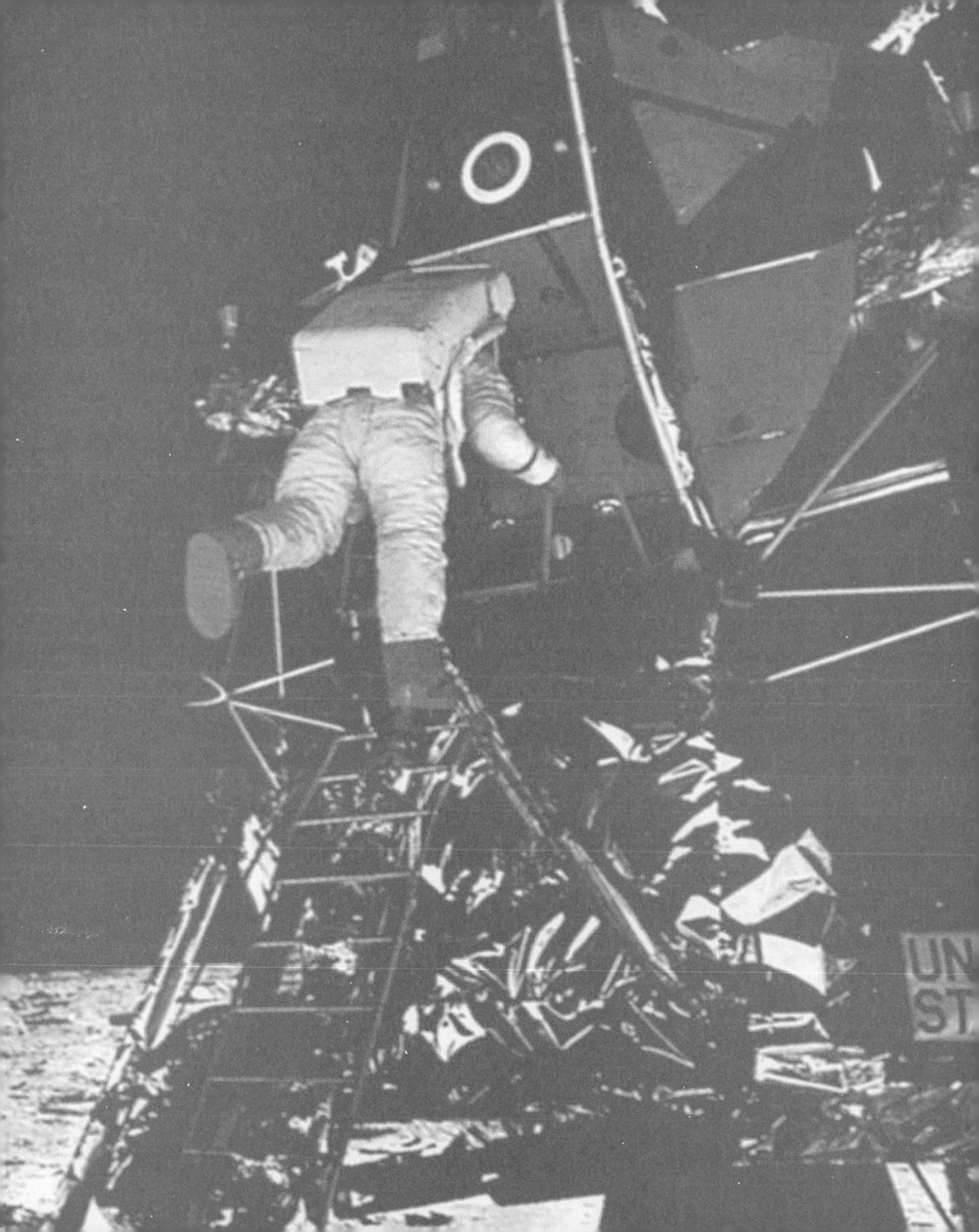
UN
ST

APPENDIX A

Secrets of the Game and Event Card List

You may find that "Buzz Aldrin's Race Into Space" is a relatively easy game to play yet a difficult game to master. Don't get desperate and start sending astronauts into space at questionable safety levels. Always try to research and develop (R&D) to the maximum levels. Sometimes, however, fate can just be against you. It's possible to have six or seven reliable rockets blow up in a row, and if that happens, your opponent will gobble up excessive prestige points. (In a situation like that, you might want to consider starting over!) Another option would be to try alternative tactics to see if you can recover and get back into the race. Quite often opponents will run into their own string of disasters. This will give you the chance to regain the lead.

Hints for Playing "Buzz Aldrin's Race Into Space"

Experiment with the game by playing both sides. You'll find that the hardware and all of the mission animations are not the same; it will seem like you're playing a different game. (The events are written in the perspective of the specific country.) If you play the Historical Model, it is yet again a different game. The U.S. player has more expensive but more reliable programs than those of the Soviet player.

Consider attempting different approaches to the Moon. There are numerous alternative animations for both sides that only pertain to specific spacecraft at certain steps of a mission or Moon landing. There are also some hidden failure mode animations that only appear in specific sequences. The more you play, the more you'll see these animations. There are also several lunar EVAs for each country and each LM performing the landing.

Mission Replay

There is an Easter Egg (surprise) during Mission replay. If you move the mouse cursor to the upper left (coordinates 0,0) and type the letter **R**, the mission animations will replay at full screen, as shown in Figure A-1.

Figure A-1. Full screen of a mission replay

Super Spacemen

Teachers and those who like to role play with their friends may like this feature. In the Preferences menu, go to the American Astronaut Roster and select the 97th name. Change the name to "von Braun" so that all of your astronauts will have each of their skills at the maximum level. All of the astronauts and cosmonauts can be Super Spacemen with a level four rating! This will let you customize individual skills for a custom roster. To keep different astronaut rosters, copy customized rosters to a separate directory for storage. When you want to play with a specific roster, copy it back into the game.

Status Box

During a mission, the status box is displayed in the bottom center of your screen, as shown in Figure A-2. This box accurately displays the reliability of the space hardware that is being utilized on the current step of the mission. The blue bar is the tolerance range (safety factor) of the hardware. The green bar that fills in represents the amount of stress on the hardware. The higher the green bar reaches in relation to the top of the blue bar, the more stress is being imposed upon the

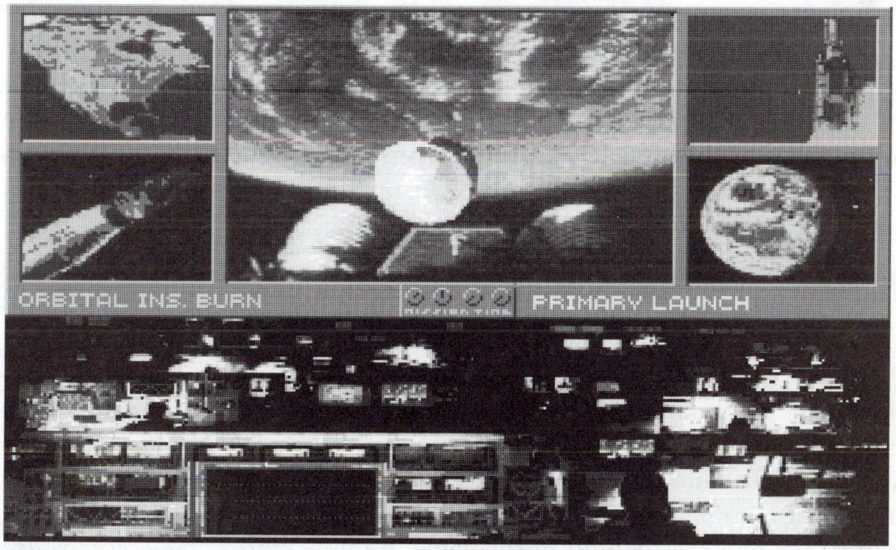

Figure A-2. The status box shows your hardware tolerance and how safe it's performing

hardware. If a red bar appears, the hardware has surpassed its tolerance limit. Failure mode goes into effect.

Failure mode affects missions differently. If the mission is unmanned, it is automatically considered a failure since there are few safety backup features. If the mission is manned, there is a chance of the failure being minor and/or repairable due to safety backup features. On manned missions, a small yellow line often appears at the top of the blue bar. This represents the astronaut skill level. This can enhance the safety range as high as an additional four percent. If you see the green bar rise quickly and cover up the blue bar and yellow line, and no red appears, it means that the astronaut's skill handled the crises. This can mean the difference between success and failure. Seat assignment on those flights, therefore, can be crucial.

Be careful to observe the safety range (blue bar). The lines across the status box denote 25%, 50%, and 75%.

Often your hardware safety factor may appear lower than you observed in R&D. This is due to the fact that you have skipped mission steps, or had suffered through failure modes on the mission. Skipping mission steps, or forging into unknown territory is risky and obviously adds additional stress tolerances on hardware. As result, your safety factor is slightly lowered.

CIA and KGB Accuracy

Okay, just how good is your country's intelligence? In real life, I figure the boys in the CIA hit the mark about 50 percent of the time. When playing Level 1 in the game, your Intelligence will be 60 percent accurate. If you play Level 2, it will be only 50 percent, and at Level 3, only 40 percent accurate. (The U.S. Intelligence Statistics screen is shown in Figure A-3.) Library and Statistics are both affected simultaneously at the same level. Some players may ignore this feature, but it can be of great assistance in gauging what your opponent is up to. (The Soviet KGB Library screen is shown in Figure A-4.)

Artificial Intelligence

The computer has the same chances on missions as the human player. The odds are identical! If you have several spacecraft programs that are at maximum safety levels but have a string of catastrophic failures, it is merely a statistical anomaly (old-fashioned bad luck!). This can easily happen to the computer as well, but you won't hear it screaming in frustration as a human might.

If the computer is behind in the game, it might start to take more risks and skip milestones. Nevertheless, it still has safety levels that must be met before it can attempt such high risk missions. But just like a human player in the same situation,

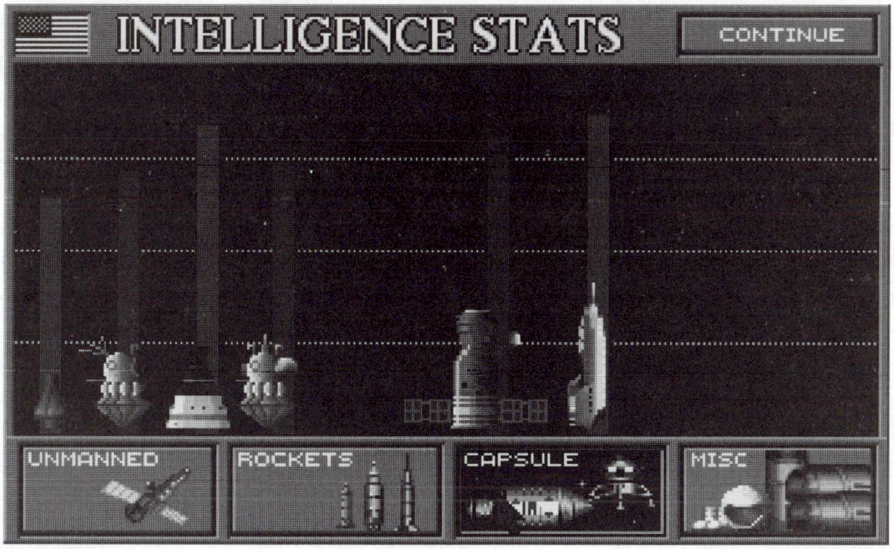

Figure A-3. US Intelligence Statistics screen

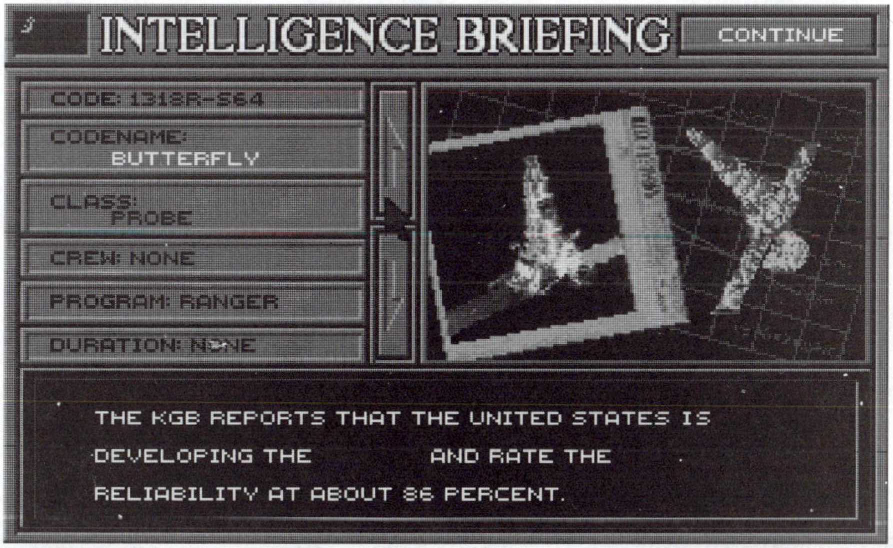

Figure A-4. The Soviet KGB Library screen

the computer will take on a more aggressive campaign if its opponent is nearing a manned Moon landing.

A human player can micromanage better than a computer can. For example, you can save two megabucks (m.b.'s) every season so that you can eventually purchase a program a few years later. The computer opponent, on the other hand, will always try to spend everything. It doesn't try to save money for future programs.

Human players can also manage the astronauts more effectively than the computer can. You can manipulate flight crew selection in order to have compatibility among individuals. You can also give more astronauts the opportunity for space missions. The computer, however, is ruthless with its astronauts. It doesn't care about astronaut morale or compatibility. It will generally burn out astronauts at a quicker rate than most humans players.

Tragedy at the Spaceport

There is an event card in the News that tells of an astronaut who died in training. This grisly event happened several times to both sides. Several U.S. astronauts nearly died in the "Flying Bedstead," or LM simulator, shown in Figure A-5. This accident is shown as an animation at the spaceport. If you see this event in News, you can go to the spaceport to view it "in person." (See Figure A-6.)

Figure A-5. The real Flying Bedstead was a difficult and often unstable craft to fly

Figure A-6. A tragic accident in the game

Lower Difficulty Level Means Budget Advantage

The way your budget is affected depends on what level of the game you're playing. The lower the level of play, the more your budget will increase (if it's at a low amount) in comparison to play at a higher level. In addition, at low difficulty levels, your budget (if it is greater than 120 megabucks) will not drop as quickly as it would at higher difficulty levels.

How to Get the Best Score

There are many strategies for achieving the highest score possible. Below are some hints which may help you increase your final point total.

Winning the Game and Level Difficulty

Winning the game will garner big points, and you can get an even higher score by playing more difficult levels. The Director Ranking screen, shown in Figure A-7, displays your progress and overall score. This is reflected by a series of formulas, but generally, the higher the level of difficulty, the more points you'll get on a given successful mission. If your opponent is playing an easier level than you are, you'll get even more points as compensation for winning at a higher difficulty level.

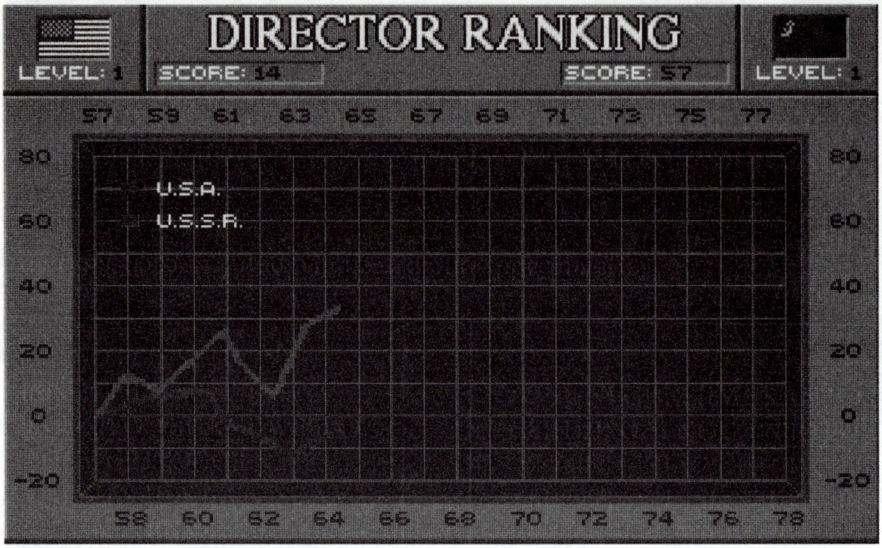

Figure A-7. Director Ranking screen

If you play Level 3 and your opponent plays Level 1, you'll be playing the best model for getting the highest score. It's also very challenging at this level, I might add! Even if you lose and your opponent lands on the Moon, it's quite possible that your Director Ranking score would be higher than in an equivalent but victorious Level 1 game.

Early/Late Landings and Overall Points

By landing earlier than the actual *Apollo 11* landing, you'll earn bonus points. If your landing is pre-1968, you'll get even more points. By the same token, if your landing is after the end of the decade deadline, you'll get fewer points. A Moon landing after 1972 will cause points to be deducted. There is one seemingly contradictory observation to remember: The earlier the Moon landing (the more streamlined the program, i.e., fewer missions), the less *overall* points you're likely to get. The converse is true as well: The later the Moon landing (the more time to complete a variety of missions for more prestige points), the larger your *overall* score is likely to be. However, keep in mind that a landing made too late can negate this advantage. I don't know the optimum date for points, but I suspect it's 1969!

Level of Difficulty and Penalties

The level of difficulty in play affects points in terms of penalties for skipping milestones. In future missions, the penalty for skipping milestone steps is least severe at the lowest level. For example, at level 1, the loss for each milestone skipped is assessed at –1%. At level 3, each new milestone step starts at a –3%. I recommend caution when skipping milestones at this level!

Event Cards

Historically, there were many things that were happening around the world at the time of the space race. Some of these events had a direct influence on the space race. Some had an indirect effect but added fuel to the fire of the cold war. A couple of events were outright bizarre and others were a stroke of luck.

The Space event news in "Buzz Aldrin's Race Into Space" is a large list based on actual historical occurrences. This feature was also in the board game "LIFTOFF!" and was expanded into a larger list of unique U.S. and Soviet events. This gives the game greater variety and a more interesting replay. The small crises and breakthroughs you'll witness are brought up randomly and offer additional challenges like those which occur in real life.

The following is a list of all of the event cards that are in "Buzz Aldrin's Race Into Space," along with some historical anecdotes. Each card has an American (U.S.) and a Soviet (U.S.S.R.) version. Following some of the cards in the list are "EFFECT ON GAME" notes which describe the effects on space hardware or other data caused by these events.

Since some events are time-specific, certain cards won't appear onscreen until after their dates have passed in the play of the game. The list that follows indicates when cards will be introduced into play.

Cards Introduced at the Beginning of the Game

U.S.

PRESIDENT EISENHOWER ANNOUNCED THE NAME OF THE DIRECTOR OF AMERICA'S SPACE PROGRAM. DIRECTOR ______ WILL LEAD THE UNITED STATES IN THE EXPLORATION OF SPACE.

U.S.S.R.

THE GENERAL SECRETARY ANNOUNCED TO THE POLITBURO THE NEW HEAD OF THE SOVIET SPACE PROGRAM. COMRADE ______ WILL LEAD THE SOVIET UNION TO GREATNESS IN THE EXPLORATION AND CONQUEST OF SPACE.

HISTORICAL NOTE: Webb and Korolev were the actual heads of the American and Soviet space programs, respectively. Korolev died on the operating table in 1966, and Webb resigned shortly after the *Apollo 1* fire.

Cards Introduced in 1957

U.S.

PRESIDENT EISENHOWER ANNOUNCED THERE WILL BE A FREEZE ON THE MANNED SPACE PROGRAM AFTER ONE OR TWO MANNED SPACE TESTS. THE PRESIDENT WARNS OF ESCALATING COSTS AND THE LACK OF GOALS IN SPACE EXPLORATION.

U.S.S.R.

MINISTERS HAVE ANNOUNCED THE INCREASE IN SPENDING FOR CONVENTIONAL GROUND FORCES AND SHORT RANGE MISSILES.

EFFECT ON GAME: No effect

HISTORICAL NOTE: Eisenhower was generally not enthusiastic about the early space race. He failed to see how it captured the public's imagination.

HISTORICAL NOTE: In the 1950s the Americans held the lead in all missile categories; all Soviet production was sacrificed to close this gap!

U.S.

OPERATION PAPER CLIP HAS BROUGHT IN A TOP GERMAN ROCKET SCIENTIST TO AID IN THE SPACE PROGRAM. R&D WILL IMPROVE FROM THIS DEVELOPMENT.

U.S.S.R.

A GERMAN SCIENTIST IS RESCUED BY THE KGB AND WILL WORK FOR THE SUPREME SPACE PROGRAM. THERE WILL BE GREAT ADVANCES IN R&D.

EFFECT ON GAME: Program adds one percentage point to all R&D attempts this turn.

HISTORICAL NOTE: German rocket scientists and engineers helped advance both American and Soviet space programs. The Soviets were a little more forceful in convincing Germans to work for the them. After the occupation of East Germany, the Soviets transported several hundred Germans deep into the Soviet Union. Many died, and some were not able to return to Germany until the mid-1950s.

U.S.

DUE TO MISMANAGEMENT AND CORRUPTION BY SOME CONTRACTORS, RESULTS HAVE BEEN FABRICATED. CONGRESS CONDUCTS AN INQUIRY. RESULTS IN R&D WILL BE WEAKER FOR ONE YEAR.

U.S.S.R.

DUE TO CORRUPTION AND POOR PLANNING, RESULTS IN R&D WILL BE WEAKER FOR ONE YEAR. SEVERAL DISHONEST DISTRICT MANAGERS HAVE BEEN EXECUTED.

EFFECT ON GAME: Program subtracts one from all R&D rolls for one year!

HISTORICAL NOTE: From time to time, contractors have been caught fabricating results.

HISTORICAL NOTE: In the 1950s and even into the '60s and '70s, Soviet managers and corrupt officials were executed. Whether the charges were true or whether it was politically expedient to execute them is not known.

Cards Introduced in 1958

U.S.

IN GOVERNMENT NEWS TODAY, THE PRESIDENT HAS INSISTED THAT THE NEXT SPACE LAUNCH WILL BE A MANNED MISSION. HE HAS AUTHORIZED AN INCREASE OF 10 M.B.'S TO SUPPORT THIS EFFORT.

U.S.S.R.

THE GENERAL SECRETARY DEMANDS THAT THE NEXT LAUNCH MUST BE MANNED. THE BUDGET IS INCREASED BY 10 M.B.'S.

EFFECT ON GAME: Program will not accept any other future launches until you successfully launch a manned mission.

HISTORICAL NOTE: The pressures of the space race caused a great demand for results.

U.S.

SOURCES IN WASHINGTON TODAY HAVE DISCLOSED INFORMATION THAT THE CIA HAS OBTAINED TOP SECRET RUSSIAN TECHNOLOGY. THIS ________ TECHNOLOGY WILL ASSIST THE UNITED STATES BY IMPROVING THE SAFETY FACTOR OF THE PROGRAM BY ___%.

U.S.S.R.

THE BRILLIANT KGB HAS UNCOVERED SECRET AMERICAN TECHNOLOGY. THE SAFETY FACTOR FOR ________ IS NOW INCREASED BY ___%.

This U.S.S.R. card, along with the screen in which it appears, is shown in Figure A-8.

HISTORICAL NOTE: The Soviets did hold and still hold the lead in several areas of space technology. Both sides had large intelligence apparatuses with numerous agents and spies collecting data.

Figure A-8. The KGB was very aggressive in obtaining American space secrets

U.S.

INTELLIGENCE SOURCES TODAY HAVE REVEALED THAT THE KGB HAS CLEVERLY FOOLED THE CIA AND THE UNITED STATES' TOP SCIENTISTS WITH FALSE DATA. THE SAFETY FACTOR FOR THE ________ IS NOW LOWERED BY ___%.

U.S.S.R.

THE EVIL CIA HAS CLEVERLY FOOLED THE KGB AND TOP SCIENTISTS WITH FALSE INFORMATION. THE SAFETY FACTOR FOR ________ IS NOW LOWERED BY ___%.

EFFECT ON GAME: Program finds space hardware similar to the opponent's, picks the most expensive hardware with a high safety factor for the U.S. (or U.S.S.R.) and the lowest safety factor for the Soviets, and lowers the U.S. (or Soviet) program to the level of the opponent's program, then fills in the program name and safety factor.

HISTORICAL NOTE: This is another possible ruse: develop or pretend to develop an expensive technological program that forces your opponent to waste valuable resources and finances. Nearly every American defense project was shadowed and also developed by the Soviets. Neither side wanted the other to have an advantage. The Americans also studied Soviet space and defense projects and developed similar programs. In the late 1950s, both sides studied the idea of space planes. The Stars Wars program is a classic example of enormous energies and funds spent on a program that is not likely to succeed.

U.S.

INTELLIGENCE SOURCES TODAY HAVE REVEALED THAT THE KGB HAS CLEVERLY FOOLED THE CIA AND THE UNITED STATES' TOP SCIENTISTS WITH FALSE DATA. THIS HAS DEGRADED THE SAFETY FACTOR OF THE ________ DOWN BY ___%.

U.S.S.R.

AN INVESTIGATION HAS REVEALED THAT THE EVIL CAPITALISTS HAVE DUPED THE KGB AND SOME MINISTRY OFFICIALS WITH MISINFORMATION. THE SAFETY FACTOR FOR ________ IS NOW LOWERED BY ___%.

EFFECT ON GAME: Program finds space hardware similar to the opponent's, picks the most expensive hardware with a high safety factor for the U.S. (or U.S.S.R.) and the lowest safety factor for the Soviets, and lowers the U.S. (or Soviet) program to the level of the opponent's program, then fills in the program name and safety factor.

U.S.	U.S.S.R.
THE PRESIDENT AND CONGRESS HAVE AUTHORIZED THE USE OF FEMALE ASTRONAUTS. THE SPACE PROGRAM MUST RECRUIT AT LEAST ONE FEMALE IN THE NEXT ASTRONAUT GROUP.	THE GENERAL SECRETARY HAS DECIDED TO SHOW THE WORLD THE SUPERIORITY OF SOVIET TECHNOLOGY. AT LEAST ONE FEMALE MUST BE RECRUITED IN THE NEXT COSMONAUT GROUP.

EFFECT ON GAME: Program makes female astronauts available in the next astronaut group unless all groups have already been selected, in which case, program ignores this card.

> **HISTORICAL NOTE:** Both sides secretly considered the use of women in space. The Americans quietly ended the FLATs program, and Khrushchev used Tereshkova only for propaganda purposes.

U.S.	U.S.S.R.
PRESIDENT _______ GAVE A STIRRING SPEECH TODAY AND A PUBLIC REQUEST FOR A MORE EARNEST EFFORT TOWARDS LANDING A MAN ON THE MOON. NATIONAL PRIDE SOARS! THE BUDGET IS INCREASED 15 M.B.'S TO ASSIST IN THIS GOAL.	GENERAL SECRETARY _______ GIVES A FORCEFUL SPEECH TO THE POLITBURO IN FAVOR OF THE SPACE EFFORT AND DEMANDS A MORE EARNEST EFFORT TOWARDS THE MOON. THIS RESULTS IN A BUDGET INCREASE OF 15 M.B.'S.

EFFECT ON GAME: When this card comes up, program checks what year you're in and fills in the correct president. (U.S.: Eisenhower, 1957-1960; Kennedy, 1961-1963; Johnson, 1964-1967; Nixon, 1968-1973; Ford, 1974-1975; Carter, 1976-1979. U.S.S.R.: Khrushchev, 1956-Spring 1964; Brezhnev, Fall 1964-1980.)

> **HISTORICAL NOTE:** Kennedy gave the Moon speech out of political necessity and survival; any other president or general secretary in the circumstances would have had the same response.

U.S.
A STORM FRONT BLOCKS THE LAUNCH WINDOW FOR THE ________ LAUNCH FACILITY. THIS LAUNCH IS CANCELED.

U.S.S.R.
A STORM FRONT BLOCKS THE LAUNCH WINDOW FOR THE ________ LAUNCH FACILITY. ANY SCHEDULED MISSIONS ARE CANCELED.

HISTORICAL NOTE: Florida squalls and Soviet blizzards caused havoc on the scheduling of space missions.

U.S.
SPACE ENGINEERS HAVE ANNOUNCED TODAY THAT A POTENTIAL FAILURE MODE HAS BEEN DISCOVERED ON THE ________ PROGRAM. THIS WAS NOTICED BY TESTING AND ANALYSIS AND WILL HELP AVOID THIS FAILURE DURING A MISSION.

U.S.S.R.
BRILLIANT SOVIET ENGINEERS HAVE, BY ANALYSIS AND ENGINE TESTS, FORESEEN A POTENTIAL FAILURE MODE ON THE ________ PROGRAM. THE NEXT FAILURE IN THAT PROGRAM WILL BE AVOIDED.

EFFECT ON GAME: This card affects rockets in the following order of priority, depending on what is in player's inventory: one-stage rockets, strap-on boosters, two-stage rockets, kickers, three-stage rockets, and four-stage rockets. Program avoids the next failure that would otherwise occur during a launch for the specified affected rocket hardware.

U.S.
THE SPACE PROGRAM HAS ANNOUNCED THAT THIS YEAR, THERE HAS BEEN A PROMISING CROP OF COLLEGE GRADUATES. R&D WILL IMPROVE FROM THIS DEVELOPMENT.

U.S.S.R.
THE GULAGS HAVE RE-EDUCATED SOME SCIENTISTS. THERE WILL BE IMPROVEMENTS THIS SEASON IN R&D.

EFFECT ON GAME: Program adds one percentage point to every R&D attempt this season.

U.S.

A NOBEL LAUREATE HAS DECIDED TO COME OUT OF RETIREMENT TO AID IN THE SPACE EFFORT. R&D WILL IMPROVE AS A RESULT.

U.S.S.R.

A GERMAN ROCKET SCIENTIST 'VOLUNTEERS' ASSISTANCE TO THE GREAT SOVIET SPACE EFFORT. THIS WILL ASSIST IN R&D.

EFFECT ON GAME: Program adds one percentage point to each R&D attempt this season.

U.S.

THE SPACE PROGRAM ANNOUNCED TODAY THAT IT HAS MANAGED TO RECRUIT A TOP SCIENTIST FROM A MAJOR CORPORATION. R&D WILL IMPROVE FROM THIS DEVELOPMENT.

U.S.S.R.

THE SPACE PROGRAM HAS MANAGED TO TRANSFER A TOP SCIENTIST FROM AN EASTERN BLOCK COUNTRY. R&D WILL IMPROVE THIS SEASON.

EFFECT ON GAME: Program adds one percentage point to each R&D attempt this season.

U.S.

A TOP SCIENTIST HAS EMIGRATED FROM AN ALLIED COUNTRY TO WORK FOR THE SPACE PROGRAM. R&D WILL IMPROVE DUE TO THIS DEVELOPMENT.

U.S.S.R.

INTERNAL SECURITY DELIVERS A TOP SCIENTIST TO WORK FOR THE SPACE PROGRAM. THIS WILL GREATLY AID R&D.

EFFECT ON GAME: Program adds one percentage point to each R&D attempt this season.

U.S.

LAST WEEK A GRADUATE STUDENT AT THE UNIVERSITY OF FLORIDA STUMBLED UPON A MAJOR SCIENTIFIC DISCOVERY. THIS ADVANCE IN MICRO CIRCUITRY WILL IMPROVE R&D AND ASSIST ALL OF AMERICA'S SPACE PROGRAMS.

U.S.S.R.

THE GREAT SOCIALIST SCIENCE PROGRAM PRODUCES A MAJOR SCIENTIFIC DISCOVERY. THIS WILL ASSIST IN R&D.

(Okay, I admit it: the University of Florida is my alma mater.)

EFFECT ON GAME: Program adds one percentage point to each R&D attempt this season.

U.S.

A FORMER GERMAN ROCKET SCIENTIST HAS ESCAPED FROM THE IRON CURTAIN. R&D WILL IMPROVE WITH THE AID OF THIS SCIENTIST.

U.S.S.R.

A SOVIET SCIENTIST WHO WENT TO THE WEST HAS BEEN RESCUED BY SOVIET AGENTS AND IS BROUGHT BACK TO WORK FOR THE SPACE PROGRAM. R&D WILL ACCELERATE THIS SEASON.

EFFECT ON GAME: Program adds one percentage point to each R&D attempt this season.

U.S.

A KEY SCIENTIST HAS ESCAPED FROM THE SOVIET UNION. INFORMATION FROM THIS SCIENTIST WILL AID IN R&D.

U.S.S.R.

AN EX-NAZI SCIENTIST HAS BEEN EDUCATED AND EMBRACES THE COMMUNIST SYSTEM. THIS WILL ASSIST R&D THIS SEASON.

EFFECT ON GAME: Program adds one percentage point to each R&D attempt this turn.

HISTORICAL NOTE: Many lower echelon German rocket technicians were kept against their will for over ten years after World War II.

U.S.

THE SPACE PROGRAM HAS LOST SEVERAL KEY RESEARCH SCIENTISTS IN A VERY SHORT PERIOD OF TIME. THIS SETBACK WILL MEAN WEAKER RESULTS IN R&D.

U.S.S.R.

THE SOVIET SPACE PROGRAM HAS LOST SEVERAL KEY RESEARCH SCIENTISTS. THEY HAVE BEEN RELOCATED TO SIBERIA. THIS SETBACK WILL MEAN WEAKER RESULTS IN R&D THIS SEASON.

EFFECT ON GAME: Program subtracts two from each R&D attempt this turn.

U.S.

SPACE SUFFERED A SEVERE SETBACK WHEN THE ________ PROGRAM HAD A MAJOR ENGINE TEST FAILURE YESTERDAY. THIS WILL RESULT IN A LOSS OF SAFETY, DOWN TO ___%.

U.S.S.R.

THE ________ PROGRAM HAS HAD A MAJOR ENGINE TEST FAILURE. THIS RESULTS IN A LOSS OF SAFETY, DOWN TO ___%.

EFFECT ON GAME: Program drops the safety factor in half, then fills in the program name and safety factor for the most advanced rocket program.

> **HISTORICAL NOTE:** Tough break, but this happened to both countries. The Atlas rocket went through a troublesome test phase, and the N-1 was a total failure for the Soviets.

U.S.

A BREAKTHROUGH IN COMPUTER AUTOMATION HAS LED TO INCREASED RELIABILITY OF THE SPACE AGENCY'S REMOTE PROBES AND SATELLITES. THE MAX R&D AND MAX SAFETY FACTOR OF THIS HARDWARE WILL BE RAISED BY 5% EACH.

U.S.S.R.

A BREAKTHROUGH IN COMPUTER AUTOMATION HAS LED TO INCREASED RELIABILITY OF THE NATION'S REMOTE PROBES AND SATELLITES. THE MAX R&D AND MAX SAFETY FACTOR OF THIS HARDWARE WILL BE RAISED BY 5% EACH.

U.S.
THERE HAS BEEN A SCIENTIFIC BREAKTHROUGH IN THE _______ PROGRAM. THE SAFETY FACTOR HAS BEEN INCREASED BY ___%.

U.S.S.R.
THERE HAS BEEN A SCIENTIFIC BREAKTHROUGH IN THE _______ PROGRAM. THE SAFETY FACTOR HAS BEEN INCREASED BY ___%.

EFFECT ON GAME: Program rolls six six-sided dice and adds the result to the current safety factor for the most advanced manned spacecraft.

HISTORICAL NOTE: Breakthroughs happened in real life after hard work and sometimes luck.

U.S.
THERE HAS BEEN A SCIENTIFIC BREAKTHROUGH IN THE _______ PROGRAM. THE SAFETY FACTOR HAS BEEN INCREASED BY ___%.

U.S.S.R.
THERE HAS BEEN A SCIENTIFIC BREAKTHROUGH IN THE _______ PROGRAM. THE SAFETY FACTOR HAS BEEN INCREASED BY ___%.

EFFECT ON GAME: Program rolls six six-sided dice and adds to the current safety factor for the most advanced rocket program.

U.S.
THERE HAS BEEN A MAJOR BREAKTHROUGH ON THE _______ PROGRAM. ITS SAFETY FACTOR HAS BEEN INCREASED BY ___%.

U.S.S.R.
THERE HAS BEEN A MAJOR BREAKTHROUGH ON THE _______ PROGRAM. ITS SAFETY FACTOR HAS BEEN INCREASED BY ___%.

EFFECT ON GAME: Program rolls four six-sided dice and adds the result to the safety factor for the most advanced satellite.

U.S.	U.S.S.R.
THE SPACE PROGRAM HAS SUFFERED A SEVERE SETBACK IN ITS _______ ROCKET DIVISION. REPAIRS CAN BE MADE FOR 15 M.B.'S OR SUFFER A 20% SAFETY FACTOR ON ITS NEXT LAUNCH.	THE SOVIET SPACE PROGRAM HAS SUFFERED A SEVERE SETBACK IN ITS _______ ROCKET DEVELOPMENT. REPAIRS CAN BE MADE FOR 15 M.B.'S OR SUFFER A 20% SAFETY FACTOR ON ITS NEXT LAUNCH.

EFFECT ON GAME: Program randomly selects any of the rockets in the player's inventory. The options are to repair the rocket, launch it with this setback, or cancel the launch. R&D will not solve this problem.

HISTORICAL NOTE: Individual rockets varied in reliability, some required additional repairs.

U.S.	U.S.S.R.
LAST WEEK, THE SPACE AGENCY DOWNGRADED THE _______ PROGRAM SAFETY FACTOR BY 25% DUE TO A TEST FAILURE.	DUE TO A TEST FAILURE, 25% SAFETY IS LOST FROM THE _______ PROGRAM.

HISTORICAL NOTE: An omission of a safety feature or flaw could easily downgrade the safety of a spacecraft, and has happened on numerous occasions in American and Soviet space programs.

U.S.	U.S.S.R.
DUE TO A TEST FAILURE, ENGINEERS HAVE DOWNGRADED THE SAFETY FACTOR OF THE _______ PROGRAM 15%.	DUE TO A TEST FAILURE, THERE IS A 15% LOSS FROM THE SAFETY FACTOR OF THE _______ PROGRAM.

U.S.

THERE HAS BEEN AN IMPORTANT ADVANCE IN ENGINEERING FOR THE SPACE PROGRAM. THIS WILL CUT THE COST IN ALL PROGRAMS BY 50% THIS SEASON.

U.S.S.R.

THERE HAS BEEN AN IMPORTANT ADVANCE IN ENGINEERING FOR THE SPACE PROGRAM. ANY HARDWARE PURCHASES WILL BE CREDITED AT HALF PRICE FOR THIS SEASON.

U.S.

THE PRIME CONTRACTOR FOR THE ROCKET PROGRAM WILL NOT BE ABLE TO DELIVER ANY ROCKETS THIS YEAR. A SPOKESMAN FOR THE COMPANY WOULD OFFER NO FURTHER COMMENT.

U.S.S.R.

THE MINISTRY OF PRODUCTION ANNOUNCES THAT DUE TO SHORTAGES, A SLOWDOWN HAS OCCURRED AND THE FACTORIES WILL NOT BE ABLE TO DELIVER ANY ROCKETS THIS YEAR.

HISTORICAL NOTE: By mismanagement or shortages, contractors often cause delays in deliveries of space hardware. The American LM was six months behind schedule; the Soviet Soyuz took two years longer.

U.S.

A MAJOR CORPORATION HAS DECIDED TO CREDIT THE SPACE AGENCY ________ ADDITIONAL FUNDS FOR THIS SEASON.

U.S.S.R.

TEN MILLION WORKERS WORK OVERTIME IN ORDER TO DONATE ___ M.B.'S TO THE SPACE PROGRAM.

EFFECT ON GAME: Program rolls one ten-sided die to determine the number of megabucks.

HISTORICAL NOTE: The Soviet economy and workers made huge sacrifices for space and military technology.

U.S.

A MAJOR CORPORATION HAS DECIDED TO CREDIT THE SPACE AGENCY ________ ADDITIONAL FUNDS FOR THIS SEASON.

U.S.S.R.

THE SUPREME SOVIET VOTES TO GIVE ___ M.B.'S TO THE SPACE PROGRAM. SOCIAL PROGRAMS WILL BE LOWERED IN ORDER TO SUPPORT THIS MEASURE.

EFFECT ON GAME: Program rolls three six-sided dice and totals them to determine the number of megabucks.

U.S.

A MAJOR CORPORATION HAS DECIDED TO CREDIT THE SPACE AGENCY ________ ADDITIONAL FUNDS FOR THIS SEASON.

U.S.S.R.

THE POLITBURO HAS VOTED ADDITIONAL FUNDING OF ___ M.B.'S FOR THE SPACE PROGRAM.

EFFECT ON GAME: Program rolls one ten-sided die to determine the number of megabucks.

U.S.

DUE TO A BUDGET DEFICIT, GOVERNMENT ACCOUNTING DECLARES A LOSS OF ___ M.B.'S IN CURRENT FUNDS FOR SPACE AGENCY.

U.S.S.R.

DUE TO A BUDGETING ERROR, AND CORRUPTION, A LARGE AMOUNT OF FUNDING IS LOST IN PAPERWORK. ___ M.B.'S IS LOST IN THE CURRENT FUNDS.

EFFECT ON GAME: Program discards 10% of the current total of megabucks.

HISTORICAL NOTE: Cost overruns, poor accounting practices, and unrealistic estimates have drained space programs' resources.

U.S.
DUE TO A TEST FAILURE, THE SPACE PROGRAM HAS SUFFERED A 20% SAFETY FACTOR LOSS ON THE ________ CAPSULE.

U.S.S.R.
THERE IS A 20% LOSS FROM THE SAFETY FACTOR OF THE ________ CAPSULE.

EFFECT ON GAME: Program deducts 20% from the safety factor of the most advanced capsule in the player's inventory. If no capsules are in the inventory, program ignores this card.

U.S.
A MAJOR EXPLOSION DESTROYED ALL ROCKET FUEL RESERVES FOR THE UNITED STATES. THIS DISASTER LEAVES ONLY ENOUGH FUEL FOR ONE LAUNCH THIS SEASON.

U.S.S.R.
A MAJOR EXPLOSION DESTROYED ALL ROCKET FUEL RESERVES FOR THE CCCP. MAY LAUNCH ONLY ONE MISSION THIS SEASON.

EFFECT ON GAME: Program allows only one launch this year; all others must be canceled. This card takes effect immediately.

> **HISTORICAL NOTE:** This happened twice in the history of America's space program. There is only one contractor for NASA and a major accident at the sole processing facility curtailed mission planning.

U.S.
SEVERE STORMS HAVE DAMAGED THE SPACE PROGRAM'S PRIMARY LAUNCH FACILITY. IT WILL COST 15 M.B.'S TO REPAIR THE LAUNCH FACILITY.

U.S.S.R.
SEVERE STORMS HAVE DAMAGED THE PRIMARY LAUNCH FACILITY. IT WILL COST 15 M.B.'S TO REPAIR THE LAUNCH FACILITY.

EFFECT ON GAME: Program selects the first launch facility in the player's inventory. It must be repaired this season; the program will close down the launch facility until it is repaired.

U.S.	U.S.S.R.
THE SPACE PROGRAM HAS MADE A DESIGN ERROR ON THE ________ SATELLITE PROGRAM. IT WILL COST 10 M.B.'S TO REDESIGN AND REPAIR THE PROBLEM, OR SUFFER A LOSS OF 10% SAFETY FACTOR ON ITS NEXT MISSION ONLY.	THERE IS A MINOR PROBLEM IN THE ________ SATELLITE PROGRAM. PAY 10 M.B.'S OR LOSE 10% ON THE SAFETY FACTOR OF THAT SATELLITE ON ITS NEXT MISSION.

EFFECT ON GAME: Program selects the most advanced satellite in the player's inventory. The options are to repair the satellite, launch it with this setback, or cancel the launch.

HISTORICAL NOTE: High technology often required additional research and development.

U.S.	U.S.S.R.
THE SPACE PROGRAM WILL HAVE TO REDESIGN THE ________ ROCKET PROGRAM. IT WILL COST 10 M.B.'S TO REDESIGN, OR A LOSS OF 5% SAFETY FACTOR ON THE NEXT ROCKET MISSION WILL RESULT.	THERE IS A MINOR PROBLEM IN THE ________ ROCKET PROGRAM. PAY 10 M.B.'S OR LOSE 5% ON THE SAFETY FACTOR OF THAT ROCKET FOR ITS NEXT MISSION.

EFFECT ON GAME: Program selects the most advanced rocket in the player's inventory. The options are to pay the repair costs, launch it with this setback, or cancel the launch.

U.S.	U.S.S.R.
THE SPACE PROGRAM HAS A MECHANICAL ERROR WITH THE ________ ROCKET PROGRAM. IT WILL COST 10 M.B.'S TO REPAIR OR A LOSS OF 5% SAFETY FACTOR WILL OCCUR ON THE NEXT ROCKET MISSION.	THERE IS A MINOR PROBLEM IN THE ________ ROCKET PROGRAM. PAY 10 M.B.'S OR LOSE 5% ON THE SAFETY FACTOR OF THAT ROCKET FOR ITS NEXT MISSION.

U.S.

THE SPACE PROGRAM HAS A MECHANICAL ERROR WITH THE ________ PROGRAM. IT WILL COST 10 M.B.'S TO REPAIR OR SUFFER A LOSS OF 10% SAFETY FACTOR ON THE NEXT MISSION.

U.S.S.R.

THERE IS A MINOR PROBLEM IN THE ________ PROGRAM. PAY 10 M.B.'S OR LOSE 5% ON THE SAFETY FACTOR ON ITS NEXT MISSION.

EFFECT ON GAME: Program selects a manned spacecraft. The options are to pay the repair costs, launch it with this setback, or cancel the launch.

U.S.

BUREAUCRATIC MISMANAGEMENT AT THE SPACE PROGRAM HAS CAUSED A LOSS OF MORALE. THIS RESULTS IN LOWER PRODUCTIVITY IN R&D FOR ONE YEAR.

U.S.S.R.

POOR PLANNING AND FAILURE TO MEET QUOTAS CAUSE A LOSS OF MORALE AND LOWER PRODUCTIVITY IN R&D FOR ONE YEAR.

EFFECT ON GAME: Program randomly subtracts either one or two percentage points on all R&D attempts for one year.

HISTORICAL NOTE: An army is only as good as the soldier in the trenches. If morale is low, effectiveness and progress is less likely in any organization. This commonly occurred when major space projects were cut back or canceled.

U.S.

A CONTRACTOR FOR THE SPACE PROGRAM HAS GROSSLY UNDERESTIMATED THE DEVELOPMENT COST FOR THE ________ PROGRAM. R&D WILL COST AN ADDITIONAL 1 M.B. PER SCIENTIST TEAM PERMANENTLY.

U.S.S.R.

THE CENTRAL COMMITTEE GROSSLY UNDERESTIMATED THE DEVELOPMENT COSTS FOR THE ________ PROGRAM. R&D COSTS AN ADDITIONAL 1 M.B. PER SCIENTIST TEAM PERMANENTLY.

EFFECT ON GAME: Program randomly selects between most advanced manned spacecraft, rocket, or module.

HISTORICAL NOTE: Contractors continually have revised the cost of programs. They always go up in price, rarely down!

U.S.	U.S.S.R.
A CONTRACTOR FOR THE SPACE PROGRAM HAS GROSSLY UNDERESTIMATED THE DEVELOPMENT COST FOR ________ PROGRAM. R&D WILL COST AN ADDITIONAL 1 M.B. PER SCIENTIST TEAM PERMANENTLY.	CENTRAL COMMITTEE GROSSLY UNDERESTIMATES THE DEVELOPMENT COSTS FOR THE ________ PROGRAM. R&D COSTS AN ADDITIONAL 1 M.B. PER SCIENTIST TEAM PERMANENTLY.

EFFECT ON GAME: Program randomly selects between the most advanced manned spacecraft, rocket, or module.

U.S.	U.S.S.R.
THE SPACE PROGRAMS OFFICIALS PLAYED HOST TO VIPS TODAY. A GROUP OF CONGRESSMEN VISITED MISSION CONTROL FACILITIES AND SEEMED PLEASED WITH OPERATIONS BUT REPORTS FROM THE WHITE HOUSE ARE NEGATIVE.	POLITBURO MEMBER VISITS THE TRAINING FACILITIES. HE APPEARS TO BE PLEASED, BUT REPORTS FROM MOSCOW ARE CRITICAL OF SPACE OPERATIONS.

HISTORICAL NOTE: Often the leadership in a country has given conflicting views and created confusion.

Cards Introduced in 1960

U.S.

SPACE ADMINISTRATORS HAVE LOBBIED SUCCESSFULLY FOR AN INCREASE OF ___ M.B.'S THIS YEAR. THE SPACE PROGRAM HAS BEEN ACTIVELY CAMPAIGNING FOR PUBLIC SUPPORT.

U.S.S.R.

PRAVDA HAS STAGED A MAJOR MEDIA EVENT. THE WORLD IS IN A FRENZY OVER THE SOVIET SPACE PROGRAM. THE POLITBURO DECIDES TO ADD ___ M.B.'S TO THE PROGRAM THIS YEAR.

EFFECT ON GAME: Program rolls two ten-sided dice and totals them to determine the number of megabucks.

U.S.

IN GOVERNMENT TODAY, THE PENTAGON HAS ISSUED AN ORDER THAT THE NEXT AVAILABLE LAUNCH WILL BE A SPY SATELLITE. A FUNDING APPROPRIATION OF 5 M.B.'S HAS BEEN MADE TO REFLECT THIS EXPENSE.

U.S.S.R.

THE DEFENSE MINISTRY HAS ORDERED THAT THE NEXT LAUNCH MUST BE AN ORBITAL SATELLITE. BUDGET IS INCREASED BY 5 M.B.'S TO FUND THIS EFFORT.

EFFECT ON GAME: Program schedules a satellite for the next season.

U.S.

SOURCES IN WASHINGTON TODAY HAVE DISCLOSED THAT THE CIA HAS OBTAINED TOP SECRET RUSSIAN TECHNOLOGY. THIS WILL ASSIST THE UNITED STATES BY IMPROVING THE SAFETY FACTOR OF THE ________ PROGRAM BY ___%.

U.S.S.R.

THE KGB HAS SUCCESSFULLY STOLEN VALUABLE AMERICAN TECHNOLOGY. THE SAFETY FACTOR FOR ________ IS NOW INCREASED BY ___%.

EFFECT ON GAME: Program looks for common programs, finds the one that is most expensive, selects the program that has the greatest difference in safety between the U.S. and U.S.S.R., then fills in the name of the program and the new safety factor.

HISTORICAL NOTE: The CIA and KGB obtained information on their opponent's space program. The batteries on the American LM were one of the most closely guarded secrets by NASA. There were rumors of sabotage on both sides. Very few details have ever been released by NASA.

U.S.

ENGINEERS HAVE ANNOUNCED TODAY THAT A POTENTIAL FAILURE MODE HAS BEEN DISCOVERED ON THE _______ PROGRAM. THIS WAS NOTICED DURING A SYSTEMS CHECK AND WILL HELP AVOID THIS FAILURE DURING A MISSION.

U.S.S.R.

MAGNIFICENT SOVIET ENGINEERS HAVE FORESEEN A POTENTIAL FAILURE MODE ON THE _______ PROGRAM DURING A SYSTEMS CHECK. THE NEXT FAILURE IN THAT PROGRAM WILL BE AVOIDED.

EFFECT ON GAME: This card affects all types of satellites: orbital, interplanetary, and probes.

HISTORICAL NOTE: Sometimes a technician or programmer discovered a potential failure in components, hardware, and/or software, and could create a patch or use a backup sequence that avoided the problem.

U.S.

ASTRONAUT _______ HAS ANNOUNCED RETIREMENT TODAY FROM THE SPACE PROGRAM. SOURCES HAVE DISCLOSED THAT THIS IS DUE TO SAFETY CONCERNS AND THE LACK OF BACKUP SYSTEMS. NEGATIVE PUBLICITY HAS CAUSED CONGRESS TO IMPLEMENT A FUNDING CUT OF 5 M.B.'S.

U.S.S.R.

COSMONAUT _______ WILL BE RELOCATED TO NORTHERN SIBERIA AT THE END OF THIS SEASON. THE POLITBURO INSTIGATES A FUNDING CUT OF 5 M.B.'S.

EFFECT ON GAME: Program goes through the list of astronauts that are in training or are active and randomly picks one. If no astronaut groups have been selected, program ignores this card.

> **HISTORICAL NOTE:** No astronaut has ever publicly admitted this, but there has been severe internal criticism (written memos) of America's and the Soviet's space programs. This happened after the *Apollo 1* fire, the *Soyuz 1* fiasco, and the *Challenger* explosion.

U.S.
THE PRIMARY CREW OF THE ________ LAUNCH HAS CAUGHT THE COMMON COLD. SPACE OFFICIALS HAVE ANNOUNCED THAT THE BACKUP CREW WILL TAKE THEIR PLACE ON THE MISSION.

U.S.S.R.
THE PRIMARY CREW OF THE ________ LAUNCH HAS CONTRACTED INFLUENZA. THE BACKUP CREW WILL FLY THE MISSION.

EFFECT ON GAME: Program looks at the missions scheduled this turn and randomly selects one of the manned missions, then fills in the mission name. If there are no manned missions, program shows another card. To select backup crew, player must pick active astronauts that have not been selected already for missions this turn. If there are not enough astronauts for a backup crew, program cancels the mission.

> **HISTORICAL NOTE:** One of the crewmen on *Apollo 13* was scrubbed and replaced by a backup, due to the common cold.

U.S.
________ IS KILLED IN A TRAINING ACCIDENT. THE SPACE PROGRAM HAS CANCELED ALL MANNED MISSIONS THIS SEASON TO ALLOW FOR ADDITIONAL SAFETY INSPECTIONS.

U.S.S.R.
________ IS KILLED IN A TRAINING ACCIDENT. ALL MANNED MISSIONS ARE CANCELED THIS SEASON TO ALLOW FOR ADDITIONAL SAFETY INSPECTIONS.

As shown in Figure A-9, an anchorman in the News Events screen reports of a tragic accident.

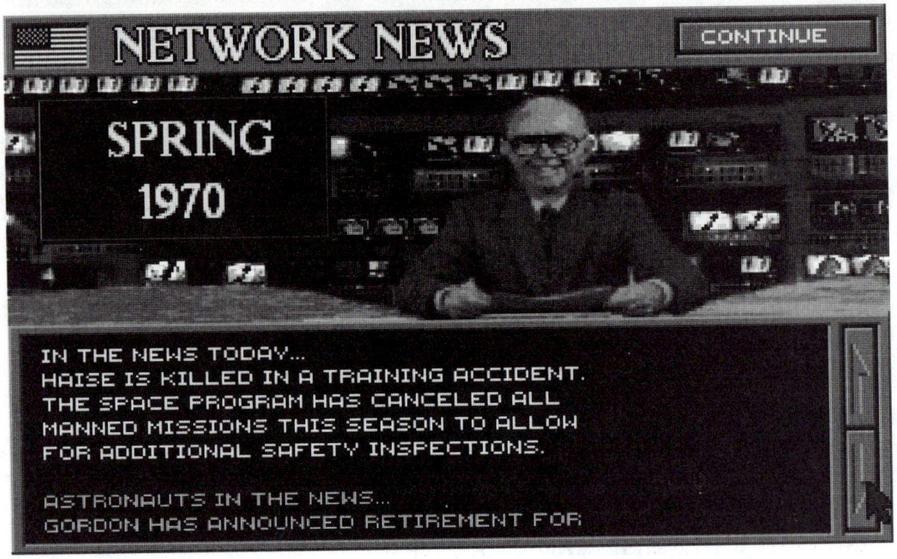

Figure A-9. Accidents unfortunately did occur in the space race

EFFECT ON GAME: Program picks an astronaut randomly from the list, and fills in the name. Obviously for this card, astronauts that are already dead or injured are out of the loop.

> **HISTORICAL NOTE:** Astronauts Bassett and See died on a routine training flight, as did cosmonaut Gagarin. This happened to several other astronauts and cosmonauts.

U.S.	U.S.S.R.
________ HAS TRAGICALLY BEEN KILLED IN AN AUTOMOBILE ACCIDENT TODAY.	________, A HERO OF THE SOVIET UNION, HAS TRAGICALLY DIED IN A TRAIN ACCIDENT.

> **HISTORICAL NOTE:** Astronauts and cosmonauts can and did die in common earth-bound car and train accidents.

U.S.	U.S.S.R.
________ FAILED A PHYSICAL AND MAY NOT PARTICIPATE IN ANY SPACE MISSIONS FOR THE DURATION OF ONE FULL YEAR.	________ FAILED A PHYSICAL AND MAY NOT PARTICIPATE IN ANY SPACE MISSIONS FOR THE DURATION OF ONE FULL YEAR.

EFFECT ON GAME: Program randomly picks an astronaut from training or active status.

HISTORICAL NOTE: Slayton was removed from flight status for 13 years due to a minor heart irregularity. Komorov was temporarily removed from flight status during training. Some astronauts and cosmonauts hid their medical ailments in order to prevent discovery and removal from flight status and from space flight.

U.S.	U.S.S.R.
FOREIGN AGENTS HAVE PENETRATED SECURITY AND DAMAGED THE ________ ROCKET PROGRAM. REPAIRS WILL COST 15 M.B.'S, OR A 20% LOSS IN SAFETY WILL RESULT.	WESTERN AGENTS HAVE PENETRATED SECURITY AND DAMAGED THE ________ ROCKET PROGRAM. REPAIRS WILL COST 15 M.B.'S, OR A 20% LOSS IN SAFETY WILL RESULT.

EFFECT ON GAME: Program fills in the most advanced rocket program. Player must pay for repairs or suffer a safety loss.

HISTORICAL NOTE: There are rumors that there was sabotage on the American LM program.

U.S.	U.S.S.R.
THE SPACE PROGRAM HAS SUFFERED A SEVERE SETBACK IN ITS ________ PROGRAM. REPAIRS CAN BE MADE FOR 20 M.B.'S, OR THE CAPSULE WILL LOSE 10% SAFETY FACTOR FOR ITS NEXT MISSION.	THE SPACE PROGRAM HAS SUFFERED A SEVERE SETBACK IN ITS ________ PROGRAM. REPAIRS WILL COST 20 M.B.'S, OR THE CAPSULE WILL LOSE 10% ON ITS SAFETY FACTOR FOR ITS NEXT MISSION.

U.S.

THERE HAS BEEN AN IMPORTANT ADVANCE IN ENGINEERING FOR THE SPACE PROGRAM. THIS WILL CUT THE COST IN ALL PROGRAMS BY 50% THIS SEASON.

U.S.S.R.

THERE HAS BEEN AN IMPORTANT ADVANCE IN ENGINEERING FOR THE SPACE PROGRAM. ALL HARDWARE HAS BEEN REDUCED IN COST 50% FOR THIS SEASON TO ACCELERATE THE PROGRAM.

U.S.

CONGRESS URGES ADDITIONAL SUPPORT FROM THE AIR FORCE AND NAVY. THREE TEST PILOTS ARE TRANSFERRED TO THE SPACE PROGRAM.

U.S.S.R.

THE MINISTRY OF DEFENSE HAS TRANSFERRED THREE TOP FIGHTER PILOTS TO THE SPACE PROGRAM.

HISTORICAL NOTE: Both sides had military space programs in operation with military pilots training for military space missions.

U.S.

SOURCES FROM THE SPACE PROGRAM HAVE DISCLOSED THAT DUE TO AN ENGINEERING ERROR, THERE IS A 50% CHANCE THAT THE NEXT MISSION WILL EXPLODE DURING LAUNCH.

U.S.S.R.

AN ENGINEERING ERROR MEANS THAT THE NEXT MISSION LAUNCHED HAS A 50% CHANCE OF EXPLODING DURING LAUNCH.

EFFECT ON GAME: This card occurs only when a launch is scheduled during the season. Program randomly determines which launch is affected.

U.S.

A MAJOR SPACE EXHIBIT OPENED LAST WEEK DISPLAYING AMERICA'S HARDWARE. THE PUBLIC HAS RESPONDED WITH GREAT INTEREST. ATTENDANCE IS REPORTED TO BE HIGH.

U.S.S.R.

A MAJOR SPACE EXHIBIT OPENS AND DISPLAYS THE GREAT SOVIET CONQUESTS OF SPACE. ALL CITIZENS ARE IN AWE.

HISTORICAL NOTE: Both space programs had enormous public relations programs in order to maintain interest and a high level of funding.

U.S.

ASTRONAUT ________ CAN NO LONGER PARTICIPATE IN THE SPACE PROGRAM. ________ WAS SERIOUSLY INJURED WHILE PLAYING TOUCH FOOTBALL.

U.S.S.R.

COSMONAUT ________ IS SERIOUSLY INJURED WHILE PLAYING ICE HOCKEY. IS OUT OF THE PROGRAM.

EFFECT ON GAME: Program randomly selects active astronauts in the program and randomly sets the length of injury to one year, two years, or makes it permanent.

HISTORICAL NOTE: Glenn was seriously injured while taking a shower.

U.S.

A TOP ROCKET DESIGNER HAS RETIRED DUE TO POOR HEALTH. R&D WILL PROBABLY SUFFER FROM THIS FOR ONE YEAR.

U.S.S.R.

A TOP ROCKET DESIGNER COLLAPSES DUE TO OVERWORK AND MUST RETIRE. THE CENTRAL COMMITTEE AWARDS THE CITIZEN'S MEDAL FOR INDIVIDUAL SACRIFICE AND BRAVERY. R&D RESULTS ARE LOWER FOR ONE YEAR.

EFFECT ON GAME: Program subtracts one percentage point from R&D attempts for all space hardware.

HISTORICAL NOTE: The pressures of the space race caused a great amount of stress for thousands of workers and probably led to numerous health problems.

U.S.	U.S.S.R.
A TOP ROCKET DESIGNER FOR THE SPACE PROGRAM HAS DIED IN SURGERY TODAY. THIS WILL CAUSE A DISRUPTION IN THE R&D OF ROCKETS FOR ONE YEAR.	A TOP ROCKET DESIGNER DIES IN SURGERY. THIS CAUSES A MAJOR DISRUPTION IN THE R&D OF ROCKETS FOR ONE YEAR.

EFFECT ON GAME: Program subtracts one percentage point from all rocket R&D attempts for one year.

HISTORICAL NOTE: The Soviets lost their Chief Designer Sergei Korolev on the operating table at the hands of an inept surgeon.

U.S.	U.S.S.R.
THE UNITED STATES SPACE PROGRAM WAS HONORED TODAY BY THE VISIT OF THE PRIME MINISTER OF ENGLAND.	THE FRENCH PRIME MINISTER VISITS THE SOVIET SPACE MUSEUM IN MOSCOW.

Cards Introduced in 1962

U.S.	U.S.S.R.
SPACE OFFICIALS ANNOUNCED TODAY THAT ONE OF ITS LAUNCH FACILITIES WILL NEED REPAIR. IT WILL CLOSE DOWN THIS SEASON AND COST 10 M.B.'S.	THE DEPARTMENT OF MAINTENANCE HAS SCHEDULED A LAUNCH FACILITY FOR REPAIR. IT WILL CLOSE DOWN THIS SEASON AND COST 10 M.B.'S.

U.S.
CONTRACTORS ANNOUNCED TODAY THAT THE GEMINI PROGRAM WILL COST AN ADDITIONAL 2 M.B.'S FOR EACH SPACECRAFT.

U.S.S.R.
THE MINISTER OF PRODUCTION ANNOUNCED TODAY THAT THE VOSKHOD PROGRAM WILL COST AN ADDITIONAL 2 M.B.'S FOR EACH SPACECRAFT.

EFFECT ON GAME: If there is no two-man spacecraft in player's inventory, program ignores this card.

U.S.
A SPOKESMAN FOR THE SPACE PROGRAM ANNOUNCED TODAY THAT THE ________ MISSION WILL BE DELAYED FOR ONE SEASON DUE TO HAIRLINE CRACKS IN THE FUEL TANKS.

U.S.S.R.
THE CHIEF MINISTER OF FLIGHT OPERATIONS HAS REPORTED THAT THE ________ MISSION WILL BE DELAYED FOR ONE SEASON DUE TO HAIRLINE CRACKS IN THE FUEL TANKS.

U.S.
SENATOR CRUTCHMIRE SHARPLY ATTACKED THE SPACE PROGRAM TODAY. SPEAKING BEFORE THE SENATE, HE ACCUSED SPACE OFFICIALS OF LAVISH SPENDING AND STRONGLY URGED FUNDING CUTS.

U.S.S.R.
POLITBURO MEMBER VOLKOVICH SHARPLY ATTACKED THE SPACE PROGRAM TODAY. SPEAKING BEFORE THE SUPREME SOVIET, HE ACCUSED MINISTERS OF LAVISH SPENDING AND STRONGLY URGED FUNDING CUTS.

HISTORICAL NOTE: Senator Proxmire actively lobbied and campaigned against the NASA's Moon program. The Soviets no doubt had their opposition and military competitors.

U.S.
THE MILITARY COMPLEX LOBBIED STRONGLY TODAY WITH SPACE OFFICIALS RECOMMENDING THE 'DIRECT ASCENT' APPROACH TO THE MOON. AIR FORCE MISSILE EXPERTS EXPLAINED THE ADVANTAGES AND SIMPLICITY OF THEIR PLAN.

U.S.S.R.
THE MILITARY COMPLEX LOBBIED STRONGLY TODAY WITH MINISTRY OF PLANNING RECOMMENDING THE 'DIRECT ASCENT' APPROACH TO THE MOON. RED ARMY MISSILE EXPERTS EXPLAINED THE SUPERIORITY OF THEIR PLAN.

HISTORICAL NOTE: Both sides seriously studied Direct Ascent. In the early 1960s it was the primary approach for the U.S. and Soviets.

U.S.
AMERICA'S TOP GERMAN SCIENTIST SPOKE TODAY ON THE ADVANTAGES OF THE 'EARTH ORBITAL RENDEZVOUS' APPROACH, IN GETTING A MAN ON THE MOON. HE SPOKE OF THE LONG RANGE ADVANTAGES OF 'EOR.'

U.S.S.R.
A SCIENTIST FROM THE SOVIET ACADEMY OF SCIENCES SPOKE TODAY ON THE ADVANTAGES OF THE 'EARTH ORBITAL RENDEZVOUS' APPROACH, IN GETTING A MAN ON THE MOON. HE SPOKE OF THE LONG RANGE ADVANTAGES OF 'EOR.'

HISTORICAL NOTE: Von Braun lobbied for EOR as the favored lunar approach.

U.S.
SEVERAL ENGINEERS SPOKE TO UNITED STATES SPACE OFFICIALS TODAY ON THE MERITS OF THE 'LUNAR ORBITAL RENDEZVOUS' APPROACH IN THE QUEST FOR GETTING MAN ON THE MOON. OFFICIALS APPEARED NONCOMMITTAL TOWARDS THIS PRESENTATION.

U.S.S.R.
A NEW PLAN WAS SUBMITTED TO THE MINISTRY OF PLANNING ON THE MERITS OF THE 'LUNAR ORBITAL RENDEZVOUS' APPROACH IN THE QUEST FOR GETTING MAN ON THE MOON.

HISTORICAL NOTE: This became the final approach of necessity for the Soviets.

U.S.	U.S.S.R.
THE VICE PRESIDENT TOURED A ROCKET CONTRACTOR'S FACILITY TODAY. HE OBSERVED A SUCCESSFUL ENGINE TEST FIRING.	A POLITBURO MEMBER TOURED A ROCKET PRODUCTION FACILITY TODAY. HE OBSERVED A SUCCESSFUL ENGINE TEST FIRING.

HISTORICAL NOTE: Politicians often took advantage of photo opportunities.

U.S.	U.S.S.R.
A TOP SCIENTIST TODAY CRITICIZED THE SPACE PROGRAM FOR THE HIGH COST OF MANNED EXPLORATION. ROBOT PLANETARY EXPLORERS WERE OFFERED AS A CHEAPER ALTERNATIVE.	A RIVAL SCIENTIST CRITICIZED THE SOVIET SPACE MINISTERS FOR THE HIGH COST OF MANNED EXPLORATION. ROBOT PLANETARY EXPLORERS WERE OFFERED AS A CHEAPER ALTERNATIVE.

HISTORICAL NOTE: Both manned space programs received criticism from their unmanned satellite counterparts, partly because of jealously and partly because they received less funding than the manned programs.

U.S.	U.S.S.R.
THE ASTRONAUTS HAVE, FOR AN UNDISCLOSED SUM OF MONEY, SIGNED A DEAL WITH A MAJOR MAGAZINE FOR THE EXCLUSIVE RIGHTS TO THEIR LIFE STORY.	COSMONAUT TRAINING HAS SUFFERED A SLOW-DOWN DUE TO POWER-OUTAGES AT FACILITIES.

HISTORICAL NOTE: The American "original seven" astronauts signed an exclusive deal for their life stories with *LIFE* magazine for $100,000.

U.S.
CHARLES LINDBERGH TOURED THE SPACE PROGRAM'S FACILITIES TODAY. THE HISTORY OF FLIGHT HAS CHANGED DRAMATICALLY SINCE 1927.

U.S.S.R.
TODAY, COMRADE FIDEL CASTRO TOURED THE SOVIET MILITARY FACILITIES. A GREAT SOCIALIST UNDER THE IMPERIALIST'S SOFT UNDERBELLY, PREMIER CASTRO COMPLIMENTED THE CCCP AS TRUE FOLLOWERS OF LENIN AND THE SOCIALIST WAY OF LIFE.

U.S.
TODAY, AN AIR FORCE COLONEL RECOMMENDED A ONE WAY MISSION TO THE MOON IN ORDER TO BEAT THE SOVIETS. OF COURSE, HE WOULD HAVE TO WAIT SEVERAL YEARS ON THE LUNAR SURFACE UNTIL THE TECHNOLOGY IS DEVELOPED TO GET HIM BACK.

U.S.S.R.
A RED ARMY GENERAL HAS RECOMMENDED A ONE WAY MISSION TO THE MOON. THIS IS ANOTHER SUPERIOR PLAN TO ALLOW A LONG-TERM SOVIET PRESENCE ON THE MOON.

HISTORICAL NOTE: The information in the U.S. card is absolutely true! The officer that proposed it volunteered for the mission. The psycho ward would have been more suitable! A couple of engineers also studied the logistics of this approach, and found that at least 20 support flights would have been necessary to keep the sole inhabitant alive.

U.S.
THE GOVERNMENT TODAY HAS ANNOUNCED THAT A CIVILIAN WILL BE LAUNCHED IN A MULTI-MANNED CAPSULE ON THE NEXT AVAILABLE MISSION. THE SPACE PROGRAM'S BUDGET WILL BE INCREASED BY 15 M.B.'S.

U.S.S.R.
THE POLITBURO HAS ORDERS TO LAUNCH A CIVILIAN IN A MULTI-MANNED CAPSULE ON THE NEXT MISSION. BUDGET IS INCREASED BY 15 M.B.'S.

U.S.

THE SPACE PROGRAM'S ENGINEERS HAVE ANNOUNCED TODAY THAT A POTENTIAL FAILURE MODE HAS BEEN DISCOVERED ON THE ________ PROGRAM. THIS WAS NOTICED DURING A SYSTEMS CHECK AND WILL HELP AVOID THIS FAILURE DURING A MISSION.

U.S.S.R.

SUPERIOR SOVIET ENGINEERS HAVE FORESEEN A POTENTIAL FAILURE MODE IN THE ________ PROGRAM. THE NEXT FAILURE IN THAT PROGRAM WILL BE AVOIDED.

U.S.

A MANUFACTURING ERROR HAS CAUSED A DEFECT IN THE ________ CAPSULE. THE SPACE AGENCY MUST PAY 10 M.B.'S OR SUFFER 10% ON THE CAPSULE'S SAFETY FACTOR ON ITS NEXT MISSION.

U.S.S.R.

DRUNKEN WORKMANSHIP HAS CAUSED A DEFECT IN THE ________ CAPSULE. PAY 10 M.B.'S OR SUFFER 10% ON THE CAPSULE'S SAFETY FACTOR ON ITS NEXT MISSION.

EFFECT ON GAME: Program selects the most advanced manned spacecraft in the player's inventory.

U.S.

A DESIGN ERROR HAS CAUSED A DEDUCTION OF 20% FROM THE SAFETY FACTOR OF THE ________ PROGRAM. SPACE OFFICIALS ARE CONFIDENT THAT THIS ERROR CAN BE QUICKLY CORRECTED.

U.S.S.R.

A DESIGN ERROR HAS CAUSED A REDUCTION OF 20% IN THE SAFETY FACTOR OF THE ________ PROGRAM.

EFFECT ON GAME: Program randomly selects the most advanced manned spacecraft or the most advanced module.

U.S.
THE PRIME CONTRACTOR FOR THE CAPSULE PROGRAM ANNOUNCED TODAY THAT THEY CANNOT DELIVER ANY CAPSULES OR 4-PERSON CAPSULE/MODULES THIS YEAR.

U.S.S.R.
THE PRODUCTION FACILITY FOR THE CAPSULE PROGRAM CANNOT DELIVER ANY CAPSULES OR 4-PERSON CAPSULE/MODULES THIS YEAR.

U.S.
THE GOVERNMENT ANNOUNCED TODAY THAT THE NSA AND THE PRESIDENT HAVE ORDERED THE SPACE AGENCY TO LAUNCH A SPY SATELLITE NEXT SEASON.

U.S.S.R.
THE KGB ORDERS THE GREAT SOVIET SPACE PROGRAM TO LAUNCH A SPY SATELLITE NEXT SEASON.

EFFECT ON GAME: If there are no satellites in the player's inventory, program ignores this card.

Cards Introduced in 1965

U.S.
MEDICAL DOCTORS AT THE UNITED STATES SPACE PROGRAM ANNOUNCED TODAY THAT AN ADDITIONAL 'DURATION MISSION E' WILL BE REQUIRED BEFORE ANY LUNAR LANDINGS WILL BE ATTEMPTED.

U.S.S.R.
THE MINISTRY OF SCIENCE HAS ANNOUNCED THAT AN ADDITIONAL 'DURATION MISSION E' WILL BE REQUIRED BEFORE ANY LUNAR LANDINGS WILL BE ATTEMPTED.

U.S.
CONTRACTORS ANNOUNCED TODAY THAT THE APOLLO PROGRAM WILL COST AN ADDITIONAL 3 M.B.'S FOR EACH SPACECRAFT.

U.S.S.R.
A RESTRUCTURING OF THE FIVE-YEAR PLAN HAS RAISED THE COST OF THE SOYUZ PROGRAM BY AN ADDITIONAL 3 M.B.'S FOR EACH SPACECRAFT.

EFFECT ON GAME: If there are no three-man capsules in player's inventory, program ignores this card.

U.S.	U.S.S.R.
THE PENTAGON ANNOUNCED TODAY THAT THE AIR FORCE HAS SHARED ADVANCED TECHNOLOGY FOR THE _______ PROGRAM. THE SAFETY FACTOR IS NOW INCREASED BY ___%.	THE MINISTRY OF DEFENSE HAS SHARED ADVANCED TECHNOLOGY; IT WILL BENEFIT THE _______ PROGRAM. THE SAFETY FACTOR IS NOW INCREASED BY ___%.

U.S.	U.S.S.R.
THE PRIMARY CREW FOR THE _______ LAUNCH THIS SEASON HAS BEEN REMOVED DUE TO THE COMMON COLD. THE BACKUP CREW WILL FLY THE SCHEDULED MISSION.	THE PRIMARY CREW FOR THE _______ LAUNCH THIS SEASON HAS BEEN REMOVED DUE TO THE COMMON COLD. THE BACKUP CREW WILL FLY THE SCHEDULED MISSION.

U.S.	U.S.S.R.
THE PRESIDENT TOURED SPACE FACILITIES IN HOUSTON. ENGINEERS AND SPACE OFFICIALS EXPLAINED TO THE PRESIDENT THE VARIOUS APPROACHES OF GETTING MAN ON THE MOON. AT SOME POINTS, THE DISCUSSION BETWEEN THE TWO GROUPS APPEARED HEATED AND QUITE LIVELY.	THE PREMIER TOURED SPACE FACILITIES IN PLESETSK. ENGINEERS AND SPACE OFFICIALS EXPLAINED TO THE PREMIER THE VARIOUS APPROACHES OF GETTING MAN ON THE MOON. ALL THE OPERATIONS ARE PROCEEDING SMOOTHLY.

HISTORICAL NOTE: President Kennedy actually did visit NASA facilities.

U.S.

TODAY, AN AGREEMENT AT THE UNITED NATIONS WAS SIGNED BY THE UNITED STATES AND THE USSR WHICH BANS HIGH ALTITUDE ATOMIC TESTS.

U.S.S.R.

AN AGREEMENT AT THE UNITED NATIONS WAS SIGNED BY THE UNITED STATES AND THE CCCP, PROHIBITING HIGH ALTITUDE ATOMIC TESTS.

U.S.

A NEW ROCKET ENGINE WAS ROLLED OUT TODAY FOR A PUBLIC VIEWING. THIS PROTOTYPE WILL HELP THE UNITED STATES IN GETTING MAN TO THE MOON.

U.S.S.R.

A NEW ROCKET ENGINE WAS ROLLED OUT TODAY. THIS PROTOTYPE WILL HELP THE MOTHERLAND IN GETTING A SOCIALIST ON THE MOON FIRST.

U.S.

ASTRONAUT ________ ANNOUNCED A TRANSFER TO THE AIR FORCE TEST PILOT PROGRAM.

U.S.S.R.

COSMONAUT ________ HAS BEEN DISCIPLINED AND TRANSFERRED FOR RE-EDUCATION.

U.S.

INTELLIGENCE EXPERTS RELEASED INFORMATION ON A NEW SOVIET ROCKET CAPABLE OF LIFTING MORE THAN ANYTHING IN THE US INVENTORY.

U.S.S.R.

THE KGB HAS OBTAINED INFORMATION ON A NEW AMERICAN MISSILE CAPABLE OF LIFTING MORE THAN ANYTHING IN THE SOVIET INVENTORY.

U.S.	U.S.S.R.
IN THE NEWS TODAY, THE PRESIDENT HAS ANNOUNCED THAT THE SPACE PROGRAM WILL PUT A MAN ON THE MOON WITHIN THREE YEARS. FUNDING HAS INCREASED BY 20 M.B.'S TO SUPPORT THIS EFFORT.	THE GENERAL SECRETARY HAS ORDERS FOR A COSMONAUT TO BE ON THE MOON WITHIN THE NEXT THREE YEARS. THERE IS A 20 M.B. INCREASE TO SHOW ADDITIONAL SUPPORT.

U.S.	U.S.S.R.
ANTI-SPACE ACTIVISTS HAVE BEEN SUCCESSFUL IN THEIR DRIVE TO CUT BACK THE SPACE PROGRAM BUDGET. CONGRESSIONAL LEADERS HAVE VOTED TO DECREASE THE FUNDING BY ___ M.B.'S.	THE OLD GUARD DECIDES TO FLEX ITS MUSCLES AND FORCES THE POLITBURO TO RECONSIDER ITS POSITION. ___ M.B.'S WILL BE DIVERTED TO THE MILITARY.

EFFECT ON GAME: Program lowers the budget by 20 megabucks or 20%, whichever is less.

U.S.	U.S.S.R.
UNITED STATES SPACE ENGINEERS HAVE ANNOUNCED TODAY THAT A POTENTIAL FAILURE MODE HAS BEEN DISCOVERED ON THE ________ PROGRAM. THIS WAS NOTICED BY TESTING AND ANALYSIS AND WILL HELP AVOID THIS FAILURE DURING A MISSION.	GLORIOUS SOVIET ENGINEERS HAVE, BY TESTING AND ANALYSIS, FORESEEN A POTENTIAL FAILURE ON THE ________ PROGRAM. THE NEXT FAILURE IN THAT PROGRAM WILL BE AVOIDED.

U.S.

THERE HAS BEEN A SCIENTIFIC BREAKTHROUGH IN THE ________ PROGRAM. THE SAFETY FACTOR HAS BEEN INCREASED TO ___%.

U.S.S.R.

THERE HAS BEEN A SCIENTIFIC BREAKTHROUGH IN THE ________ PROGRAM. THE SAFETY FACTOR HAS BEEN INCREASED TO ___%.

EFFECT ON GAME: Program rolls four six-sided dice to determine the new safety factor for the four-manned Cap-Module or LM. If these programs are not in the player's inventory, program ignores this card.

U.S.

THE ________ ADMINISTRATION HAS DECIDED TO CUT FUNDING FOR THE SPACE PROGRAM IN FAVOR OF INCREASED FUNDING FOR SOCIAL PROGRAMS. THE SPACE AGENCY'S BUDGET HAS BEEN REDUCED TO ___ M.B.'S.

U.S.S.R.

THE MILITARY AND KGB HAVE DECIDED TO CUT FUNDING FOR THE SPACE PROGRAM IN FAVOR OF MORE CONVENTIONAL WEAPONS AND INTERNAL SECURITY. THE SPACE BUDGET HAS BEEN REDUCED TO ___ M.B.'S.

EFFECT ON GAME: When this card comes up, program checks what year you're in and fills in the correct president. (U.S.: Eisenhower, 1957-1960; Kennedy, 1961-1963; Johnson, 1964-1967; Nixon, 1968-1973; Ford, 1974-1975; Carter, 1976-1979. U.S.S.R.: Khrushchev, 1956-Spring 1964; Brezhnev, Fall 1964-1980. Program decreases budget by 30 megabucks or 30%, whichever is less.

U.S.

DUE TO THE VIETNAM CONFLICT, THE PRESIDENT DECREASES THE SPACE PROGRAM'S BUDGET TO ALLOW FOR A MILITARY BUILDUP. THE SPACE PROGRAM'S BUDGET IS ___ M.B.'S.

U.S.S.R.

THE GOVERNMENT ORDERS THAT THE SPACE PROGRAM'S BUDGET BE CUT TO ALLOW FOR WHAT IT FEELS IS A MUCH-NEEDED MILITARY BUILDUP. THE BUDGET IS NOW ___ M.B.'S.

EFFECT ON GAME: Program reduces budget by 25 Megabucks or 25%, whichever is less.

U.S.
TODAY, AN X-15 AIR FORCE PILOT FLEW TO A NEW ALTITUDE RECORD OF 53 MILES. AFTER REACHING THE ALTITUDE RECORD, THE AIR FORCE MAJOR SAFELY GLIDED THE X-15 TO A LANDING AT EDWARDS AIR FORCE BASE.

U.S.S.R.
A NEW SURFACE TO AIR MISSILE WAS TESTED TODAY AND SUCCESSFULLY SHOT DOWN ITS TARGET.

HISTORICAL NOTE: Those X-15 sled drivers earned their astronaut wings by technically reaching the limits of space.

U.S.
UNITED STATES ACTIVISTS PROTESTED IN FRONT OF THE SPACE PROGRAM HEADQUARTERS IN WASHINGTON TODAY. THE STAGED SIT-IN CALLED FOR AN INCREASE IN SPENDING FOR SOCIAL PROGRAMS.

U.S.S.R.
DISSIDENTS WERE ROUNDED UP TODAY AFTER THE ORGANIZED DISRUPTION AT THE MAY-DAY PARADE.

HISTORICAL NOTE: Yes, there are actually people who are against space exploration—some of them actually believe we really never landed on the Moon!

U.S.
THE GOVERNMENT OFFICE OF ACCOUNTING AUDITORS ANNOUNCED TODAY THAT THE UNITED STATES SPACE PROGRAM WENT OVER BUDGET LAST YEAR AND WILL DEDUCT 5 M.B.'S THIS YEAR FROM THE SPACE PROGRAM'S BUDGET.

U.S.S.R.
THE MINISTRY OF ACCOUNTING HAS DISCOVERED WASTE AND MISMANAGEMENT IN THE SPACE PROGRAM AND WILL DEDUCT 5 M.B.'S FROM FUNDING THIS YEAR.

U.S.

TODAY, THE NATION'S LARGEST ROCKET CONTRACTOR ANNOUNCED THE SUCCESSFUL TEST OF THEIR NEWEST ROCKET ENGINE.

U.S.S.R.

SOVIET TECHNOLOGY TRIUMPHS AGAIN! THE U.S.S.R HAS SUCCESSFULLY TESTED THE LARGEST ROCKET ENGINE IN THE WORLD!

U.S.

THE AUSTRALIAN GOVERNMENT HAS ANNOUNCED THAT THEY WILL GENEROUSLY PROVIDE LAND TO BE USED FOR AN AMERICAN SPACE TRACKING STATION.

U.S.S.R.

THE SOVIET UNION HAS CONSTRUCTED A TRACKING SHIP TO ASSIST IN RELAY AND COMMUNICATIONS WITH SATELLITES AND DEEP SPACE MISSIONS.

U.S.

TODAY THE WHITE HOUSE ANNOUNCED THE FORMATION OF THE COMSAT CORPORATION, WHICH WILL BE RESPONSIBLE FOR THE LAUNCHING OF COMMUNICATION SATELLITES.

U.S.S.R.

THE KREMLIN ANNOUNCED TODAY THE FORMATION OF GLAVCOSMOS, A STATE AGENCY RESPONSIBLE FOR THE LAUNCHING OF COMMUNICATION SATELLITES. NOW THE GLORY OF SOCIALISM CAN BE HEARD BY THE OPPRESSED EVERYWHERE.

U.S.

A STUDY BY THE RAND CORPORATION, RELEASED TODAY, WARNS OF THE DANGERS OF A SOVIET LEAD IN THE SPACE RACE AND THE POTENTIAL IMPACT ON WORLD AFFAIRS. CONGRESSIONAL LEADERS EXPRESS CONCERN THAT THE U.S. IS NOT MAINTAINING THE LEAD.

U.S.S.R.

THE POLITBURO WARNS ITS CITIZENS OF THE DANGERS OF THE CAPITALISTS LEADING THE SPACE RACE. ALL CITIZENS ARE URGED TO WORK HARDER IN DEFENDING THE MOTHERLAND.

U.S.
THIS MORNING, A CORPORATION SUBMITTED TO SPACE LEADERS A PROPOSAL FOR A SMALL SPACE SCOOTER. THIS LIGHTWEIGHT HOPPER COULD FERRY ONE ASTRONAUT DOWN TO THE LUNAR SURFACE FOR A SIX-HOUR DURATION.

U.S.S.R.
A SOVIET SCIENTIST HAS SUBMITTED A PROPOSAL FOR A LUNAR HABITAT CAPABLE OF SUSTAINING COSMONAUTS FOR UP TO THIRTY DAYS ON THE MOON.

As shown in Figure A-10, an anchorman in the news events screen reports of alternate LM proposals.

U.S.
A CONGRESSMAN SPOKE TODAY ON HOW ADVANCES IN SPACE TECHNOLOGY ASSIST IN MEDICAL SCIENCE. SPIN-OFFS FROM THE SPACE PROGRAM WILL HELP SAVE LIVES.

U.S.S.R.
GLORIOUS SPACE SPIN-OFFS! THE MINISTRY OF AGRICULTURE ANNOUNCED TODAY THAT SUPERIOR SOVIET SPACE TECHNOLOGY WILL ASSIST IN IMPROVING CROP YIELDS.

Figure A-10. This was a real proposal that was considered

U.S.

THE UNITED STATES ANNOUNCED THE SUCCESSFUL TEST-FIRING OF ITS FIRST NUCLEAR ROCKET ENGINE, KIWI. NUCLEAR ROCKETS WILL NOT AFFECT THE LUNAR PROGRAM, BUT IT IS EXPECTED TO MAKE MANNED MARS FLIGHTS FASTER AND SAFER.

U.S.S.R.

SOVIET ENGINEERS HAVE SUCCESSFULLY TEST-FIRED A NUCLEAR ROCKET ENGINE. SOMEDAY, THIS SUPERIOR TECHNOLOGY WILL ENSURE THAT THE FIRST MEN ON MARS WILL BE COSMONAUTS.

U.S.

THE UNITED STATES ANNOUNCED TODAY THE CONSTRUCTION OF A SMALL LAUNCH FACILITY FOR ICBM TESTS AT WALLOPS ISLAND, VIRGINIA.

U.S.S.R.

THE SOVIET UNION HAS SUCCESSFULLY LAUNCHED A NEW, MORE POWERFUL ICBM. THIS IS ANOTHER WEAPON TO STOP AMERICAN AGGRESSION.

U.S.

DOD OFFICIALS ARGUED BEFORE A CONGRESSIONAL HEARING THAT THE U.S. SHOULD BEGIN AN ANTI-SATELLITE WEAPONS PROGRAM. THE SUGGESTION WAS NOT WELL RECEIVED.

U.S.S.R.

MARSHALL KAMINOV PROPOSED TO THE SUPREME SOVIET THAT SPACE FUNDING SHOULD BE DIVERTED TO SPACE WEAPONS TO SHOOT DOWN THE CAPITALIST AGGRESSORS.

U.S.

A NEW SPACE GAME MAKES ITS DEBUT IN HOLLYWOOD. BOTH REALISTIC AND ENTERTAINING, IT IS EXPECTED TO INSPIRE THE PUBLIC WHILE EDUCATING THEM.

U.S.S.R.

THE MINISTRY OF INFORMATION HAS PRODUCED AN INSPIRING GAME ABOUT OUR SOVIET HERO'S IN THE SPACE PROGRAM.

(Okay! It's an inside joke!)

U.S.

PROPOSALS HAVE BEEN SUBMITTED REGARDING THE CONCEPT OF A REUSABLE 'SHUTTLE-BUS' FOR ROUTINE ACCESS TO SPACE. INTENDED FOR DEVELOPMENT AFTER THE LUNAR PROGRAM, IT IS EXPECTED TO MAKE ACCESS TO SPACE COST MUCH LESS.

U.S.S.R.

THE SOVIET MINISTRY OF AEROSPACE HAS SUBMITTED A PROPOSAL FOR A LARGE REUSABLE SPACE SHUTTLE FOR USE IN THE POST-LUNAR PROGRAM.

U.S.

THE AIR FORCE HAS SUBMITTED TO CONGRESS A PROPOSAL FOR A MANNED ORBITING LABORATORY, OR MOL. THE SMALL TWO-MAN MILITARY OBSERVATION PLATFORM WOULD HAVE A DURATION OF THIRTY DAYS.

U.S.S.R.

THE SOVIET MILITARY HAS DEVELOPED PLANS FOR A PERMANENT ORBITAL STATION FOR MONITORING AMERICAN AGGRESSION.

U.S.

STUDIES HAVE BEEN RELEASED ON THE RELIABILITY OF CAPSULE LANDINGS ON THE PLAINS OF TEXAS AFTER RETURNING FROM THE MOON. THE IMPACT OF A GROUND LANDING WILL BE SLIGHTLY HARDER.

U.S.S.R.

THE SOVIET NAVY HAS SUCCESSFULLY COMPLETED CAPSULE WATER-RECOVERY TRIALS, AND IS CONFIDENT OF ITS ABILITIES.

HISTORICAL NOTE: NASA actually had plans to have returning lunar missions land in the plains of Texas. It wasn't very romantic, but technically feasible. The Navy probably objected to this in fear of a smaller Navy. Would have eliminated the Carrier Task Force needed for water recoveries.

U.S.

THE AIR FORCE HAS SUBMITTED A PROPOSAL FOR THE LAUNCHING OF AN X-15 ON A NAVAHO ROCKET INTO A LOW EARTH ORBIT. THE PILOT WOULD PARACHUTE BACK TO THE EARTH AFTER THE EXPENDABLE VEHICLE SAFELY RE-ENTERS THE ATMOSPHERE.

U.S.S.R.

THE SOVIET MINISTRY EVALUATES THE APPLICATIONS OF THE SANGER ANTIPODAL BOMBER. A PROTOTYPE IS EXPECTED TO BE READY FOR DROP TESTS NEXT YEAR.

HISTORICAL NOTE: The X-15B mission actually was proposed, but Project Mercury superseded it.

U.S.

THE PENTAGON HAS LOBBIED FOR A MANNED MILITARY SPACE FLIGHT. CONGRESSIONAL LEADERS HAVE VOICED OPPOSITION TO THE IDEA.

U.S.S.R.

POLITBURO MEMBERS ARE EXCITED BY THE PROSPECT OF A FUTURE SOVIET MILITARY SPACE MISSION. AMERICAN AGGRESSION CAN BE MORE CLOSELY MONITORED ON SUCH MISSIONS.

APPENDIX B

History of the Game: How It Really Came About!

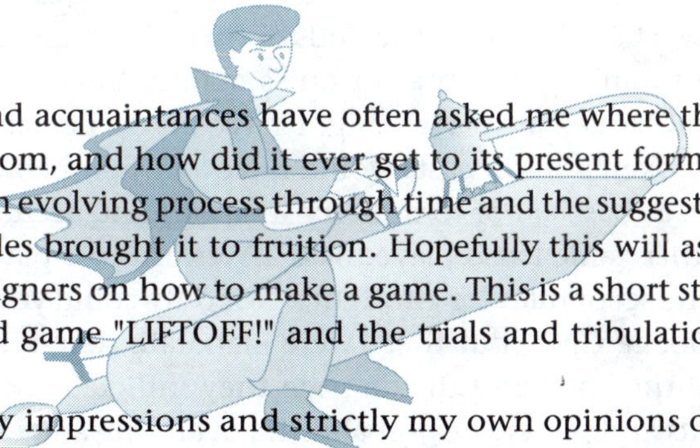

Several friends and acquaintances have often asked me where this idea of a space race game came from, and how did it ever get to its present form. Looking back, the game has been an evolving process through time and the suggestions from many friends and comrades brought it to fruition. Hopefully this will assist some future potential game designers on how to make a game. This is a short story of the development of the board game "LIFTOFF!" and the trials and tribulations of the computer game.

What follows are my impressions and strictly my own opinions on games, dates, individuals, and the industry. Many are likely to disagree with me or my opinions. My being an "outsider," however, might bring a different perspective to the matter.

The Foundation

In the summer of 1982, I can fondly remember playing numerous board games with several friends. We would start with World War I dogfights and then jump to "Diplomacy" and then medieval combat, often all in a single evening. After work, four or five of us would get together and enthusiastically play these games late into the night.

It was around that time that the role-playing game "Traveler" appeared. This science fiction game opened a richly-textured universe to explore, defend, and plunder. It was a text-based game system relying equally on the imaginations of a game master and players. In our minds we described technological wonders and alien worlds that could never be adequately portrayed in film or television. This beautiful, open design permitted individual creation and the ability for innovation. It encouraged all of us to research for more material, ideas, and concepts in other books and magazines to add more flavor to the game. I remember reading through a wide variety of magazines, ripping out photographs of an Arctic base in National Geographic, pictures of terrorists from *Newsweek*, and weird fashions out of some glamour magazine. All of these were part of the mosaic and background tapestry of the game universe. As game players, we were hooked: so many worlds and creatures, fantastically detailed charts, and aids for character development!

"Traveler" also introduced me to the ten-sided die, a molded die that has ten sides for counting one to ten. With two ten-sided dice (one for the first decimal and one for the second decimal), you have numbers ranging from 1 to 100, perfect random percentiles! With these dice, I began to calculate the percentages for completing a task. What are the odds of throwing paper into the waste basket from across the room? Depending on one's ability, the chance of success can very; a professional basketball player might have an 85 percent chance, while someone with two broken arms in casts would have a low probability. This can be applied to sports such as baseball or to military weapons and accuracy.

Another game I also played was a Western role-playing game that included a very realistic set of rules for gunfights. Not only could you maneuver your character around "Dodge City," but there were elaborate descriptions of where the bullets penetrated the body and the damage they inflicted. The rules determined movement points and percentile probabilities of aiming and hitting an object. There was a base chance to hit, say 50 percent, depending on the character's ability and type of weapon. If the target was running and dodging or concealed, the chance of hitting the target decreased. If the character firing the weapon was also moving, the chance of hitting the target decreased further. While the subject matter may be a bit morbid to some, it remains a mathematical engine to simulate an actual event.

This actually led us to read and study more on the anatomy of the human body so that we could accurately reconstruct the path of the bullet. In our minds during play, we could see how the bullet punctured the epidermis, would violently tear through vital organs and ricochet off one bone and splinter another as it finally exited out the other side. Using mathematics, physics, and a little imagination it was perfect virtual reality. In our minds we vividly saw all the texture and detail.

The Planting of the Seed

I remember I was at Steve Stipp's house and I had just lost another gunfight in the Western game with Steve. Once again my whole team of characters was annihilated. I decided to take a break and watch the local news. Still frustrated with the results of that last encounter, I was replaying in my mind what mistakes I might have made, when a rocket launch appeared on the news. Now, I must be honest in saying off the top, that I'm a space fanatic. The game player in me suddenly wondered what the odds were for a successful launch. The next thought I had was the chance of failure. Immediately I formulated in my mind a guess on the total number of launches verses failures. I quickly concluded that out of ten previous launches, nine of them were successful. Just before liftoff, I rolled the percentile dice and rolled below the range, which indicated to me that the launch would be successful. A few moments later, another satellite reached orbit. I was elated that I had come up with pseudo-model for launch success. The thought struck me that there was a lot of data available on spaceflight to help form a statistical base. Immediately I wondered how a manned launch would work. I told Steve what had just happened and started to play with some rough mathematical figures. I selected a one-stage rocket and a two-stage rocket and then realized that I would have to devise a simple system for payload weights and lift capabilities. I also needed to devise a safety factor for a capsule. I think I came up with around 85 percent for the capsule. Then I plunged into what mission steps would occur in spaceflight. I rolled the dice on a three-step suborbital flight and to my excitement it worked! Suddenly, each step of the mission was monumentally important. I became tense as I rolled the dice. It reminded me of the flavor of the early spaceflights. I called Steve over and told him of my successful suborbital flight. After his own successful flight, we both gleefully started scribbling notes on possible payload weights and additional mission steps. Soon we had scraps of paper filled with my horribly drawn stick figures of capsules that were lofting astronauts into space.

At this point, it was success or total failure on a mission step. We both realized that it was too crude and unrealistic for a rocket to always blow up on the pad. There were cancelled launches and aborts that should be considered. We laughed and played and scribbled more notes and sketched drawings for several hours and then folded it up and forgot about it for several years. There was no game really, it was only percentages for missions. There were no goals or objectives other than completing a mission. It was just one little scenario that I had devised among hundreds of other amusing little games that our group had momentarily toyed with. We moved on to new games that caught our fancy, but for some reason this game stayed in the back of my mind. I thought there was more to it.

More Board Games

There are two more games that had a strong influence on me and provided lots of hours of enjoyment. The first was the World War II beer and pretzel game, "Axis and Allies." This was the paper counter version before Milton Bradley bought the game and made the wonderful plastic molded pieces. "A&A" is a simplified world strategy view of the war based on the premise that territory is worth production points and the military pieces cost production points. Capture territory and you can buy more units. It works beautifully! It also teaches young people logistics, planning, geography, and the strategic difficulties faced by each opponent.

The second game that I loved was "Civilization." This was not the fine computer game published by Microprose in 1992, but a board game that had been published several years earlier with the same name. It was an abstract game about the cradle of civilization. You led your nomadic tribe through history, making it grow and prosper, become nations, and eventually become empires or perish. One of the nice features was the way your civilization achieved sophistication and progressed through various epochs.

Out of the Wilderness...

A few years passed and it was 1985 and I was living in New York City. On the weekends, my future wife Margie was away flying for Eastern Airlines. That left me with the company of our dog Saba and lots of solitude. I remember watching an inspiring space documentary narrated by Martin Sheen. It was an account of a close race for the conquest of space with the mythical finish line as the Moon. The idea came to me that it was the perfect background to apply all of the space routines that I had mentally thought of while taking Saba on long walks in Central Park. Full of energy, I started working on potential hardware and the mission steps to the

Moon. Now, like the space program, I had a goal, a direction: the Moon! There was also a competitor, the Soviets, which would add excitement to the game.

In one long weekend (with little sleep), I completed most of the U.S. and Soviet hardware and mission steps. The costs of the programs were roughly worked out and prestige points were developed. The prestige points became the "epochs" of space achievements.

The following weekend, I decided to improve the look of the game with color-coded counters and a game map. That took most of the weekend, leaving me just a little time for additional play testing and tuning.

Now I was ready to find "victims" to play my game. I was determined to find anyone! I recall that for several weekends I lured all of my friends into playing at least a few turns. Some of them were obviously being very polite. Surprisingly, most of them were enthusiastic. I started to think about marketing the game.

For those of you who have the ambition of marketing a game idea, hopefully you'll learn from my blunders. I made some right choices, some wrong ones, and had some luck. My first idea was that I needed a prototype of the game. People generally need to physically see your proposal or they won't properly understand your ideas. A game publisher also likes to see what kind of components are needed for manufacturing costs, and they need to be able to play test the game.

My first bit of luck was knowing Steve Stipp and his artistic talents. Steve has a great eye for technology and is excellent at line art. I called Steve, who was now an air traffic controller in the Air Force in Germany and begged and pleaded with him to draw the hardware pieces and come up with a montage of the space race for cover art. It worked out well for both of us (Steve needed funds because the US dollar was taking a beating at that time) and in a few months he delivered a beautiful cover and line art. I decided to call the game "Race to the Moon." The cover art by Steve Stipp is shown in Figure B-1. Originally, up to four players could play, representing the United States, the Soviet Union, Asia, and Europe. The game took place from 1956 to the year 2020. Besides the Moon race, I included the Space Shuttle, Skylab, Salut, Mir, and the U.S.-proposed space station. The game ended with construction of a permanent lunar base. It was four games in one, the original Moon race, the race to build a space station, the race to build a lunar base, and an overall campaign game. It was huge, with lots of counters, but it was fun to play!

Submissions

I came up with a list of about 25 game companies, large and small, to which I decided to send letters inquiring about submissions. Let me tell you, the board game

Figure B-1. Game cover art for the prototype of "Race to the Moon"

industry in general is a funny lot. With strategy games in a holding position, you'd think that they would be more active and open in looking for new products. Not only was response slow, but a great majority refused to even open the initial letter. (Often they were returned sealed.)

Over the next few months, a few companies did respond. A couple had "killer" submission/confidentiality forms that basically would have given the company the right to steal my idea for a measly fifty bucks. I stored those in the circular file! In the fall of 1985, a company in New York responded with a very fair submission form. I sent them a complete photocopy of my prototype with game cover, rules, counters, and game board.

After a few months of nail biting, I called the game company to gauge the response. Nobody had looked at it. They were, however, having open gaming nights on submissions, and invited me over with the possibility of the group playing my game on one of those nights. The folks at the company were very friendly and offered encouraging words. The office was smaller than I expected, as was the size of the staff, for a company that had produced so many board games.

I walked into the game room where there were four or five "gamers" involved in a game in the center of the room. They were laughing and giggling hysterically, obviously enjoying this game. In the corner I saw a pile of odd-looking prototype games. I found mine in the bottom third and pulled it out. I looked through the

contents and noticed not a piece had been moved. My eyes scanned the room and, to my horror, on one of the other tables I saw another space game! I couldn't believe it—it was a game that covered the same exact subject matter, sitting for submission at the very same company. All of my enthusiasm began to crumble.

I casually asked the people in the room what they thought of this game that was laid out on the table.

Only one person responded and said it was just set up for the first time and that no one had played it.

I asked if it was okay to take a look at it. He said, "Sure, go ahead." To my relief, after a quick look, it was a totally different approach. These were two completely novel designs on the space race. I began to feel a little better.

I decided to join in on the game that was causing everyone to roll on the floor. Unfortunately, I don't recall the title. It was a great beer and pretzel game and was definitely a unanimous hit among us, but to my knowledge it never got published.

I called the game company three weeks later to asked if a decision had been made on "Race to the Moon." They told me that no one has played it yet and they were going to send it back.

I basically just went on to other activities and new interests. I decided I would wait a few months and then consider whether I would go through the whole process again.

Once More into the Breach!

In 1987, I moved to California. After six months of settling in and exploring Los Angeles, I discovered a couple of board game stores. I was determined to clear the cobwebs and think about submitting again. I also decided to spiff up the appearance of the game board. My theory on this is that you have to convince the game company that your product is ready to ship. You still have to let them know that you're open to redesigning the game, but the more professional your prototype is, the more respect they'll give the product.

In early 1988, while at a game store, I discovered that Task Force Games was practically down the street.

I immediately went in search of one of their games on the shelf and wrote down the address. The next day, I called up Task Force and demanded to talk to the president. Within minutes, I had the president, John Olsen, on the phone and I told him that he had to see this game and that I would be over in twenty minutes.

I was determined to get this game at least one chance at bat. I showed up at Task Forces games and introduced myself. I met John Olsen, who was a very cheerful and likable Englishman. I was then introduced to Robert Sassone and I began to explain the game. After about twenty minutes, I was able to cover all of the main points of the game. I had made my pitch and at that point was happy to have gotten my game an audience.

John then said, "Aren't we going to play it?" I told him that it might take a couple of hours if he and Robert were up for it. To my delight, they enthusiastically said yes. Ironically, all of my early launches were successful and all of John's and Robert's missions were catastrophic failures. This statistical irregularity began to concern me, but soon it all balanced out. On a long shot mission, John came from behind and was victorious in landing on the Moon. After playing the game, John told me that they were definitely interested.

I was thrilled! After nearly five years, this game had a publisher! In the next few weeks we went through several rounds of contract negotiations. There wasn't as much money involved as I thought there would be. I had gotten an attorney who specialized in merchandising. He had never done a game before, but his area of specialty was an asset. His price wasn't so great, though, but in the long run, his expertise and his willingness to add my suggestions of specific limitations, and so on, greatly helped me.

After about a month and a half, I signed the contract, giving Task Force Games for the rights to my board game, "Race To the Moon." My entire advance on signing went to paying for the attorney's fees. I didn't really care, though, because it was going to be published and I was looking forward to the royalties!

Publishing a Board Game

The next several months were spent on play testing and modifying and tuning the game. John and Robert let me have a lot of input and control of the game. I came up with a list of shorter titles for the game. "LIFTOFF!" was my favorite choice, and everyone else agreed on it. They also agreed to my suggestion of acquiring a top space artist for the box cover art. I went through *The Black Book,* a graphic industry reference guide listing illustrators and photographers. There was one artist who immediately stood out. His name was Ken Hodges. I found out that he not only had done extensive work for NASA and JPL (Jet Propulsion Laboratory), but he also had done top secret work for nearly all of the major aerospace contractors. More recently, he had completed a series of U.S. postage stamps on space exploration. We called Ken in to see if he would be interested in the layout of the cover. Ken is an instantly likable person. The concept and subject matter interested him enough to give us a break from his normal rate. What amazed me was not only how skilled Ken is, but how fast he works. He was finished in less than two weeks. The game cover artwork of "LIFTOFF!" is shown in Figure B-2. The good work paid off for Ken, too—it got him an award and a lot more clients!

The game has a lot of components, so costs had to be closely watched. We also were trying to make a release deadline to correspond with the Origins National Convention (for board gamers) in Los Angeles, and it was the 20th anniversary of the first Moon landing. I managed to get Steve Stipp again for drawing the hardware counters. It took him a long time, but his work was worth the nervous wait. Robert assembled the rules. Miraculously it all fell into place the last week in June and was sent off to the printer. On the first day of the Origins Convention, the initial batch of games was finished!

Release

Origins is a huge board gamer's convention with thousands of people and booths, filled with lots of games and products. I strongly recommend everyone to give these conventions a try. They are a ton of fun, and you can meet a lot of people with similar interests. There are several in the Los Angeles area throughout the year and numerous conventions around the country. Your local game store and magazines can provide you with information about upcoming game conventions.

Figure B-2. The "LIFTOFF!" board game, cover by Ken Hodges, published by Task Force Games

At Origins the game sold fairly well. People who played it during the convention came back to the booth and complimented the variety of game play and novel design. This was music to my ears! Satisfied gamers enjoying my game was more important to me than all of the money in the world! (That was fine, since the overall profits were minimal.)

In the next few months, orders were moving well. What I didn't know was that the hobby shop industry is a mom and pop industry. Their attitude is that they take a risk and order one game from the distributor, and if it sells they made their money. Only on the megahits will they ever take the risk of reordering. (Generally megahits are the games that have lots of plastic pieces and have 6 pages or less of rules.) They assume that if they sold their one copy of "LIFTOFF!," why take any more risk. Overall that makes sense, too. I soon learned that the board game business shows small profits. Some other observations:

- Over half of all games ever bought are never opened. (They are bought on impulse or as gifts.)
- Board gamers are as loyal to their genre of game as cigarette smokers are to their brand of cigarettes. (Most Civil War buffs are not going to buy sci-fi or, for that matter, space or role-playing games.)

The average board gamer is not as affluent as the computer player. The computer player has cultivated a sense of experimentation, since the computer requires occasional patience and experimenting. I think this is why you'll find a wider range of games in a computer player's library.

In general, board gamer wallets are stretched and they are not willing to risk experimenting on different game genres. Understandably, they're going to save their hard earned money for the next published game on the Civil War or on whatever subject matter interests them. This is why the board game industry is semi-stagnant, with little growth. It also makes it very difficult for the industry to take risks on new game ideas. It is safer to imitate proven lines of product. Now I understood why my submission responses were so slow. If board gamers were a little more adventurous, distributors more imaginative, and hobby stores willing to risk one more game on the shelf, there would be lot more new game entries on the market.

Reaction and Sales

A couple of reviews came out on "LIFTOFF!" in the board game magazines. (The back cover of the board game is shown in Figure B-3.) The reviews were favorable, but that didn't help sales, due to the reordering policy of the small hobby stores.

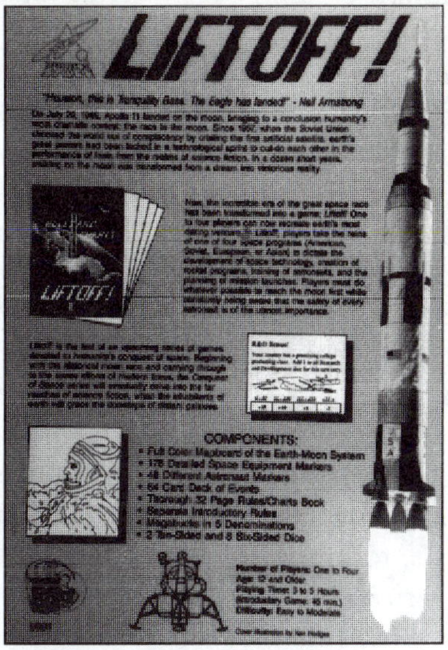

Figure B-3. Reverse view of the box of the board game, "LIFTOFF!"

Sales were also adversely affected by the economic situation in Task Force. Task Force was going through some budget cutting measures by its parent company, a computer game organization that closely monitored operations. Task Force had zero budget for marketing any of its projects. In a few months things were getting worse, and it looked like Task Force would close permanently.

This was sad, because John Olsen worked very hard and cared about Task Force. Fortunately, he managed to work out funding and made a deal with the parent company and bought out Task Force Games. In order to cut down costs, John moved the company to Texas. It was a struggle for a few years, but happily it worked out for John and Task Force Games. I've heard that sales have been very good for them recently.

The Next Phase: The Computer Version of the Game

One provision that I had in my contract with Task Force and the parent company was that they had one year to option the computer rights of the game. This was what made the attorney's fees worthwhile.

During the development of "LIFTOFF!," I approached several people in Task Force's parent company about how "LIFTOFF!" would work as a computer game. Since they made computer games, I thought they had taken a look at "LIFTOFF!" and would be interested. Unfortunately my questions fell upon deaf ears. Not only were they not interested, but I was given the impression that it was a foolish idea.

I decided to regroup and evaluate the computer game idea. I wanted to see "LIFTOFF!" presented in a different form. Most board games that went to computer were still two-dimensional transplants. Slightly improved graphics and components re-drawn over the same game board. This was very uninspiring to me.

I wanted more. I thought about the images that I remembered seeing as a child on television of early space launches. The thrill of not knowing what was going to happen added tension. I started to go to computer game stores and study what the medium was capable of. I also looked at all of the computer systems: IBM, Amiga, Atari ST, Apple, and Macintosh. For games, it appeared that the Atari ST and Amiga were the dominant players, followed by Apple. On the horizon, though, the IBM PC and Macintosh were growing. This made it challenging to determine which format was going to survive and which would disappear.

Most successful games have simple concepts that are cost effective (or keep costs low) for production purposes. The simpler the design is, the easier it is to program. The fewer the exceptions there are in a design, the less of a nightmare for programming. All of this saves time and money. This is an extremely successful method in the computer gaming industry; it works and it makes companies very rich. This is why there are so many spin-offs using the same engine on a different theme. But for the consumers, over a period of time games start to blur. I won't say that it stifles creativity, but occasionally the wheel should be reinvented.

If a game is marketed properly, the consumer quickly understands the concept and is either interested in buying the product or is turned off. A good game idea sitting on the store shelf can be successful, but often these games have a one-note theme or interface. Too often, after 20 minutes into the game, all of the wrinkles, bells, and whistles are seen by the player. For many people this is fine. I was curious to see whether a computer simulation could be made that had complex variables and a multimedia presentation with all the bells, whistles, and rocket explosions.

As a consumer myself, I wanted to see more in games. I desired more variety and exceptions. I also wanted more flexibility in game options with more graphics. I know there are different views on whether graphics are necessary for strategy games. For so long the general public has seen strategy games as nothing more than arcane chores. If they were presented as crisis management in a multimedia presentation, even the addicted joystick jockey might give strategy games a second look.

I wanted to recreate the feeling of being in the mission control room with all the multiple screens showing the different aspects of a space mission. At this time, there were no computer games that had even digitized graphics. The concept of multiple animations being assembled to tell numerous mini-stories had never been done. I had no idea if these constantly-updating multiple images would even work on a computer. (A Mission Control screen from the computer game is shown in Figure B-4.) I ruled out the Atari ST, because it offered only sixteen colors, and had a limited market and growth potential. I bought a used Amiga 500 to study its capabilities. The Amiga and the growing VGA IBM formats looked the most promising. In my head, some of the interface of the game began to jell.

Words Into Action

In the summer of 1990, I decided that I needed to find a computer programmer. I went to a couple of computer stores and asked if anyone knew of a good programmer for the Amiga and IBM. It took a few months before I had any leads.

Michael K. McCarty had no idea who I was when I called him in November 1990. I had gotten his name from a salesman at an Amiga computer game store. I called Michael and told him about "LIFTOFF!" and about some of my plans for a computer game. I briefly told him of my background as an actor and asked Michael about his programming skills. He was 22 years old at the time, was currently attending college, and was in the process of completing his degree in computer science. He had worked on computers since he was nine. I asked him if he had any

Figure B-4. Mission Control launch countdown

recollection of the Moon landings. He told me that he was born within a few weeks of *Apollo 11*. Well, that certainly dated me and what era I came from. I also asked him if he had any interest in space. He said that he did and mentioned that his father had designed numerous satellites for Hughes Spacecraft. That certainly qualified as having some interest in space. I decided to meet Michael in person and pursue this further.

After telling Michael of this bold adventure, I asked him if he was up to the task. He said yes, but I'm sure during the long haul he had second thoughts about my sanity and his own.

The rest of November and December, Michael and I had numerous discussions on computer technology—about whether my ideas of multiscreens would work and about our business relationship. We decided to form a partnership and started work on putting a demo of the game together. The name of our partnership was called Strategic Visions. We looked into the possibility of publishing the game ourselves but quickly discovered the enormous expenses were well beyond our means.

Early Designs and a Demo

In late December we decided to put together a demo of the game. Since a prototype of the board game got a publisher, perhaps a slick demo of the computer game would generate some interest as well. Michael had more experience on the Amiga, so we started on that platform. We decided to produce a video tape demo, rather than a disk-based demo, and our release estimates were three months for the demo and six more months for the game itself.

Michael and I went to the Consumer Electronics Show (CES) in Las Vegas in January 1991. The section on entertainment software was overwhelming. With only a business card in hand, we started to make contacts with various computer game publishers and distributors. Looking at the company booths and the products gave us a good idea of who we wanted to talk to.

We worked on interface screens. We both had distinct views on interface, but fortunately we were able to incorporate them well. Michael wanted to use a style of screens similar to Next computer screens, including gray borders and buttons. I wanted a *USA Today* and CNN style of graphic presentation for statistics: a splash of colors with bold shapes and icons. This was a good merging of ideas.

Work on the demo continued into late May. Twice we had hard drive failures and had to redo major sections. Lots of photos and animations needed to be digitized or drawn from scratch. I can recall the frustration of four repetitions of digitizing several thousand still photos. The support hardware was sometimes very temperamental. During that time, over 30 companies from CES agreed to see our demo and sign our confidentiality forms. In June we assembled a brochure with literature about the game and had 35 video packets with little space trinkets ready to ship to publishers. An early screen shot of the demo is shown in Figure B-5.

Within a few days we started receiving faxes from publishers. Some even sent contracts and offers. A couple of the offers were realistic, others were not. Many said it was one of the best demos that they had seen. Some said that it was a great demo but wondered if the technology existed for the game that we had in mind. One or two companies couldn't understand how a game could be made out of old film footage. They basically said, "Nice demo, but where's the game?" For two months we sorted through various publishers and studied their company profiles and offers. In the end, over 30 publishers had seriously responded to our demo. Now we had to pick one.

We also had to be realistic about our position. We were very small and we had no track record. Why would a publisher be willing to risk money on a relatively unknown game with a virtually unknown creator? Publishers love to emphasize this point, too! Fortunately, "LIFTOFF!" had gained respect within the computer industry as a well-designed game, and our demo displayed a strong and focused design concept.

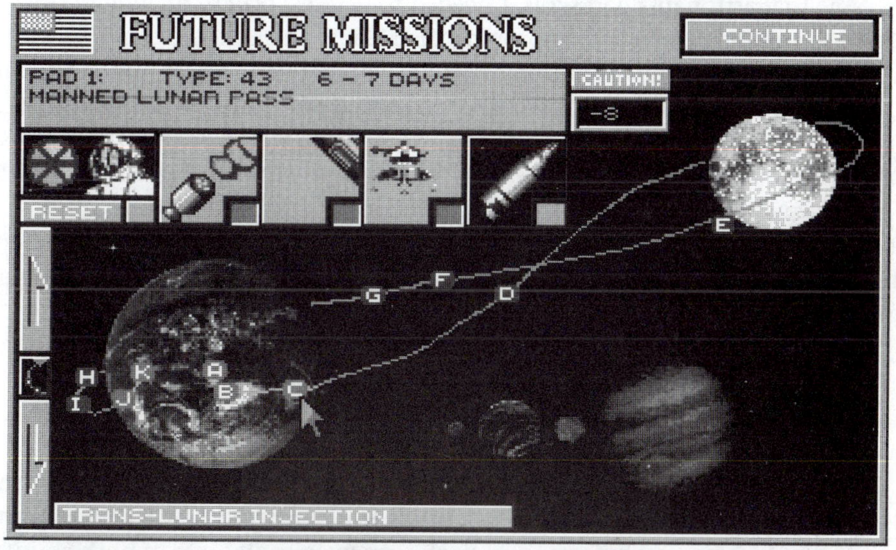

Figure B-5. One of the first screen designs for the game

It's a Deal!

We decided to go with Interplay, a Los Angeles area publisher that had a good track record and tremendous growth potential. We felt that a local publisher was the right choice for our first game because there were advantages to having local support and assistance. There were other larger companies with larger offers, but we would have had less control on the outcome of the game. Another strength was that Brian Fargo, president of Interplay, liked the suggestion of getting Buzz Aldrin involved. I had seen Buzz interviewed on several television shows and knew he would be interested in an educational simulation. He is very active in new space proposals and in proposals for getting back to the Moon and manned landings on Mars. Another important feature was that Brian Fargo was willing to give us autonomy and control on the project.

We went over contract negotiations for about a month and finally signed a deal in August 1991. I was very excited. We hoped to have it released in May 1992. With Buzz on board, I wanted the game to be more involved with astronauts. In the board game astronauts had no effect on the missions; they were nothing more than Spam in the can, just going along for the ride. This is how astronaut skills and, later on, morale, came to be added to the game. Suddenly those factors had a purpose and value in the game. Having real astronauts gave us the added dimension of using skills and morale to assign certain astronauts to specific missions to utilize their skills. (Of course, I randomized the skills of the astronauts from game to game so that there would be variety with each game.)

Now that astronauts had skills, I needed to devise a way to include in the game a feature for training them and improving their skills. This is how I came up with the various buildings for different training sessions. Earlier I had broken up the structured board game sequence into free-flowing form. I asked Michael if having buildings and offices that the player could enter and send astronauts to would be technically possible or present any memory problems. He said, "Not only will it work, but why not include the turn sequences?" The dam burst! This flooded me with ideas to create a whole spaceport of program buildings, museums, and the development of the presidential review (see Figure B-6) and CIA intelligence (see Figure B-7) features of the game. As soon as I designed a building layout, I knew I had to have equipment and rockets moving around the spaceport. I also wanted this "living" spaceport to grow and change over time and events. I went overboard with it. I probably scared Michael with the immensity of it. With astronaut skills and spaceports, it would be a better game and, consequently, much more complicated for Michael to program.

I called Steve Stipp and had him start work on the rough sketches I had made of the spaceport. I told him to give it a bird's eye perspective, with the capitol in the distance. In a few weeks Steve sent back a marvelous diagram of both countries' spaceports. Steve Stipp's sketch of the U.S. Spaceport and the final computer version by Dave Mosher are shown in Figure B-8.

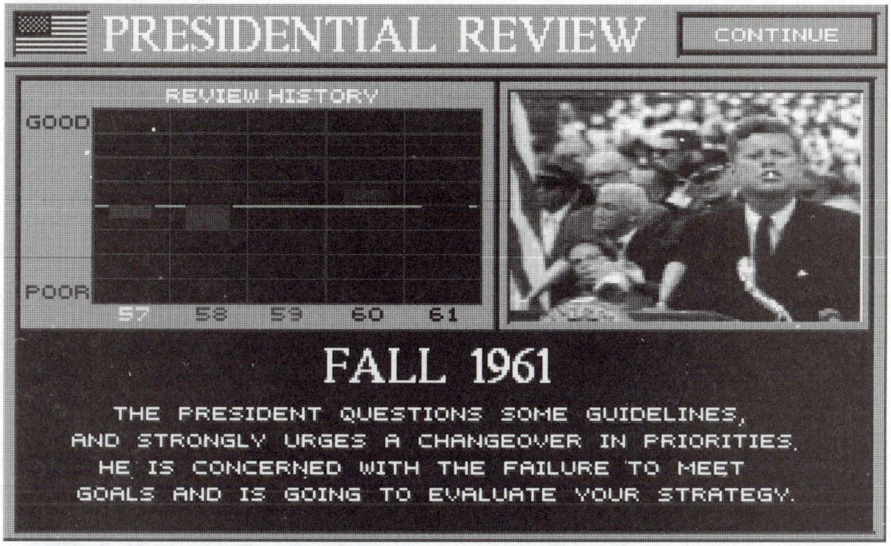

Figure B-6. Presidential review

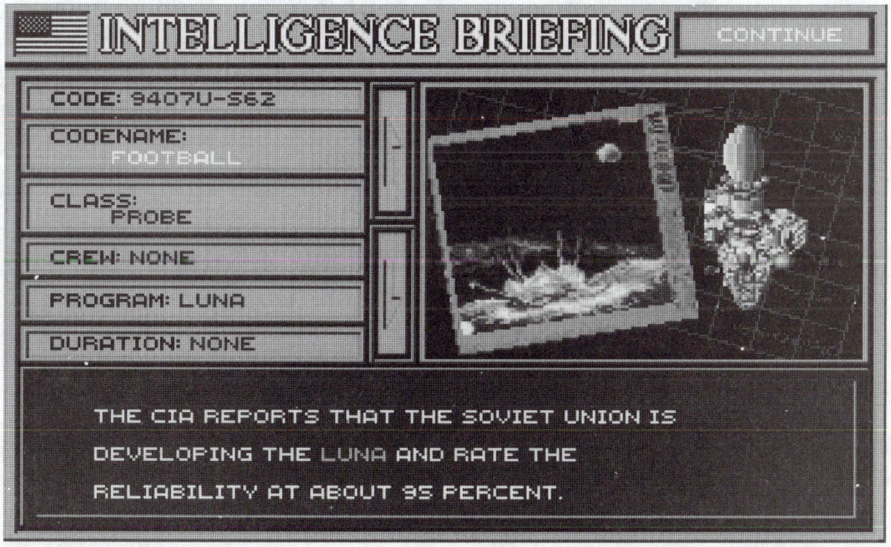

Figure B-7. CIA intelligence

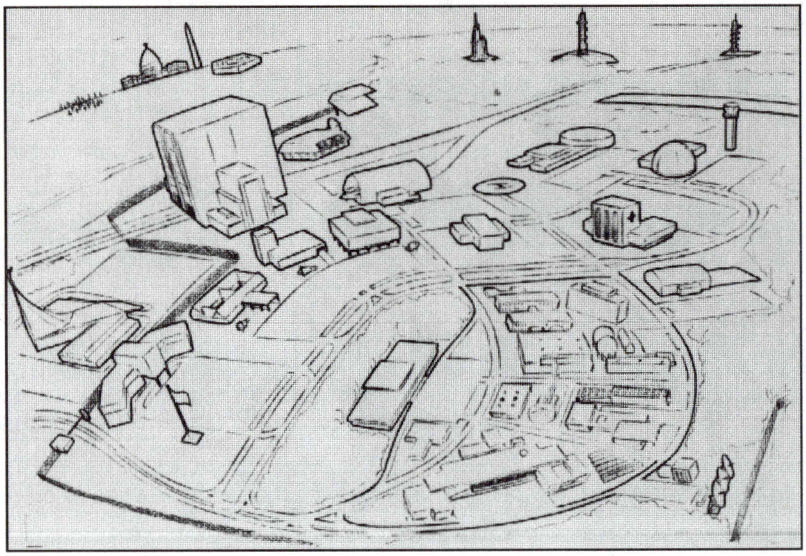

Figure B-8. Sketch of the U.S. Spaceport (top), Final art in the game (bottom)

The People at Interplay

The very first person that I met at Interplay was Rusty Buchert. At the time, Rusty was in charge of reviewing product submissions. He told me that not only had his playing of the board game "LIFTOFF!" made him interested in the demo, but also that the game had received favorable reviews. Rusty is a jack-of-all-trades and was thoroughly involved with "Buzz Aldrin's Race Into Space" from start to finish.

Next, Mike and I met Bruce Schlickbernd, who was assigned as our Interplay producer. At the time, he was on the home stretch on another game for Interplay, "Star Trek—The 25th Anniversary," which became a huge success. Bruce would be our liaison for support or technical needs. This is a challenging function at any company, and Interplay was growing enormously at this time. At times it took perseverance and a battle manual to meet our needs. Bruce went all the way in getting the support.

One of my requirements with Interplay was to have Ken Hodges and his artwork involved again. He drew a more enhanced painting that included more features of the space race. His work was magnificent! Unfortunately it ended up only on the cover of the historical game manual. This is my opinion only, but I thought it should have been on the box cover, too! I was overruled, but I understand the reasons. The photo of Buzz Aldrin is appropriate as a cover.

The Salt Mines

In September we switched the design platform from Amiga to IBM DOS because the Amiga market was shrinking. This meant Michael would have to redo all of his code, but a 256-color palette gave us a greater range in graphics and, more importantly, IBM became the dominant consumer computer platform. With these changes and new design features, our target release date of May 1992 began to look unrealistic. This being our first product and because we were reinventing the wheel, Michael had to develop every routine from scratch: routines for Future Missions, Spaceport, Mission Animations, and all of the data and statistics, which would further delay the release date.

For several months digitizing had to be done all over again, since we could use more colors on the IBM. Also, in late August 1991, Mike brought in a friend who worked part-time to take over this growing media task. His name was Tom Chamberlain. Poor Tom had no idea what he was getting into and how much he would learn about the space program. (Interestingly enough, Tom's father is also in the space business. Employed at McDonnell Douglas, he is one of the key individuals involved with Space Station Freedom.) Tom started the long project of grabbing images and cleaning up frames.

By cleaning up frames and taking the detail out of the background, less information or data is used. This saves disk space, which is at a premium with this game. Saving space meant more room for additional animations. As it turned out, we ended up using over 250 different mission step animations. That's a lot of frames. We originally had additional hardware and other animations that were longer, but they were cut due to the an obvious disk limitation.

So now there were three of us busily banging away on various functions of the game. This is when I began to spend half of my time on the road going to Interplay or Michael's.

After hearing numerous demo tapes, I settled on two contrasting musical composers for the musical score. Brian Langsbard had a strong classical background and Michael McNerny a more modern style. I designated sections that best suited their styles. I was very pleased with their results.

Christmas started to creep up so we needed to start work on a *new* demo for CES. It was an extra project, and for over a month we had even less sleep than before. Charles Deenen, Interplay's composer, was very helpful in the final music mix. He has a knack for tinkering with and enhancing music and sound.

The 1992 spring CES in Las Vegas arrived. We finished the demo for the press just a day before the deadline! Whew! Interplay provided a gracious reception at Caesar's Palace for the press. Buzz Aldrin shared his personal moments of the Moon landing with the audience. It was a very moving experience.

Deeper in the Salt Mines

After CES, we went back to our chores. Margie Beeson came aboard and began doing a lot of the detail artwork. In February, David Mosher, one of Interplay's artists, began coloring in Steve Stipp's diagrams of the spaceports. David is a dedicated and talented computer artist. His talents brought the spaceports to life with color, texture, and depth. The Soviet Spaceport screen is shown in Figure B-9.

I was determined to have a historical manual included with the game that followed the likes of Lucas Art's "Battle of Britain" game. I started gathering hundreds of books and sources in late 1991. I also contacted Robert Reeves, a dedicated independent Soviet space researcher located in Texas. I had met Robert through GEnie and asked him to assist in this endeavor. It had grown too huge for one person to complete.

By March, all of the components of the historical manual were finished. After carrying a three-pound brick of pages to Interplay, I made a discovery. I miscalculated the size of the manuscript—it was a little large. I made the mistake of thinking that the 200-page limit was two hundred physical pages instead of one hundred pages front and back. In the font size that I used, it would have converted into about

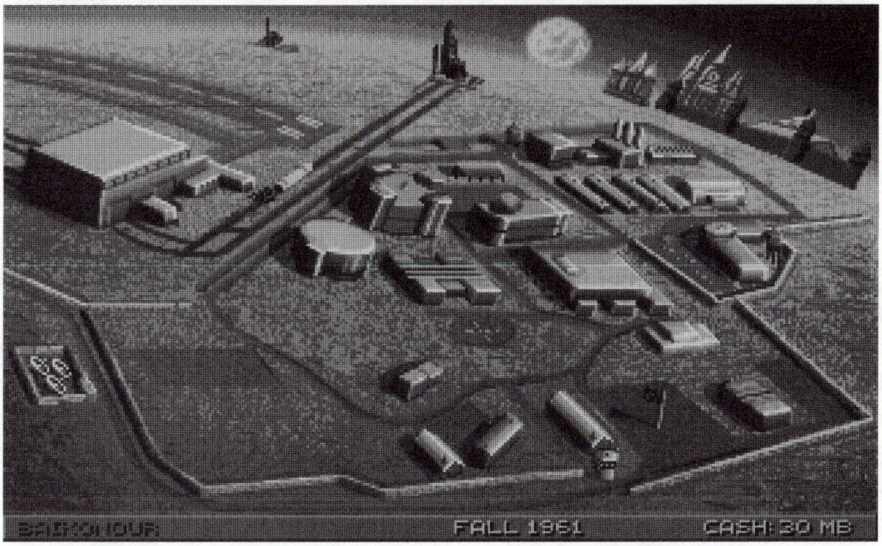

Figure B-9. Soviet Spaceport: Baikonour

500 pages after formatting and artwork was added. This didn't include any of the rules at this point, so obviously major sections had to be deleted.

After several months of persistence, Interplay hired someone to assemble the manual. After a series of meetings, we felt that in a few months all of the manual would be assembled and completed. None of us expected it to take as long as it did.

The immensity of the game was taking its toll. We couldn't even maintain a schedule, let alone estimate how long sections of the game would take. We were extremely tight on budget and undermanned. We needed a second programmer. Fortunately, Morgan Roarty, one of Michael's college classmates, joined the team at this time. Getting him on the project was like getting reinforcements, and it rejuvenated our morale. Now Michael and Morgan battled the Artificial Intelligence and Failure Mode portions of the game.

Spring turned into summer and then, suddenly, fall was upon us. Where did the time go? The game was close to completion, but there were still numerous bugs in September. Play testing went into full force, but it was a daunting task. One of the play testers, Mike Packard, calculated that there were over 12,000 different types of space missions when all of the various space hardware combinations were included. When Failure Mode was incorporated, it raised the number to around 50 million potential outcomes. Understandably, this didn't make the folks in play testing very happy, but they continued to plug away.

Trapped in the Salt Mines

We all worked feverishly to make the Christmas deadline, but there were additional snafus. The manual was even further behind schedule.

October passed, as did November and, frustratingly, Christmas. At this point, changes were made in order to get things moving again. In January, Jerry Friedman was assigned to assemble the manual and the rules.

Jerry is a fast and meticulous graphic artist and is very easy going. He put up with all of my miserable demands and constant changes of photo and art placement. Even under such an accelerated schedule, the finished manual was quite satisfying. In about two and a half weeks the manual was sent off to the printers. Now it was the printers who held us up. Instead of a few weeks, it would turn into over six weeks before the manual and rules were printed.

Light at the End of the Tunnel

Finally, the game shipped. After 28 months, the project was complete! I never thought it would take over ten years from the conception of the game idea to its completion. It was hard work, and I'm very grateful to Michael McCarty for his tireless, around the clock programming. He not only worked long hours with very little sleep for over two years, but also contributed enormously to the game. Morgan Roarty, Tom Chamberlain, and Margie Beeson also spent a tremendous amount of time, offered enthusiasm, and made many fine contributions to the game.

The "Buzz Aldrin's Race Into Space" CD-ROM Version

Immediately following the completion of the DOS version of the game, we started work on the CD-ROM version (with a planned release date of Christmas 1993).

Due to the limitations of disks for the IBM DOS version, many photographs and animations had been cut from the release product. They also would have taken up over 100 Megabytes on a hard drive! Modem play is another feature that I dearly wanted to include, and it also had to be deleted.

The CD-ROM will have a lot more animations that will be longer and more varied. These animations will also have a faster, smoother frame rate. Modem play will finally be incorporated, which will allow you to leave saved turns on boards so that you can play with friends across the country. There will be a few surprises as well!

Other Plans

There are further plans for the board game as well. Regarding a "LIFTOFF!" expansion, I have done numerous designs and play tested several modules. There is the "JPL Scenario: Unmanned Exploration of the Planets," a campaign game from 1957 to 2025, which includes the Moon race up through the shuttle and space stations, lunar bases, and a manned landing on Mars. It includes mini-games in which you can build the space station and a lunar base, and there are additional player countries, Brazil and India. There is also a multiplayer individual space goal scenario with each country having a defined combination of missions and hardware to complete. They all work quite well. Fortunately, the rights to the board game "LIFTOFF!" are back in my possession. If there are any board game publishers out there who are interested, don't hesitate to call. I'd like to see a deluxe version of "LIFTOFF!" with an enhanced game board and color components in time for the 25th anniversary of *Apollo 11*'s Moon landing on July 20, 1994.

All of these expansions would be ideally suited for a computer sequel to "Buzz Aldrin's Race into Space." Imagine a hundred years of realistic space exploration. It would be a lot of fun.

There are other projects looming in the heads of Michael and myself. Hopefully, we'll be able to do justice to those subjects as we did to "Buzz Aldrin's Race into Space." Look for us, hopefully we'll be around for awhile.

Eleven Years...

My main purpose and motivation in working on this project for the last ten years was to spark interest in the space program. The current space program is in the doldrums, with little interest from the media and government. The general public is unaware of the sacrifices and wonderful accomplishments that have been made in space exploration. Sadly, few can remember man's greatest technological achievement—landing on the Moon—nor can many name the individuals who took part in that great endeavor and walked on another world. Hopefully this obsession of mine will interest and spur on the von Brauns, Korolevs, Gagarins, Armstrongs, Aldrins, and Sally Rides of tomorrow.

I hope this simulation will be a helpful aid for teachers and students. Students will learn that they can create their own history. This is just the first step!

Reach for the stars and you'll achieve nothing less!

Bibliography

Aldrin, Buzz. *Men From Earth.* New York: Bantam Books, 1989.

Aldrin, Buzz, and Wayne Warga. *Return To Earth.* Random House, 1973.

Alexander, Tom. *Project Apollo.* Harper & Row, 1963.

Allen, Joseph P., and Russell Martin. *Entering Space.* Stewart, Tabori & Chang, Publishers, 1986.

Allward, Maurice. *The Encyclopedia of Space.* Hamlyn House, 1969.

Always Aim to Be the First. Moscow: Novosti Publishers, 1991.

Apollo-Soyuz: Mission Profile. Rollins Press, Inc., 1975.

Appel, Fred. *The Moon and Beyond.* London: Aldus Books, 1971.

Baker, David. *Conquest: A History of Space Achievements.* 1984.

Bedwell, Theodore, and Hubertus Strughold. *Bioastronautics and the Exploration of Space.* Alexandria, VA: Defense Document Center, 1965.

Bell, Joseph H. *Man Into Orbit.* Avon Books, 1961.

Benford, Timothy, and Brian Wilkes. *The Space Program Quiz and Fact Book.* Harper & Row, 1985.

Bixby, William, and Robert Scott. *Antarctic Pioneer.* J.B. Lippincott Company, 1970.

Blaine, B.C.D. *The End of an Era in Space Exploration.* American Astronautical Society Publication, Univelt Inc., 1976.

Bono, Philip, and Kenneth Gatland. *Frontiers of Space.* New York: Macmillan Publishing Co., Inc., 1976.

Borman, Frank, and Robert J. Serling. *Countdown.* Silver Arrow Books, 1988.

Braun, Wernher von, and Frederick I. Ordway. *History of Rocketry and Space Travel.* New York: Thomas Y. Crowell Company, 1969.

Burgess, Eric. *Pioneer: First to Jupiter, Saturn and Beyond.* NASA SP, 1980.

Caidan, Martin. *The Greatest Challenge.* E.P. Dutton & Co., 1965.

---. *Man into Space*. New York: Pyramid Books, 1961.

---. *Rendezvous in Space*. E.P. Dutton & Co., 1962.

Canby, Courtlandt. *A History of Rockets and Space*. Hawthorn Books Inc., 1963.

Cassutt, Michael. *Who's Who in Space: The First Twenty-Five Years*. G.K. Hall & Company, 1987.

Cipriano, Anthony J. *America's Journeys into Space: The Astronauts of the United States*. Wanderer Books, 1979.

Clark, Phillip. *The Soviet Manned Space Program*. New York: Orion Books, 1988.

Collins, Michael. *Liftoff*. New York: Grove Press, 1988.

Cortright, Edgar M. *Apollo Expeditions to the Moon*. NASA, U.S. Government Printing Office, 1975.

Cox, Donald W. *The Space Race from Sputnik to Apollo...And Beyond*. Philadelphia: Chilton Books, 1962.

Cunningham, Walter. *The All-American Boys*. Macmillan Publishing Co., 1977.

Curtis, Anthony R. *Space Almanac*. Woodsboro, MD: Arcsoft Publishers, 1990.

Editors of Fortune. *The Space Industry*. Prentice-Hall, 1962.

Ertel, Ivan. *Apollo Spacecraft,* vol. 1. NASA SP-4009, 1969.

Ezell, Edward. *On Mars: Exploration of the Red Planet, 1958-1978*. NASA SP-4212, 1984.

Ezell, Edward Clinton, and Linda Nueman. *The Partnership: A History of the Apollo-Soyuz Test Project*. Washington D.C.: U.S. Government Printing Office, 1978.

First to Venus: The Story of Mariner II. McGraw-Hill, 1963.

French, Dr. Bevan. *What's New on the Moon*. Langley Research Center.

Furniss, Tim. *Manned Spaceflight Log*. London: Jane's Publishing Company, 1983.

Gartmann, Heinz. *The Men Behind the Space Rockets*. Weidenfeld & Nicolson, 1955.

Gatland, Kenneth. *Fact Finder Space Diary*. Crescent Books, 1989.

---. *Manned Spacecraft*. New York: Macmillan Publishing Co., Inc., 1976.

---. *Missiles and Rockets*. New York: Macmillan Publishing Co., Inc., 1975.

---. *Robot Explorers*. London: Blandford Press, 1972.

---. *Space Technology*. Harmony Books, Crown Publishers Inc., 1981.

Glushko, Valentin, ed. *Soviet Cosmonautics: Questions and Answers*. Novosti Press Agency, 1988.

Golovanov, Yaroslav. *Our Gagarin*. Moscow: Progress Publishers, 1978.

Gordon, Yefim, and Bill Sweetman. *Soviet X-Planes*. Motorbooks International, 1992.

Gurney, Gene. *Rocket and Missile Technology*. New York: Franklin Watts, Inc., 1964.

Haley, Andrew G. *Rocketry and Space Exploration*. Princeton: D. Van Nostrand Company, Inc., 1958.

Hall, Al. *Peterson's Book of Man in Space: The First Small Step*. Los Angeles: Petersen Publishing Co., 1974.

Hall, Al, ed. *Man in Space*, vols. 1, 2, 3, and 4. Peterson Publication Company, 1974.

Hart, Douglas. *The Encyclopedia of Soviet Spacecraft*. Bison Books Corp., 1987.

Harvey, Brian. *Race into Space: The Soviet Space Programme*. Halsted Press, 1988.

Hendrickson, Walter B. *Reach for the Moon*. Bobbs-Merrill Company, 1962.

Highland, Harold Joseph. *Encyclopedia of Space Science*, vols. 1 and 2. New York: Theodore Audel & Company, 1963.

Isakowitz, Steven J. *International Reference Guide to Space Launch Systems*, American Institute of Aeronautics and Astronautics, 1991.

James, Peter N. *Soviet Conquest from Space*. New Rochelle, NY: Arlington House Publishers, 1974.

Jane's Spaceflight Directory 1986. Jane's Publishing Company, 1986.

Johnson, Nicholas. *Handbook of Soviet Manned Space Flight*. San Diego: Univelt Inc., 1980.

---. *Soviet Space Programs 1980-1985*. San Diego: Univelt Inc., 1987.

---. *The Soviet Year in Space-1987*. Colorado Springs: Teledyne Brown Engineering, 1988.

---. *The Soviet Year in Space-1988*. Colorado Springs: Teledyne Brown Engineering, 1989.

---. *The Soviet Year in Space-1989*. Colorado Springs: Teledyne Brown Engineering, 1990.

---. *The Soviet Year in Space-1990*. Colorado Springs: Teledyne Brown Engineering, 1991.

Kerrod, Robin. *The Journey of Voyager*. Mallard Press, 1990.

Kidron, Michael, and Dan Smith. *The War Atlas: Armed Conflict-Armed Peace*. Simon & Schuster, 1983.

Koppes, Clayton R. *JPL and the American Space Program*. New Haven: Yale University Press, 1982.

Larson, Stephen. "Saturn". *Academic American Encyclopedia*. Grolier Electronic Publishing, Inc., 1991.

Lebedev, V. et al. *Sons of the Blue Planet*. Moscow: Political Literature Press, 1973.

Lebedev, Vladimir, and Yuri Gagarin. *Survival in Space*. Bantam Books, 1969.

MaGill, Frank N., ed. *MaGill's Survey of Science, Space Exploration Series*, vols. 1, 2, 3, 4, and 5. Pasadena: Salem Press, 1989.

Mailer, Norman. *Of a Fire on the Moon*. Signet Books, 1971.

Manno, Jack. *Arming the Heavens: The Hidden Military Agenda for Space, 1945-1995.* Dodd, Mead & Company, 1984.

McAleer, Neil. *The Omni Space Almanac.* Ballantine Books, 1987.

McDonough, Thomas R. *Space: The Next Twenty-Five Years.* Wiley & Sons, Inc., 1987.

McDougall, Walter A. *...the Heavens and the Earth.* New York: Basic Books, Inc., Publishers, 1985.

Medaris, Major General John B. *Countdown for Decision.* New York: G.P. Putman's Sons, 1960.

Miner, Ellis. "Uranus". *Academic American Encyclopedia.* Grolier Electronic Publishing, Inc., 1991.

Murray, Bruce. *Journey into Space.* New York: W.W. Norton & Company, 1989.

Murray, Charles, and Catherine Bly Cox. *Apollo: The Race to the Moon.* Simon & Schuster Inc., 1989.

National Aeronautics and Space Administration (NASA). *Space: The New Frontier.* Washington, D.C.: U.S. Government Printing Office, 1962.

Newkirk, Denis. *Almanac of Soviet Manned Space Flight.* Houston: Gulf Publishing Company, 1990.

Oberg, James E. *Red Star in Orbit.* New York: Random House, 1981.

---. *Uncovering Soviet Disasters.* Random House, 1988.

Ordway, Frederick. *The Rocket Team.* New York: Thomas Y. Crowell Company, 1979.

Ordway, Frederick, and Randy Liebermann. *Blueprint for Space.* Smithsonian Institution, 1992.

Peebles, Curtis. *Guardians Strategic Reconnaissance Satellites.* Novato, CA: Presidio Press, 1987.

Pellegrino, Charles R. *Chariots for Apollo.* New York: Atheneum, 1985.

Powers, Robert R. *The World's First Spaceship Shuttle.* Stackpole Books, 1979.

Riabchikov, Evgeny. *Russians in Space.* Moscow: Novosti Press Agency Publishing House, 1971.

Ridpath, Ian, ed. *The Illustrated Encyclopedia of Astronomy and Space.* New York: Thomas Y. Crowell Company, 1976.

Rockwell International. *Twentieth Anniversary of Apollo Program.* (Rockwell International), 1989.

Rockwell International, Space Division. *Man on the Moon.* (Rockwell International), August 1969.

Ryan, Peter. *The Invasion of the Moon: The Story of Apollo 11.* Penguin Books, 1969.

Rycroft, Michael. *The Cambridge Encyclopedia of Space.* Cambridge University Press, 1990.

Sheldon, Charles S. *Soviet Space Programs, 1966-1970.* Washington D.C.: U.S. Government Printing Office, 1971.

Shelton, William R. *Man's Conquest of Space.* National Geographic (publishers), 1968.

Smith, Arthur. *Planetary Exploration: Thirty Years of Unmanned Space Probes.* Richard Clay Ltd., 1988.

Smith, Marcia et al. *Soviet Space Programs: 1976-80.* Washington D.C.: U.S. Goverment Printing Office, 1984.

Smolders, Peter. *Soviets in Space.* New York: Taplinger Publishing Co., Inc., 1974.

Stanford, Walter H. *Artificial Earth Satellites and Successful Solar Probes.* Huntsville, AL: George C. Marshall Space Flight Center, 1961.

Sternfeld, Ari. *Soviet Writings on Earth Satellites and Space Travel.* New York: The Citadel Press, 1958.

Stoiko, Michael. *Soviet Rocketry: Past, Present, and Future.* New York: Holt, Rinehart & Winston, 1970.

Swenson, Loyd S., James M. Grimwood, and Charles C. Alexander. *This New Ocean, A History of Project Mercury.* NASA, 1966.

Thomas, Shirley. *Men of Space.* New York: Hillman Books, 1961.

TRW. *Study of an RTG Version of Pioneer F and G: Volume 1, Final Technical Report.* Document 11339-6005-R0-00. (TRW), 1969.

Turnill, Reginald. *The Observer's Space Flight Directory.* London: Frederick Warne Ltd., 1978.

U.S. Congress. Office of Technology Assessment. *Salyut: A Technical Memorandum.* Washington, D.C.: U.S. Government Printing Office, December 1983.

U.S. Department of Defense. *The Soviet Military Power: 1990.* Washington, D.C.: U.S. Government Printing Office.

---. *The Soviet Space Challenge.* Washington, D.C.: U.S. Government Printing Office, November 1987.

Wilson, Andrew. *Solar System Log.* London: Jane's Publishing Company, Ltd., 1987.

Winter, Frank H. *Prelude to the Space Age, The Rocket Societies: 1924-1940.* Washington D.C.: Smithsonian Institution Press, 1983.

Wulforst, Harry. *The Rocket Makers.* Orion Books, 1990.

Young, Hugo, Bryan Silcock, and Peter Dunn. *Journey to Tranquility.* Doubleday & Company, 1970.

Other References

Aviation Week and Space Technology. Weekly issues. McGraw-Hill Publications, 1990-1992.

Final Frontier. Bi-monthly issues, 1989-1992.

Grumman Aerospace Corporation. *Apollo Lunar Module* (pamphlet and transparencies).

The McGraw-Hill Encyclopedia of Space. Paris: Editions Rombaldi, 1967.

Smithsonian, Air and Space. Bi-monthly issues, 1989-1992.

Space Flight. Monthly issues, 1988-1991.

Spaceworld, The National Space Society. Monthly issues, 1986-1988.

Photo Credits

Front Matter Illustrations: Steve Stipp, Margie Beeson

Foldout Illustration: Margie Beeson, Fritz Bronner

NASA: xxvi, 6, 11, 12, 13, 18, 19, 22, 26, 38, 40, 42 (both), 45, 46, 47 (both), 53, 56, 57, 67, 68, 73, 74, 75, 78, 80, 82, 83, 84, 86, 87, 89, 99 (both), 101, 107, 110, 112, 113, 115, 117, 119, 121, 124 (both), 127, 129, 134, 139, 158, 159, 161, 162, 168, 172, 175, 177, 178, 180, 182, 184, 185, 272, 273, 280

USAF: 137 (both)

DOD: 8, 25, 32, 36

Grumman Aerospace: 142, 144, 145, 146

US Army: 24, 173

Frederick Ordway III: 1

Randy Liebermann: 20

Fritz Bronner: (Foldout), 38, 50, 61, 148, 164, 260

Bronner/Beeson: (Foldout), 122

Margie Beeson: 147

Steve Stipp: 150, 153, 330, 342

John Olsen, Task Force Games: 333, 335

Strategic Visions: 194, 196, 199, 205 (both), 206, 207, 208, 209, 210, 211, 212, 213, 214, 215, 216, 224, 226, 227, 228, 229, 230, 231, 234, 235, 236, 237, 238, 239 (both), 241, 243, 244, 245, 246, 248, 249, 250, 251, 252, 254, 255 (both), 256, 258, 262, 263, 268, 269, 270, 271, 276, 277, 279, 281, 282, 286, 304, 321, 322, 337, 339, 341 (both), 342, 345

Send your comments, space records, and winning strategies on "Buzz Aldrin's Race into Space" to this address. I'd like to hear suggestions for BARIS 1.5 the Silver Anniversary Edition, scheduled for July 1994, the twenty-fifth anniversary of the *Apollo 11* Moon landing. Aslo include comments for "LIFTOFF EXPANSIONS," "LIFTOFF II," and the sequel to BARIS. Tell me what your earliest Moon landing was and what your winning strategies are. I look forward to your comments!

—Fritz Bronner

NOTE: By sending comments to this address, you are waiving all rights. Your comments may be re-edited, altered, and reprinted by Fritz Bronner and/or Strategic Visions Inc., in other material and media. All rights will be reserved by Fritz Bronner and Strategic Visions.

Send to: Fritz Bronner
P.O. Box 3241
North Hollywood, CA
91609